The Design of Foundations for Buildings

McGRAW-HILL SERIES IN MODERN STRUCTURES:
Systems and Management
THOMAS C. KAVANAGH *Consulting Editor*

BAKER, KOVALEVSKY, AND RISH *Structural Analysis of Shells*
FORSYTH *Unified Design of Reinforced Concrete Members*
FOSTER *Construction Estimates from Take-Off to Bid*
GILL *Systems Management Techniques for Builders and Contractors*
JOHNSON *Deterioration, Maintenance and Repair of Structures*
JOHNSON AND KAVANAGH *The Design of Foundations for Buildings*
KOMENDANT *Contemporary Concrete Structures*
OPPENHEIMER *Directing Construction for Profit*
PARKER *Planning and Estimating Dam Construction*
PARKER *Planning and Estimating Underground Construction*
PULVER *Constructions Estimates and Costs*
RAMASWAMY *Design and Construction of Concrete Shell Roofs*
SCHWARTZ *Civil Engineering for the Plant Engineer*
TSCHEBOTARIOFF *Foundations, Retaining and Earth Structures*
WOODWARD, GARDNER, AND GREER *Drilled Pier Foundations*
YANG *Design of Functional Pavements*
YU *Cold-formed Steel Structures*

The Design of Foundations for Buildings

SIDNEY M. JOHNSON and THOMAS C. KAVANAGH

McGRAW-HILL BOOK COMPANY

New York San Francisco Toronto London Sydney

THE DESIGN OF FOUNDATIONS FOR BUILDINGS

 Library of Congress Catalog Card Number 68-18576

ISBN 07-032583-9

567890 MAMM 7543

Preface

Beginning with the pioneer efforts of Terzaghi and others in the 1920s and 1930s, the design of foundations has become a highly developed field of specialization. An enormous amount of analytical and experimental work has been done. Nevertheless, foundation design, at least insofar as the action of the soil is concerned, remains a field in which experience and judgment are the primary determinants.

Unfortunately, despite the vast technical literature on foundation engineering, data on experience and value judgments are few. Some descriptions of actual projects even present a distorted picture of the design decision process, as noted by Casagrande in his 1964 Terzaghi Lecture:

> Unfortunately it is common practice in publications describing projects in applied soil mechanics to present a rationalized picture that has little resemblance to the truth about the actual approach followed in arriving at major design decisions. If judgment based on empirical knowledge played an important role in such decisions, why not admit it? If unknown risks or calculated risks were involved because of the larger gaps in our understanding of the mechanics of soils, why not admit it? Authors who, a posteriori, try to rationalize their decisions or make theories fit the facts, merely reveal their limited grasp of the realities of applied soil mechanics. With distorted presentation of important case records they hinder rather than promote progress; and they mislead their younger and less experienced colleagues.*

One reason for this situation is that value judgment (call it "practice") often is based on empirical procedures (experience) or on gross simplifi-

* A. Casagrande, *Journal of the Soil Mechanics and Foundations Division, Proceedings of the American Society of Civil Engineers*, vol. 91, no. SM4, part 1, July, 1965.

cations which, almost without exception, can be controverted in particular cases, are subject to differences of opinion, and while considered by some as the best practice, are widely contested by others and thus not readily defended.

This reluctance to publicize the true nature of value judgments in one's everyday practice constitutes a prime deficiency in the profession of foundations engineering. Bits and pieces of information are scattered sporadically in the literature. However, the information is not available to the extent that it is in other fields of practice. For example, in steel or concrete design, there are comprehensive, well-publicized, and nationally accepted design standards, such as the Specifications for the Design, Fabrication, and Erection of Structural Steel for Buildings (American Institute of Steel Construction), the work of Committee 318 of the American Concrete Institute, and the design specification for bridges of the American Association of State Highway Officials. As those who use these standards are aware, their provisions are based in part on theory and in part on "good practice," which includes empirical and simplified procedures based on past satisfactory performance.

Perhaps, someday, there will be a nationally accepted design standard for foundations as well as for steel, concrete, and other fabricated materials. The obstacles to development of such a standard are formidable, however, and the primary purpose of this book is to summarize and present current standards of practice to the extent that they exist. There are large gaps in these standards, of course, some of which the authors have attempted to fill. Naturally, the suggested procedures and value judgments contained herein represent the personal views and practices of the authors, for which they anticipate and accept the inevitable controversy.

The theoretical and empirical bases for both the accepted and the suggested standards are presented, where they can be inferred, and the validity of conventional assumptions and approximations is discussed. The original basis for many of these assumptions and approximations is but poorly understood and could stand a good airing.

The book is not intended as a textbook, however. Its purpose is to acquaint architects and engineers in their day-to-day work with the practices and opinions of some of their peers in the common routine aspects of the design of foundations for buildings and to suggest standards for the application of such practice. That such practice must be supplemented by large doses of individual interpretation is axiomatic. In addition, the book delineates potential problem areas and, where such areas occur, directs the reader to other sources or to the soils specialist for assistance.

As background for this work, the authors have enjoyed a rare opportunity to wrestle with the problem of formulating standards for the design and construction of foundations. For a period of nearly four years the firm of Praeger-Kavanagh-Waterbury was engaged in preparing the structural and foundations provisions of a new building code for the City of New York. During the course of this effort, proposed code provisions relating to foundation design and construction were discussed and reviewed by the authors with numerous prominent foundations engineers and specialists in soil mechanics, and with contractors, materials suppliers, and manufacturers, many of them recognized experts in their respective fields.

Many detailed commentaries were prepared, presenting the principles underlying the various proposed provisions and summarizing the various views presented during the many discussion periods. Several reviewers suggested that a record of these commentaries and discussions would be of value to the practicing designer, and this book has been prepared in response to those suggestions. It summarizes much of the pertinent discussion and is supplemented by gleanings on practice found in the technical literature, plus the authors' personal experience in the design of numerous foundation projects.

The authors gratefully acknowledge the many helpful suggestions and the frank discussions and reviews which were made by the several committees, professional associations and societies, and individuals who were engaged in review of the authors' previous work; the authors have drawn freely from these suggestions. Particular acknowledgment is made to the following organizations:

Code Committee of the New York Association of Consulting Engineers
Code Committee of the Metropolitan Section, American Society of Civil Engineers
Code Committee of the New York Chapter, American Institute of Architects
Code Committee of the Architects Council of New York City

and to the following individuals:

Mr. William Eipel, Mr. Gordon Fletcher, Mr. Leon Fraikin, Mr. Sidney Gilbert, Mr. David Greer, Mr. Leo Nalen, Mr. James Parsons, Dr. Paul Rongved, Mr. Reuben Samuels, Mr. Cove Sullivan, Dr. Gregory Tschebotarioff, Mr. Joseph Ward, Mr. Edward White, and many others.

The preparation of the building code above noted was executed under the immediate supervision of Dr. Robert B. B. Moorman, Department of Civil Engineering, Polytechnic Institute of Brooklyn, to whom the authors express their sincere gratitude for his help and patient encouragement.

Sidney M. Johnson
Thomas C. Kavanagh

Contents

The Design of Foundations for Buildings

CHAPTER 1
Introduction

The engineering aspects of building design require knowledge derived from both theory and empiricism. This applies whether the element being designed is the structural frame, the foundation, or the service equipment and utilities.

The designer's theoretical knowledge is provided by his education in the basic sciences as well as in such engineering sciences as mechanics, strength of materials, fluids, etc. His empirical knowledge is derived from experience, which may be his own, the reported or related experiences of others, or reported test or performance data. It also includes judgment, which is the extrapolation of experience to new situations, this extrapolation being based on theoretically or empirically established principles or on related experiences.

The theoretical aspects of design have an essentially fixed, albeit not necessarily scientific, basis and, if applied without error and in accordance with accepted procedures to problems with similar background conditions (assumptions), give certain and consistent answers.

Empiricism is different. Ostensibly an empirical relationship is established by compiling test data and extrapolating these data to other related conditions. In practice, the extrapolation may not be at all clear, simply because the field conditions to which it is desired to apply the empirical data are not precisely related in the way that laboratory conditions can be duplicated. As a result, interpretation and application of empirical data in practice involve a large measure of experience and judgment and become a subjective matter wherein the conclusions drawn depend not only on the facts but on the individual making the conclusions, on his background of experience, and to a surprising extent, on his age, personality, and professional position.

In structural design and in the design of service equipment and facilities, the subjective aspects of the empirical factors have been largely eliminated by the promulgation of design specifications such as the "Specification for the Design, Fabrication, and Erection of Structural Steel for Buildings" (AISC), the "Building Code Requirements for Reinforced Concrete" (ACI 318), and the "American Safety Code for Elevators" (USAS A17.1). These standards are prepared by committees. As those who are familiar with the activities of these committees are aware, the standards contain much that is of an empirical nature. Moreover, they very often do not represent unanimity of opinion by the committee members, but rather a compromise or perhaps a consensus of differing and sometimes diametrically opposite opinions.

Still, having been codified, these opinions become the accepted standard of practice, and designers follow the standard (often unaware of the empiricism or controversy underlying the procedures which they are following). Despite these problems, the system works well. The committees are staffed by knowledgeable persons, the standards are reasonably conservative, the structures (and systems) designed to the standards function satisfactorily, and a condition of order is enjoyed by the design professions.

This situation does not prevail in foundation design. There is no accepted design standard, other than local building codes plus reliance on the aforementioned AISC and ACI standards, for detailing the steel and concrete elements. The problem is compounded by the fact that because of the variable and heterogeneous nature of soil, a purely theoretical approach to foundation engineering is of limited use, and experience is even more important than it is in other facets of building design. The result is that it is very common in foundation work to find that two engineers, both capable and experienced, will have had entirely different experiences with a given type of design, method of construction, or material and will accordingly have entirely different views on its use or may recommend radically different safety factors for a given design or entirely different solutions to a given problem.

Casagrande has written particularly well on this matter in the 1964 Terzaghi Lecture on the "Role of the 'Calculated Risk' in Earthwork and Foundation Engineering."* His frank discussions† of the controversies underlying the reports of several consulting boards and boards of review will provide many readers with their first insight into the functioning of such boards or committees.

* A. Casagrande, *Journal of the Soil Mechanics and Foundations Division, Proceedings of the American Society of Civil Engineers,* vol. 91, no. SM4, July, 1965, part 1.

† See quotation in Preface.

This lack of accepted standards in foundation design is largely the consequence of the enormous number of variables involved in soils engineering. Its effect has been the development of an interesting field of specialization, whose practitioners are keenly enthusiastic and have developed a sort of fraternal organization—which is well and makes for good fun. But for the general practitioner, the engineer or architect in general practice who must design moderately simple foundations as a part of a routine service which does not warrant the cost (or the time) of retaining the services of a specialist (or of making extensive investigations), the lack of standards is a considerable annoyance and a decided risk. He is left almost entirely to his own devices to evaluate and apply the experience and judgment factors necessary for proper design, and he must do so based on what often includes very limited data and often within a very limited time schedule.

As stated in the Preface, the primary purpose of this book is to mitigate this lack of generally accepted standards by presenting, summarizing, and reviewing those standards of practice in foundation design which do exist, and by suggesting additional procedures which may be used to fill some of the many gaps. It is emphasized that these standards and the suggested procedures apply only to the design of routine or moderately simple foundations. This implies that the user of the text can recognize when he is dealing with a foundation problem which is *not* routine or moderately simple—in itself not an easy judgment to make. Accordingly, the authors have indicated a considerable number of conditions which could be troublesome and which must be treated with caution. However, the text cannot hope to describe all such conditions, and so the authors advise a generally cautious approach to foundation design particularly if there is any question as to an unusual circumstance in the proposed work.

The discussions in the text also presume that the reader has had sufficient background in foundation design and construction to be conversant with basic soils engineering and construction techniques and that the building code (if any) under which the designer is working affords some flexibility for the exercise of judgment. It is further assumed, of course, that the designer will operate within the limitations imposed by the building law.

Before proceeding to the technical portions of the text, the reader should consider a few points of basic philosophy and approach on which a number of the recommendations in the text are based:

1. *The Need for Caution.* The need for caution in foundation design has been stated and is repeated. One specific point is emphasized: the tendency to assume that if the loads are light and/or the building is small, the foundation is simple. This approach has some theoretical justification and also the very practical justification in that, for a small

building, so little time and money are available for foundation design that one is forced to treat it simply. Unfortunately, experience seems to indicate that this approach is fallacious and that, in fact, the majority of problems with completed foundations (as measured by the number of instances where remedial underpinning is required) occur with relatively less-important buildings of one and two stories. It is suggested that no foundation design should be treated cavalierly, particularly since it usually is possible to incorporate considerably extra strength into a structure in the design phase at relatively little cost, whereas the cost of repairing or strengthening that same structure because of some deficiency in the foundation is likely to be very high and the results unsightly.

This conservative approach will be evident in the text.

2. *The Need for Inspection.* The text presumes that inspection will be provided for certain key elements of the construction work in progress and for the operations incident to the soils exploration program, as follows:

a. Borings

Inspection or supervision of borings is required, for several reasons. As an absolute minimum, such inspection would assure that the borings are made—it is not unknown for boring logs to be submitted for borings which were never made, or for samples to be presented for locations where samples were never taken.

More important, the discussions and suggestions in the text about allowable soil bearing capacity and pile loads rely heavily on empirical relations between performance of the construction and such data as the standard penetration resistance determined from borings. Therefore it is important that the borings be made using proper techniques so that the results of the penetration test will be meaningful and consistent.

Lastly, the presence of relatively thin (say 2 to 3 ft) layers or lenses of soft materials in an otherwise firm deposit often is the governing factor in the design of a foundation. Detection of these layers is too important a matter to be left to the driller, except of course where local soil conditions are known to be uniform or consistent. While it is generally agreed that the inspection of borings can be carried out adequately by subprofessional personnel, as long as some training and orientation are provided, the proper classification of soil samples requires considerable experience.

b. Subgrade for Footings

The design drawings for the proposed construction (the plans) will show the levels of the bottoms of the footings. In general, these levels are established by the requirement to bear on, or in, a specific type or density of soil material which is capable of developing the design bearing pressure. The levels at which such type or density of material occurs are determined by interpretation and interpolation of the borings. This determination is approximate at best and should be confirmed in the field by visual inspection of the subgrade for each footing after excavation.

c. Piling

Opinions as to the need for and the required degree of inspection of piling differ. It is the authors' opinion that, except for relatively low-capacity piles driven in areas of uniform soil conditions by a known and reliable contractor and under the supervision of an experienced superintendent, such inspection is required, and the discussion in the text on the subject of piling presumes that such inspection will be provided for all operations incident to the installation of piles.

3. *Test Data.* An attempt has been made in developing the text to limit the field or laboratory test data required for the development of the design to those tests which can be conducted with rudimentary equipment. In general, these include visual inspection, sieve analysis, Atterberg limits, moisture content, and the standard penetration test. The text presumes that the reader is familiar with these tests and with the significance of the results derived therefrom. If the reader is not familiar therewith, he is referred to the excellent texts by Tschebotarioff [Chap. 5, Ref. 17], Terzaghi and Peck [Chap. 5, Ref. 5], or other standard works on soil mechanics.

In some cases, more sophisticated test data are required. Where such is the case, the required tests and data are indicated. The methods and procedures and equipment incident to conducting these tests are not described. It is assumed that the tests will be performed by one of the many reliable laboratories specializing in this service and that it is sufficient merely to designate the test to be performed in terms which the laboratory will understand.

4. *Coordination with Standards for Design of the Superstructure.* The authors note the current tendency for the development of soil mechanics and foundations engineering to progress as a separate field of specialization, independent of the practice of structural design. In point of fact, however, the two fields of specialization are based on the same physical laws and basic engineering sciences, and a basic concept of this text is that foundation engineering and structural engineering are tandem fields of specialization, that what is good practice or a proper approach in one probably is equally applicable in the other, and that practice, approach, and recent developments in structural engineering may be considered to constitute guidance for a similar practice and approach in foundation design.

5. *The Need for Experience.* Lastly, the text presumes that the reader has had some experience and/or education in foundation design.

As a matter of advisement, he should make every effort to broaden this experience. Reading of the technical journals is one way. Field observations and laboratory work are even better. This does not necessarily imply a lot of time-consuming or sophisticated effort. For example, some time spent in observing borings in progress or just the simple expedient of inspecting the sides of newly opened cuts and excavations, wherever opportunity offers, will be valuable.

Perhaps most valuable of all for the relative novice in soils work would be some practice in the visual and manual methods of soil identification and some familiarity with the tentative criteria for performance and properties of soils. These are described by Burmister [Chap. 2, Ref. 7], and his paper is recommended for study.

CHAPTER 2
Soil Exploration

The first of the problems in design of building foundations for which theoretical guidance is lacking, and the designer must rely primarily on experience and judgment is that of establishing the type, number, location, and depth of explorations required to determine the soil conditions at the site.

There are no accepted standards. Except where the pertinent requirements are established by code, the required decisions are entirely predicated on the designer's judgment, which in turn must consider that too detailed an exploration program imposes a penalty of excess cost and delay while an insufficient program may fail to provide the information necessary for a safe and adequate design.

This chapter is intended to provide criteria for guidance in selecting the number, depth, locations, and methods of soil explorations to provide the information required for design of the foundations for types of buildings which are encountered in general practice.

Section 2A METHODS AND TECHNIQUES

The principal data sought in any program of soils exploration for a proposed building foundation are:

1. *The Levels at Which the Different Strata Occur.* This is determined from the penetration resistance, from the samples recovered, or from the wash from the boring operation.

2. *The Classification of the Soil or Rock Strata Encountered* (i.e., sand, silt, clay, type of rock, etc.). Classification is made by inspection of

Laboratory Classification Criteria		
6		
Use grain-size curve in identifying the fractions as given under field identification. Determine percentages of gravel and sand from grain-size curve. Depending on percentage of fines (fraction smaller than No. 200 sieve size) coarse-grained soils are classified as follows: Less than 5% GW, GP, SW, SP; More than 12% GM, GC, SM, SC; 5% to 12% Borderline cases	$C_u = \frac{D_{60}}{D_{10}}$ Greater than 4 $C_c = \frac{(D_{30})^2}{D_{10} \times D_{60}}$ Between 1 and 3	
	Not meeting all gradation requirements for GW	
	Atterberg limits below "A" line or PI less than 4	Above "A" line with PI between 4 and 7 are borderline cases.
	Atterberg limits above "A" line with PI greater than 7	
	$C_u = \frac{D_{60}}{D_{10}}$ Greater than 6 $C_c = \frac{(D_{30})^2}{D_{10} \times D_{60}}$ Between 1 and 3	
	Not meeting all gradation requirements for SW	
	Atterberg limits above "A" line or PI less than 4	Limits plotting in hatched zone with PI between 4 and 7 are borderline cases.
	Atterberg limits above "A" line with PI greater than 7	

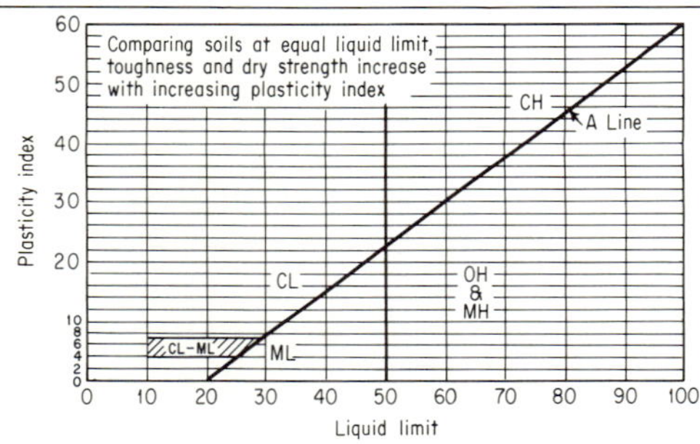

PLASTICITY CHART
For laboratory classification of fine-grained soils

FIG. 2-1 The Unified Soil Classification System. (Adopted by

Note 1 to Figure 2-1—Grain Size Classifications for Coarse-grained Soils

Gravel: (#4 sieve to 3″ size)
Coarse—¾″ to 3″
Fine—#4 sieve to ¾″

Sand: (#200 sieve to #4 sieve)
Coarse—#10 sieve to #4 sieve
Medium—#40 sieve to #10 sieve
Fine—#200 sieve to #40 sieve

Note 2 to Figure 2-1—Identification Procedures for Fine-grained Soils or Fractions. These procedures are to be performed on the minus No. 40 sieve size particles, approximately 1⁄64 in.

Dilatancy (reaction to shaking)

After removing particles larger than No. 40 sieve size, prepare a pat of moist soil with a volume of about one-half cu in. Add enough water if necessary to make the soil soft but not sticky.

Place the pat in the open palm of one hand and shake horizontally, striking vigorously against the other hand several times. A positive reaction consists of the appearance of water on the surface of the pat which changes to a livery consistency and becomes glossy. When the sample is squeezed between the fingers, the water and gloss disappear from the surface, the pat stiffens, and finally it cracks or crumbles. The rapidity of appearance of water during shaking and of its disappearance during squeezing assist in identifying the character of the fines in a soil.

Very fine clean sands give the quickest and most distinct reaction, whereas a plastic clay has no reaction. Inorganic silts, such as a typical rock flour, show a moderately quick reaction.

Dry Strength (crushing characteristics)

After removing particles larger than No. 40 sieve size, mold a pat of soil to the

Major Divisions			Group Symbols	Typical Names	Identification Procedures (excluding particles larger than 3 in. and basing fractions on estimated weights)		
1	2		3	4	5		
Coarse-grained Soils. See Note 1. More than half of material is larger than No. 200 sieve size.	Gravels. More than half of coarse fraction is larger than No. 4 sieve size.	Clean Gravels (Little or no fines)	GW	Well-graded gravels, gravel-sand mixtures, little or no fines	Wide range in grain sizes and substantial amounts of all intermediate particle sizes		
			GP	Poorly graded gravels or gravel-sand mixtures, little or no fines	Predominantly one size or range of sizes with some intermediate sizes missing		
		Gravels with Fines (Appreciable amount of fines)	GM	Silty gravels, gravel-sand-silt mixture	Nonplastic fines or fines with low plasticity (for identification procedures see ML below)		
			GC	Clayey gravels, gravel-sand-clay mixtures	Plastic fines (for identification procedures see CL below)		
	Sands. More than half of coarse fraction is smaller than No. 4 sieve size. (For visual classification, the 1/4-in. size may be used as equivalent to the No. 4 sieve size)	Clean Sands (Little or no fines)	SW	Well-graded sands, gravelly sands, little or no fines	Wide range in grain size and substantial amounts of all intermediate particle sizes		
			SP	Poorly graded sands or gravelly sands, little or no fines	Predominantly one size or a range of sizes with some intermediate sizes missing		
		Sands with Fines (Appreciable amount of fines)	SM	Silty sands, sand-silt mixtures	Nonplastic fines or fines with low plasticity (for identification procedures see ML below)		
			SC	Clayey sands, sand-clay mixtures	Plastic fines (for identification procedures see CL below)		
The No. 200 sieve size is about the smallest particle visible to the naked eye.					Identification Procedures on Fraction Smaller than No. 40 Sieve Size		
Fine-grained Soils. See Note 2. More than half of material is smaller than No. 200 sieve size.					Dry Strength (crushing characteristics)	Dilatancy (reaction to shaking)	Toughness (consistency near PL)
	Silts and Clays	Liquid limit is less than 50	ML	Inorganic silts and very fine sands, rock flour, silty or clayey fine sands or clayey silts with slight plasticity	None to slight	Quick to slow	None
			CL	Inorganic clays of low to medium plasticity, gravelly clays, sandy clays, silty clays, lean clays	Medium to high	None to very slow	Medium
			OL	Organic silts and organic silty clays of low plasticity	Slight to medium	Slow	Slight
	Silts and Clays	Liquid limit is greater than 50	MH	Inorganic silts, micaceous or diatomaceous fine sandy or silty soils, elastic silts	Slight to medium	Slow to none	Slight to medium
			CH	Inorganic clays of high plasticity, fat clays	High to very high	None	High
			OH	Organic clays of medium to high plasticity, organic silts	Medium to high	None to very slow	Slight to medium
Highly Organic Soils			Pt	Peat and other highly organic soils	Readily identified by color, odor, spongy feel, and frequently by fibrous texture		

Corps of Engineers and Bureau of Reclamation, January, 1952.)

consistency of putty, adding water if necessary. Allow the pat to dry completely by oven, sun, or air drying, and then test its strength by breaking and crumbling between the fingers. This strength is a measure of the character and quantity of the colloidal fraction contained in the soil. The dry strength increases with increasing plasticity.

High dry strength is characteristic for clays of the CH group. A typical inorganic silt possesses only very slight dry strength. Silty fine sands and silts have about the same slight dry strength, but can be distinguished by the feel when powdering the dried specimen. Fine sand feels gritty, whereas a typical silt has the smooth feel of flour.

Toughness (consistency near plastic limit)

After particles larger than the No. 40 sieve size are removed, a specimen of soil about one-half inch cube in size, is molded to the consistency of putty. If too dry, water must be added and if sticky, the specimen should be spread out in a thin layer and allowed to lose some moisture by evaporation. Then the specimen is rolled out by hand on a smooth surface or between the palms into a thread about one-eighth in. in diam. The thread is then folded and rerolled repeatedly. During this manipulation the moisture content is gradually reduced and the specimen stiffens, finally loses its plasticity, and crumbles when the plastic limit is reached.

After the thread crumbles, the pieces should be lumped together and a slight kneading action continued until the lump crumbles.

The tougher the thread near the plastic limit and the stiffer the lump when it finally crumbles, the more potent is the colloidal clay fraction in the soil. Weakness of the thread at the plastic limit and quick loss of coherence of the lump below the plastic limit indicate either inorganic clay of low plasticity or materials such as kaolin-type clays and organic clays which occur below the A-line.

Highly organic clays have a very weak and spongy feel at the plastic limit.

the samples recovered. In most cases, visual examination will suffice. In more difficult instances, some simple laboratory procedures such as sieve analysis may be required. The Unified Soil Classification System, shown in Fig. 2-1, will be the standard of soil classification used in this text. Rock will be identified in terms of its geologic classification.

3. *The In-place Density or, for Rock, the Soundness of the Material.* The density of soil strata is estimated by field tests conducted on the material, in place. The standard penetration test (SPT) is the usual criterion employed with regard to soil strata. The soundness of rock strata is estimated from the percentage of core recovery.

With regard to exploration technique, there are many methods available.* In fact, subsurface exploration has become a small science in itself. Of these many techniques, however, relatively few are in common use for the exploration of building sites and only these need be of concern

* See Ref. 1 for a detailed discussion of methods and techniques.

TABLE 2-1 ***Methods of Subsurface Exploration Commonly Used to Explore Building Sites***

Purpose	Method	Description
Soil sampling	Wash-type boring with dry samples	Hole advanced by chopping and jetting, with or without casing. Samples recovered with split-spoon or tube samplers
	Auger boring	Hole advanced by use of hand or power auger. Samples recovered by periodic withdrawal of auger or, for continuous flight auger, single withdrawal
	Test pit	Samples are recovered from open pits
Rock sampling	Core drilling	Used to recover a core sample of material being explored for direct examination
	Rotary or percussion drilling	Samples recovered consist of chips or a water slurry of material being sampled
Define contours of top of hard material (or of bearing stratum)	Probing	Executed by use of augers or by driving probe rods
	Geophysical methods	Seismic methods most common. (Based on time of travel, from source to pickup, of reflected or refracted seismic waves)

to building designers. These few are listed in Table 2-1, and the following discussion will be limited to the methods listed there.

2A.1 Soil Sampling

a. Wash-type Borings

Wash-type borings are the most common method of soils exploration. They are applicable for use in virtually any type of soil (other than rock) and may be used for the recovery of samples suitable both for classification

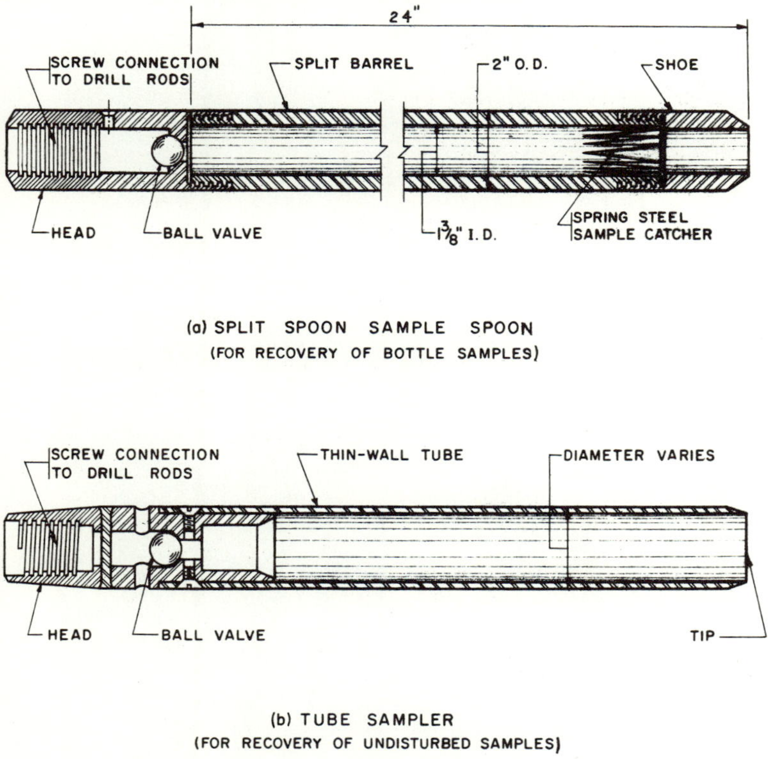

FIG. 2-2 Common Types of Soil Samplers.

and for laboratory test. In this technique, the bore hole is excavated by a combination of chopping with the wash pipe and jetting. Where required, the hole is kept open by means of a casing or a slurry of drilling mud (a thixotropic volcanic clay in water).

Samples are obtained by driving or pressing a sampling tool into the soil, withdrawing the tool, and recovering the sample, which is held in place in the sampler by the effects of wedging, friction, and vacuum. For samples which are difficult to hold (such as loose sands) special mechanical devices are available. Several of the boring and sampling tools used with wash-type borings are illustrated in Figs. 2-2 and 2-3.

There are many techniques and refinements for making wash-type borings, and a number of different types and sizes of sampler may be used. Of these latter, two are standard, the split spoon and the Shelby

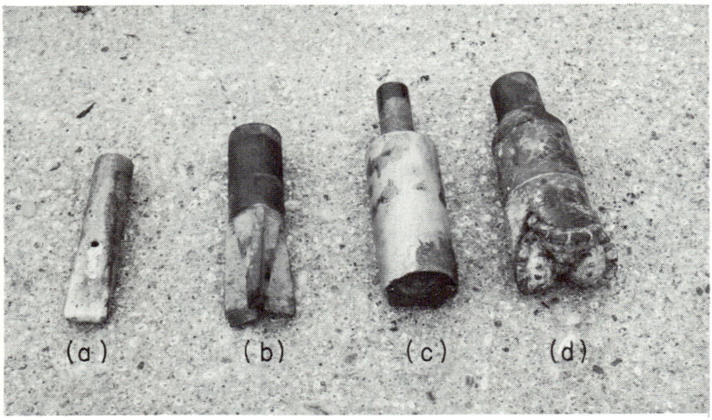

(*a*)

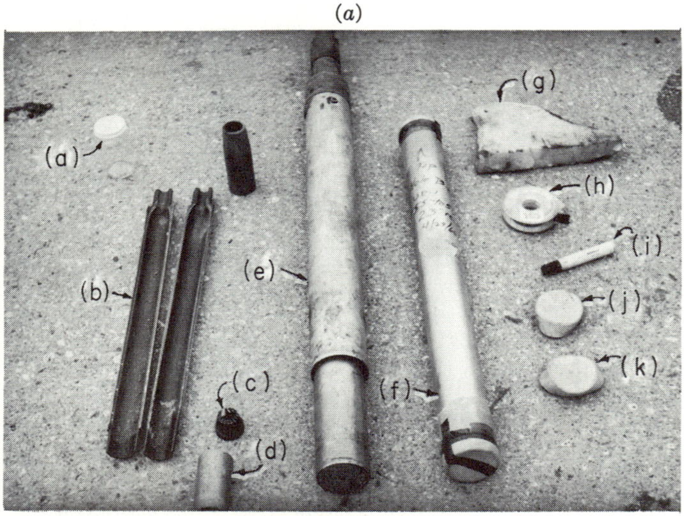

(*b*)

FIG. 2-3 Boring and Sampling Tools. (*a*) (*a*) Washing and chopping bit for use in soil. (*b*) Chopping bit for use in soil. (*c*) Cleaning bit for cleaning soil from bottom of hole. (*d*) Rotary rock bit. (*b*) (*a*) Jar for bottle samples. (*b*) 2-in.-diam standard spoon sampler. (*c*) Spring-type sample catcher. (*d*) Point for spoon sampler. (*e*) Piston sampler for undisturbed samples. (*f*) Undisturbed sample tube ready for shipment to laboratory. (*g*) Paraffin for sealing sample tube. (*h*) Tape for sealing tube. (*i*) Marker. (*j*) and (*k*) Caps for ends of sample tube. (*c*) (*a*) Washing out casing. (*b*) 300-lb weight for driving casing. (*c*) 140-lb weight for standard penetration test.

tube, and discussion will be limited to the uses and limitations of these two. A suggested specification for the making of wash-type borings is contained in Sec. 2F, Chap. Note 1.

1. *The Split-spoon Sampler.* The 2-in.-OD, $1\frac{3}{8}$-in.-ID, 24-in.-long split spoon (Fig. 2-2 and piece b in Fig. 2-3*b*) is the sampler most commonly used with wash borings. When driven with a 140-lb weight falling 30 in., the blows per foot required to drive the spoon are a measure of the standard penetration resistance which is designated by the symbol N. The operation itself is known as the standard penetration test.

Properly executed, wash-type borings provide samples which are of adequate quality to permit classification of the soil strata and the technique itself permits determination of the boundaries of the soil strata within an accuracy of a few inches. In addition, when used with the standard penetration test, such borings will provide information as to soil properties which is sufficient for the design of the foundations for most buildings. The reason is that several soil-strength parameters have been related, empirically, to the N value so that the bearing capacity of the

(*c*)

FIG. 2-3 *(Continued)*

soil material can be estimated without recourse to laboratory test data. (Some of these relations are shown in Figs. 2-4, 2-5, and 5-9.)

A. *Execution of the Standard Penetration Test.* The principle of the standard penetration test, as described in the previous paragraph and in the suggested specification (see Sec. 2F, Chap. Note 1), is exceedingly

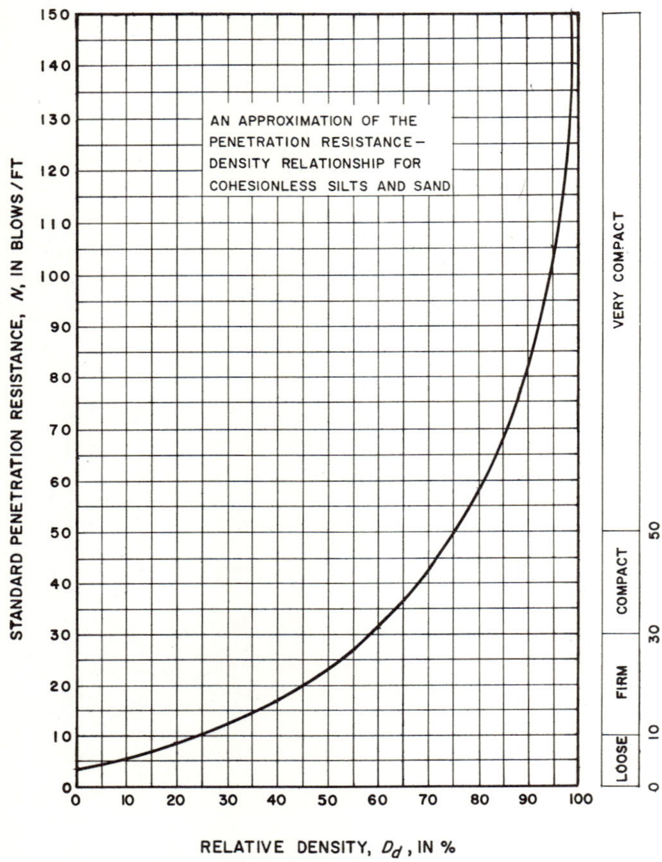

FIG. 2-4 Standard Penetration Resistance versus Relative Density—Granular Soil. *(From B. K. Hough, "Basic Soils Engineering," The Ronald Press Company, New York, 1957.)*

simple, in fact deceptively so, because execution of the test actually requires considerable care. The results are quite sensitive to variations in technique.

Gordon Fletcher, one of the pioneers in the development of this test, has indicated [3] that the following variables and points of procedure are the most common sources of error.

(1) *Fall of the Weight.* The intent of the test is for the sampler to be

driven with a 140-lb weight having a *free* fall of 30 in. The weight itself presents no problem because it usually weighs quite close to 140 lb and can be checked and adjusted if doubt exists. The exact height of fall, however, depends on the driller who may have had a bad night or a heavy lunch so that the drop of the weight becomes short or erratic. Also, if the driller takes too many wraps of the hoist rope around the drum of the winch so that the rope tends to drag, or if he otherwise fails to release the

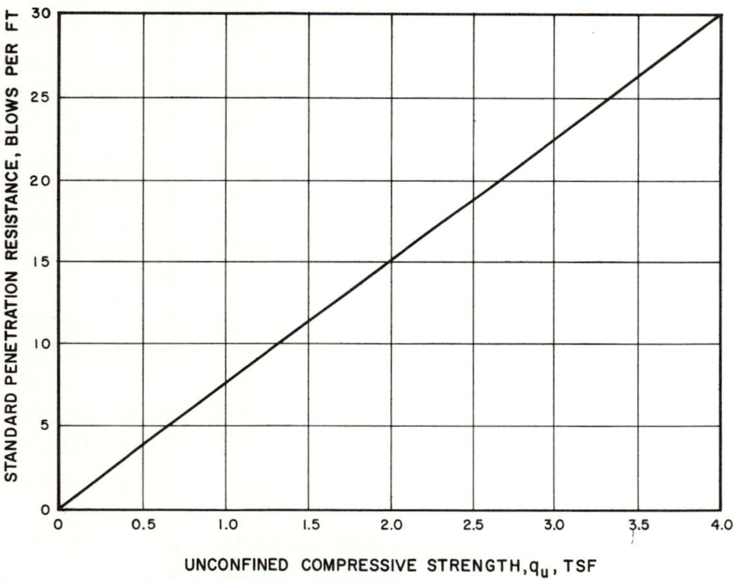

FIG. 2-5 Standard Penetration Resistance versus Unconfined Compressive Strength for Clay Soils. (*After Terzaghi and Peck, Ref. 5.*)

tension on the rope, the blow is rendered only partly effective, and the apparent penetration resistance is deceptively increased. Any other circumstance interfering with the free fall of the weight, or preventing the weight from striking concentrically on the drive cap (such as tilt of the guide for the weight, or failure to keep the striking surface of the drive cap horizontal), will have a similar effect.

Automatic tools intended to obviate the effects of these human variables are available. The Pilcon Automatic Trip Monkey, developed in Great Britain, for example, is described by Stubbings in Ref. 8. The reader is cautioned, however, to satisfy himself as to the actual operation of such devices, with regard to their efficiency and the degree of accuracy with which the drop actually can be controlled, before using same.

(2) *Depth of the Hole.* As the bore hole is deepened, the length and weight of the drill rods connecting the sampler to the drive head increases.

As this weight (mass) increases, the inertial resistance correspondingly increases, decreasing the effective energy of the blow delivered to the sampler. Similarly, the lateral "spring" of the long, slender rods under the force of the hammer blow will absorb an additional portion of the impact energy. The result is that the apparent penetration resistance of a deep-seated stratum is increased compared with the apparent resistance of an identical stratum at or near the surface. Fletcher indicates that, based on experience, this distortion of the apparent penetration resistance is not important to depths of 125 to 140 ft, but that at depths of 200 ft the results are significantly too high and unreliable. Yang, discussing the explorations for the caissons for the Narrows Bridge [10], describes a condition where the borings indicated the presence of a very dense stratum (N = 85 to 320 blows per foot) at a depth of 160 to 240 ft which, nevertheless, was relatively easily penetrated by large (30-in.-OD) open-end pipe piles. The discrepancy between apparent and actual penetration resistance was attributed to the influence of the weight and flexibility of the drill rods.

This variation of apparent resistance with variations in weight of the drill rods is the reason that standardization of size of drill rod is important. One-in. extra-heavy pipe or A rods conventionally are used to connect the sampler to the drive cap. The use of substantially heavier rods causes an apparent increase in penetration resistance due to the inertial effect. The use of substantially lighter rods would increase the energy losses due to whip.

(3) *Cleaning the Hole.* There is a tendency for sludge to collect at the bottom of the drill hole in the interval between cleaning the hole and driving of the sampler (see Fig. 2-6). The sludge accumulates largely from material settling out of suspension in the column of water in the drill hole, although some inflow of material from the bottom of the hole may occur. For an uncased hole, there is settlement of the drilling slurry, or there may be some slumping of the sides. When the sampler is driven through such material, the top section of the sample recovered is sludge and not *in situ* material, and the penetration resistance offered by the sludge is not representative of the density of the parent soil deposit.

In general, it is impractical to prevent the accumulation of sludge entirely, and precautions must be observed to detect and compensate for its presence, as follows:

1. The hole should be cleaned to the bottom, and the sampler should be lowered and driven as quickly thereafter as possible.

2. The presence and depth of the accumulation of sludge should be checked by comparing the length of wash pipe entering the hole with the length of the sampler plus drill rods lowered before sampling.

3. If an uncased hole and a slurry are not used, maintaining the over-

flow of wash water from the hole until it flows clear will help to reduce the amount of sludge which settles out.

4. If all else fails, and the sludge deposit exceeds, say, 6 in. in depth, an extra-long sampling spoon having a sludge chamber should be used.

5. Lastly, the blows on the sampler should be recorded for 6-in. intervals, and the blows for the first 6 in. of penetration should be disregarded.

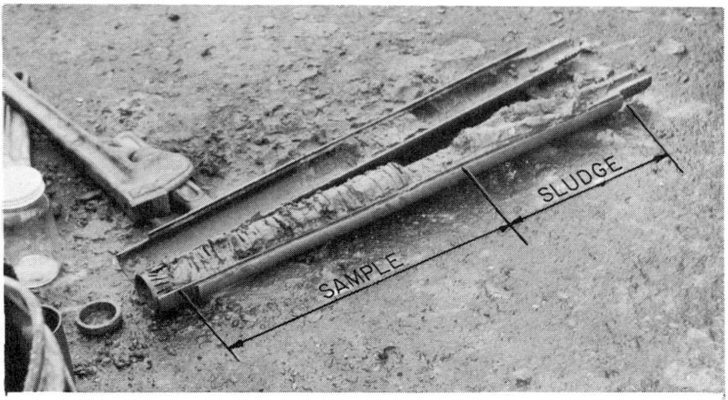

(*a*)

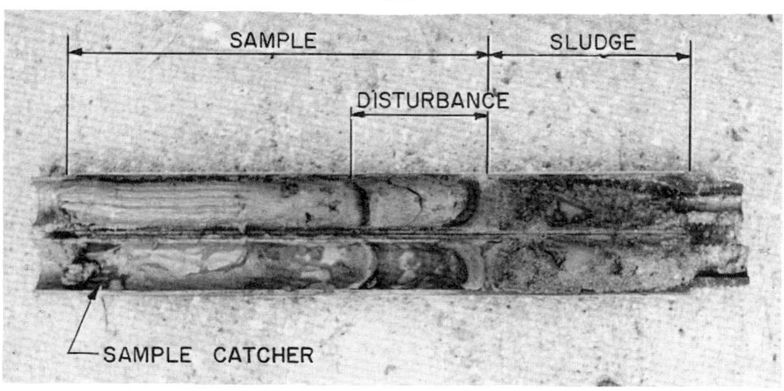

(*b*)

FIG. 2-6 Spoon Sample. (*a*) and (*b*) Spoon samples obtained using 2-in. OD standard spoon sampler. Note accumulation of sludge at top of samples and evidence of disturbance of soil layers.

It is suggested that the *N* value used for design purposes should be the sum of the two resistances for the second and third 6-in. increments (see Sec. 2F, Chap. Note 4).

A final important point is to be certain that the hole is cleaned to the bottom of the casing. Otherwise, the sampler may tend to wedge in the soil confined by the casing, giving excessively high apparent resistance.

(4) *Inflow of Material.* In making borings through granular soil below groundwater level, there is a tendency for the soil to flow into and up the

casing if the hydrostatic pressure in the soil is greater than the hydrostatic pressure of the water (or slurry) column in the casing. If this occurs, the structure of the soil being sampled is disturbed, and the penetration test values will be deceptively low. To prevent this occurrence, the level of the water column in the casing must be kept at or above the groundwater level. Where the soil at the bottom of the casing is under an artesian head, the use of a mud slurry may be required to provide additional weight.

(5) *Size of Hole.* The resistance of the soil to the penetration test will be affected by the confining pressure provided by the surrounding soil. If the test is made in a large-bore hole, some of the confining effect is lost (or there may be a tendency for material to run into the bottom of the hole). In either event, the apparent penetration resistance will be too low. Conventionally, the standard test is made in a 2½-in.-diam hole. Fletcher [3] has suggested that the test results will not be affected by the size of the hole if the soil being sampled is a granular material above the water table or a cohesive soil of adequate stiffness to preclude its being squeezed into the hole. However, as a precaution he also suggests that if the hole diameter exceeds 4 in., a casing of 4 in. or lesser diameter should be inserted, and the sampling operation conducted through the casing. The authors concur in this suggestion but would add that the confining effects should be simulated by driving and cleaning out the temporary casing for a distance below the bottom of the bore hole, a distance not less than twice the difference in diameter between the bore hole and the temporary casing.

(6) *Presence of Coarse or Very Fine Material.* The standard sampling spoon has a relatively small cross section (2-in. OD, 1⅜-in. ID). Accordingly, a relatively small piece of gravel or other coarse or cemented material can block the opening, inhibit penetration of the sampler, and drastically increase the apparent penetration resistance. Results of the test, if conducted in such materials, should be discounted.

Similarly, in very fine sands and silts, a deceptively high value of penetration resistance may be simulated at and just above the phreatic surface (water table) by the negative pore water pressures present at that location or by the dashpot effect wherein the penetration of the sampler is resisted by pressure build-up of pore water which cannot readily flow away from the point of impact. Either effect tends to produce a false indication of a dense layer which is, in fact, nonexistent.

At the opposite extreme, fine sands and silts are readily disturbed by the flow of wash water from the boring or may be rendered "quick" by the vibration incident to the test.

A stiff clay, on the other hand, may show a high blow count but have a low shear strength due to the presence of slickenslided surfaces (old failure scars).

All these factors must be considered when interpreting the boring logs. In particular, the SPT is a poor criterion of strength and density when used in soils containing gravel.

(7) *Careless Blow Count.* It is, of course, axiomatic that the blows of the drop hammer must be accurately counted. Occasionally, this is not done, and detection of such occurrence is one of the many reasons that inspection of borings was suggested in Chap. 1 as being of great importance.

To these several points the authors would add the following:

(8) *Condition of Tools.* The sampling spoon should be sharp and straight. If the tip dulls or the cutting edge is peened over, it should be sharpened or replaced. Occasionally, the spoon will be bent due to hard driving. It should be replaced.

(9) *Dropping the Drill Rods.* The sampler and drill rods conventionally are allowed to slide into the hole, the drop being controlled by the driller who grips the rods with a pipe wrench or similar tool which is alternately slacked or given purchase on the drill rods by levering the tool. Very often, under this arrangement, the sampler and rods land on the bottom of the hole with considerable impact, in effect wedging a length of sample into the spoon before the spoon is driven.

The spoon and rods should be gently "bottomed" in the hole with as little impact as possible.

(10) *Overwashing.* When cleaning the hole, the driller may tend to wash below the lower end of the casing (or this may occur as a means of reducing the driving resistance on the casing). This loosens the soil below the casing, i.e., at the sampling level, giving a deceptively low blow count.

II. *The Thin-wall Shelby Tube Sampler.* (See Figs. 2-2 and 2-3*b*.) This is a thin-wall, seamless tube sampler which is pushed or driven into the soil. Tubes in general use range in size from 2 to 4½ in. OD, with a length of 30 in. The 3-in.-tube is suggested for use because it provides (after trimming) a convenient specimen to fit the 2½-in.-consolidation test ring, and the tube fits down a conventional 4-in. drive casing. A 2-in.-diam tube sampler can be used but does not yield as satisfactory a test specimen as the larger sampler.

Except where the strength and consolidation properties of the local strata are already known from previous test data, thin-wall tube samples should be recovered whenever cohesive soil strata (silts or clays) are encountered in the boring exploration program. Such samples are needed for test in the laboratory to determine the strength and/or consolidation properties of the material. They should represent, as nearly as possible, the undisturbed condition of the soil, in place.

Recovering a good undisturbed sample suitable for test and estimating the probable effects of the inevitable disturbance which does occur during sampling are sufficiently difficult problems that if such samples are required, a reliable soils specialist or a testing laboratory should be retained to supervise or to actually make the borings and to conduct the tests. There are many firms offering this service.

The recovery of tube samples should be supplemented by recovery of companion dry samples using the standard penetration test. This costs little and provides a valuable correlation between the N value and the soil properties as indicated by laboratory test.

There is one small problem in connection with recovering tube samples. A 4-in.-diam cased hole is advisable instead of the more usual $2\frac{1}{2}$-in.-diam bore. A 4-in.-diam hole is more expensive than a $2\frac{1}{2}$-in.-diam hole and so is not to be used unless necessary. Therefore, it becomes desirable to know beforehand that undisturbed samples will be required; i.e., some precognition of the probable soil conditions is needed. This matter is discussed in greater detail in Sec. 2B, but basically it means that the boring program should be carried out in two or more stages, unless soil conditions in the area are generally well known.

A suggested specification for the recovery of undisturbed tube samples is given in Sec. 2F, Chap. Note 2.

b. Auger Borings

Auger borings also are in common use. Potentially they are an economical alternate to the use of wash-type borings because, where the soil is sufficiently open to permit penetration of the auger, the work progresses more rapidly. However, there is the problem of recovery of samples. The material which comes up on the flights of the auger is so greatly disturbed that it provides little evidence of in-place density or strength and, accordingly, is virtually worthless for providing the data necessary to determine bearing capacity and allowable pile loads.

Accordingly, where auger borings are used, the samples should be recovered using the standard penetration test and not by taking samples from the flights. When making the penetration test, if the diameter of the hole exceeds 4 in., a temporary drop-in casing should be utilized, as discussed in Sec. 2A.1*a*. In wet material, full hydrostatic head must be maintained in the hole, usually by the use of drilling mud.

The costs of these items (withdrawing the auger, drop-in casing, and drilling mud) usually offset the potential economies in the use of auger borings, which, accordingly, are most frequently used for reconnaissance explorations or for probings rather than for detailed foundation explorations.

c. Test Pits

Under some circumstances (chiefly where shallow exploration is required in dry ground above the water table as, for example, where rock is at shallow depth), test pits also are an economical alternate to the use of wash borings. They provide excellent data on subsurface conditions within the depth to which the pit is penetrated because a large section of ground is opened for inspection and because it is possible to inspect the samples in place.

The in-place density of the soil is evaluated by use of the usual cone or balloon field-density tests (see ASTM D 1556 or D 2167) in lieu of the standard penetration test.

The samples recovered from a test pit usually are "bottle" or dry samples as described in Sec. 2F, Chap. Note 1. Such samples can be recovered directly from the workman's shovel or off the bucket of the shovel or crane. However, high-quality undisturbed samples for use in the consolidation or unconfined compression test also can be recovered from test pits. An empty beer can makes an excellent sampler which is easily sealed, is just the right diameter to fit the 2½-in. consolidation test ring, has a height/diameter ratio suitable for the unconfined compression test, is readily available, and provides certain potential fringe benefits. The sample is introduced into the can by pressing and trimming a block of the soil exposed in the sides or bottom of the pit.

2A.2 Rock Sampling

As previously indicated, the basic purposes of drilling the rock at a site are (1) to ascertain at what level the rock occurs, (2) to determine its classification, i.e., the kind of rock (in geological terms), and (3) to determine its soundness (as regards cracks, seams, and degree of weathering). The classification is based on visual examination of the samples which are recovered, and the soundness is estimated from the percentage of core recovery.

a. Core Borings

In order to recover a core, drilling is required. This is conventionally accomplished by use of a diamond-bit core barrel sampler which is rotated and ground into the rock, eroding an annular hole, and leaving a pinnacle of rock which is snapped off and retrieved. (Photograph of such a core is shown in Fig. 2-7.)

The size of the core barrel sampler should be such as to provide at least a $1\frac{3}{8}$-in.-diam core (the same as the dry samples from the overlying soil), and, except in circumstances described in Sec. 2A.2*c*, it is important where weathered, soft, or fractured rock is encountered to specify the use of a twin-tube core barrel sampler in order to get the maximum recovery

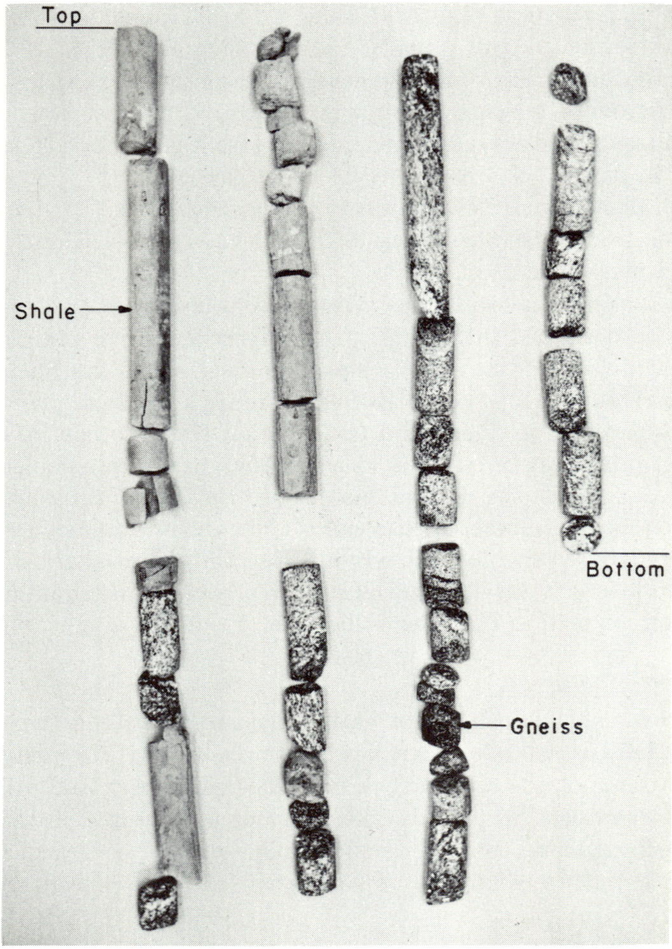

FIG. 2-7 Photograph of Rock Core.

and to minimize breaking of the core. The importance of obtaining maximum recovery is simply that of obtaining as complete and representative a sample as possible. With other types of core samplers, the breakup of the core tends to mislead as to the solidity of the stratum.

A suggested specification for core borings in rock is given in Sec. 2F, Chap. Note 3.

b. Rotary and Churn Drilling

Rotary (and occasionally churn) drilling has been used for the exploration of rock at building sites and has an advantage with regard to cost. The bore hole is advanced more economically than when using a core sampler. However, except for cases where the local rock strata are unusually uniform, since no core is recovered, it is difficult to evaluate the condition or the type of material being drilled with the accuracy needed to evaluate its load-bearing capacity. This defeats the primary purpose of exploring the rock. Further, for the depth of rock drilling usually required for the exploration of a building site (5 to 10 ft), the cost saving is not large. For these reasons, except as described in Sec. 2A.2*c*, or in cases where some circumstance of uniformity provides adequate precognition of the condition of the rock so that only its surface elevation need be determined, diamond-bit core drilling is recommended in preference to rotary or churn drilling.

c. Cases of Lightly Loaded Foundations

The preceding discussion (Secs. 2A.2*a* and *b*) relates to the type of exploration required to provide design data in cases where the rock is to be loaded to a substantial fraction of its bearing capacity. There are many cases where the foundation load is relatively so light that, for nominal-size foundations, any kind or condition of rock, or anything resembling rock, would be adequate to support the building. For such cases, it is necessary only to determine that hard material exists, and subject to the limitation described in Sec. 2C.2 relating to the need to differentiate rock from boulders, it makes little difference in design whether the rock is drilled or not, or, if it is drilled, how it is done. Accordingly, that which is cheapest is best, and the cheapest usually is rotary or churn drilling or else not to drill at all but to use probings or geophysical methods, as described in Sec. 2A.3.

2A.3 Probings and Geophysical Explorations

Probing (sometimes called sounding) consists of forcing or driving a rod, pipe, or similar device into the ground and attempting to correlate the resistance to penetration to the type or properties of soil encountered. Geophysical exploration operates similarly except that it is wave velocity (or change of velocity), electrical resistivity, gravity, or some other physical variable for which attempt is made to correlate to the type or properties of the soil. Much has been claimed for these methods of subsurface exploration, and much has been done with them. Nevertheless, the accuracy of the data which they provide is a matter of debate, and the following points of caution apply.

a. Stratification of Deposits

The techniques of probing and geophysical exploration are best limited in application to cases where it is desired to determine the contours and/or limits of some abrupt change in stratification, such as defining the top of rock or the level of firm material (say, sand) underlying a deposit of soft clay. Note: Do not attempt to use seismic refraction techniques to determine the presence of a soft deposit underlying a firm one. Also, unless the foundation loads are so light that accurate definition of strata is not important, these types of exploration are not recommended for use in cases where the strata grade into one another, for example, to define top of rock where the sound material is overlain by disintegrated rock, a boulder stratum, or a till.

b. Need for Check Borings

Two cases occur:

CASE 1 Where the loads are heavy or a consolidation problem exists, accurate definition of the contours of the bearing stratum or of the depth of compressible material is required.

In such case, if the proposed foundation is to consist of footings, piers, or walls bearing on the soil, inasmuch as the bearing stratum will be uncovered during their construction, the purpose of check borings would be only to assure that a softer stratum does not underlie the bearing stratum or that the bearing stratum does not contain pockets of softer material. For this purpose perhaps one-third the number of borings required under normal circumstances (see Sec. 2B.1*a*) would suffice as check borings with an even fewer number required where conditions are reliably known to be uniform.

On the other hand, if the foundation consists of piling, then the bearing stratum will not be uncovered during construction, and in addition to questions relating to inclusions or underlayments of softer materials, for the reasons discussed in Chap. 6, it is important (particularly for high-capacity piles bearing on rock) to define the contours of the top of the bearing stratum with accuracy. Therefore, the number of check borings should be increased to at least one-half the number of borings normally (Sec. 2B.1*a*) required. Also, particular attention should be given that the check borings confirm that the top of the bearing stratum does, indeed, occur as an *abrupt unconformity,* as assumed. If not, it is best not to rely on probings or geophysical methods at all but to carry out a conventional exploration program.

CASE 2 Where the loads are so light, any kind of firm material would be of adequate capacity to support the building.

For this case, check borings are not important in defining the top of the bearing stratum, but the problem of inclusions or underlayments of softer strata is just as critical. Accordingly, the recommended number of check borings is one-third the normal requirement. It is not necessary that they be carried as deeply as would be required for a more heavily loaded foundation (see Sec. 2C).

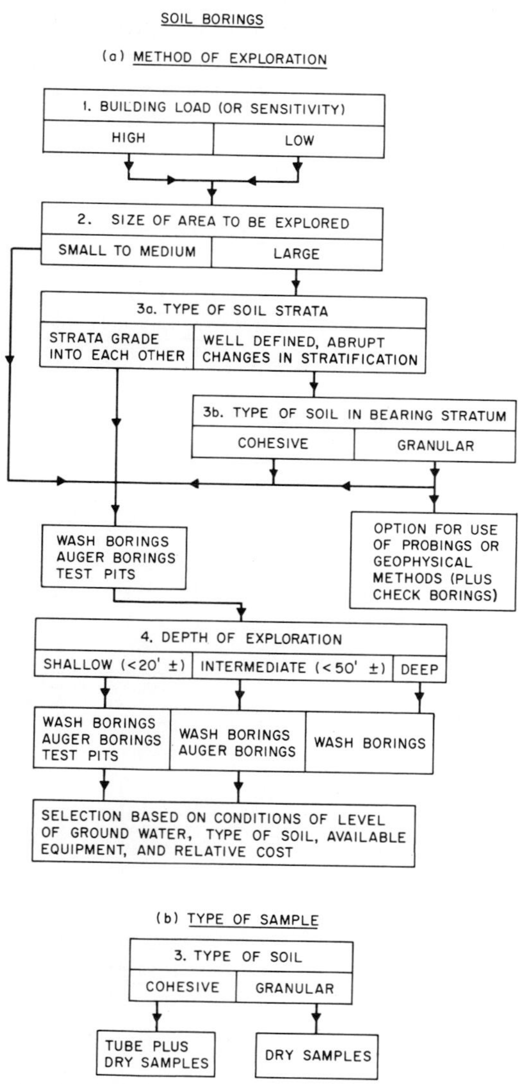

***FIG. 2-8** Selection of Methods for Exploration and Sampling of Soil.*

c. Qualified Personnel

Geophysical methods of exploration require specialized training for execution and interpretation, and the services of qualified specialists should be retained whenever such methods are to be employed.

2A.4 Summary

The process of selecting an appropriate method of exploration and sampling for use in investigating the site of a proposed building is summarized in Figs. 2-8 and 2-9. As previously discussed, of the many methods and samplers which might be utilized, the indicated scope of

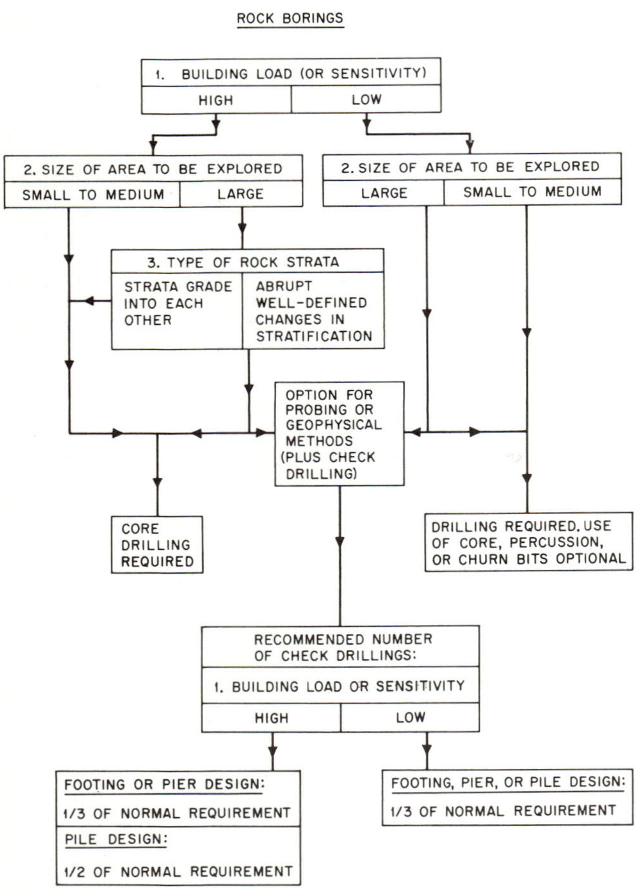

FIG. 2-9 Selection of Methods for Exploration and Sampling of Rock.

selection is limited to those relatively few (dry-sample wash borings, auger borings, test pits, probings, and geophysical explorations) which are in most common use. In connection with these charts, it is significant to note that the selection process is ultimately dependent upon only four basic considerations: (1) size of the area to be explored (large or small to moderate), (2) anticipated type of soil to be found (cohesive or granular), (3) the type of building (heavily or lightly loaded or, alternately, sensitive or insensitive to small movements of the foundation), and (4) the required depth of exploration.

Section 2B NUMBER OF EXPLORATIONS REQUIRED

In the previous section, the problems involved in selecting types and methods of subsurface exploration and sampling applicable to a given site and project were discussed. Careful selection of the method of boring and sampling was necessary in order to provide proper data as to soil classification and properties.

The next problem is to decide the number of explorations (borings) which will be required. An adequate number of borings is needed to (1) provide a reasonably accurate determination of the contours of the proposed bearing stratum and (2) locate any soft pockets in the supporting soil which would adversely affect the safety and performance of the proposed design.

It is not generally practical, because of limitations on time and cost, to carry out a sufficiently detailed exploration program to permit definitive evaluation of even these two essential points. For example, the designer cannot be certain exactly what conditions prevail in the soil between two bore holes or, in a single bore hole, between two samples. The best that can be done is to test hopefully representative locations and to recover hopefully representative specimens. As a result, the foundation design inevitably incorporates some degree of risk and the number of borings which should be made is a function of (1) the degree of risk which can be tolerated and (2) the degree of risk which is involved.

The degree of risk which can be tolerated is related to the sensitivity of the building (or of the type of foundation being contemplated) to variations or local weakness in the soil. For example, a tall slender building or a building filled with delicate instruments would be potentially more sensitive than would a low rambling warehouse. A building supported on a few high-capacity piles would be more sensitive than the same building supported on many low-capacity piles.

The degree of risk involved is related to the apparent uniformity (or lack thereof) of the soil strata at the site. For example, if the soil strata were to occur as uniform horizontal beds over the full area of the site, only one exploration would be required.*

These two variables (sensitivity of the building and apparent uniformity of the soil) will be used in Secs. 2B.1 to 2B.3 to develop some quantitative recommendations on required numbers (and spacings) of soil explorations. Sensitivity of buildings will be categorized, as follows:

1. Insensitive
   - Highly flexible constructions
   - Rigid monoliths
2. Normal
   - Frame structures
   - Statically determinate structures
   - Low buildings
3. High Sensitivity
   - Statically indeterminate with long spans
   - Tall buildings and towers
   - Buildings sheathed in frangible materials such as thin masonry veneers with large panels and few joints
   - Buildings housing equipment which requires careful and stable alignment
   - Certain types of shell structures and continuous frames

In this connection, it is noted that this same problem of required numbers of checks and samples also occurs with regard to the design of the superstructure. For example, the assumed strength and properties of the steel and concrete used in the work must be checked. There are established and accepted standards for this purpose. For steel, ASTM standard A6 stipulates the number of samples to be taken per ton of material and establishes the locations from which such samples are to be recovered. ACI 318 makes similar stipulations with regard to required numbers of test cylinders of concrete.

It is unfortunate that, for soil, such standards are lacking, and (except where stipulated by code) the determination is left solely to the judgment of the individual designer.

2B.1 Where the Building Is to Be Supported on Spread Footings Bearing on Soil

a. For "Normal" Conditions

A tabulation of requirements of several local and national building codes relating to required spacing of borings is given in Table 2-2.

* Unfortunately, nature is not often so cooperative, and even if she were, the extent of her cooperation would not be known until the necessary explorations had already been made.

TABLE 2-2 ***Some Sample Building Code Requirements Relating to Number and Depth of Soil Explorations***

Building code	Requirement: Number of explorations	Requirement: Depth	Miscellaneous
Building Officials Conference of America (BOCA)	For buildings more than 40 ft in height or where it is proposed to use a floating, mat, or deep foundation, one boring per 2,500 sq ft of building area	To rock or 50 ft below top of load-bearing strata	
Uniform Building Code	Sufficient	Sufficient	
Southern Standard Building Code	———	———	
American Standard Building Code	———	———	
Baltimore	———	———	For piers on rock, drill rock at each pier to prove that bedrock has been reached
Boston	Sufficient	Sufficient	
Cleveland	———	———	Minimum 5 ft penetration into rock
City and county of Denver	Sufficient	Sufficient	For loads exceeding 20,000 psf on rock, drill hole at each pier or footing
Los Angeles	Sufficient	Sufficient	
Miami	BOCA	BOCA	
New York City	One boring for every 2,500 sq ft of building area	Not less than 25 ft of penetration into satisfactory material. One boring in 10,000 sq ft of building area to penetrate 100 ft	
New York State	———	———	
Pittsburgh	———	———	
Portland, Ore.	———	———	
St. Louis	For buildings more than 40 ft in height, one boring per 10,000 sq ft of building area	To rock or 50 ft below top of load-bearing strata	5 ft minimum depth of core

The most notable aspect of this table is the general absence of requirements. There is a limited range to those stipulations which do appear, however. For example, the New York City code and the BOCA (which is a national code) require one boring every 2,500 sq ft of building area, i.e., a 50-ft spacing. St. Louis requires one boring every 10,000 sq ft of building area, i.e., a 100-ft spacing.

The U.S. Navy Design Manuals [4] suggest a spacing of approximately 50 ft for a large structure with separate closely spaced footings, a spacing of 100 to 200 ft with intermediate borings under the specific buildings where a large general site is to be developed, and a spacing of 25 ft where subsurface conditions are erratic.

Terzaghi and Peck [5] state that the spacing between explorations is "governed primarily by convention and not by rational considerations" and indicate that for building sites the borings are commonly spaced at about 50-ft centers.

The general range of spacing, thus, is about 50 to 100 ft. Based on this, plus their own practice, and considering the basic philosophy of conservatism expressed in Chap. 1, as well as the fact that, within limits, improved exploration tends to improve economy in the final construction, the authors suggest that a 50-ft spacing of explorations be used as an average requirement for the investigation of subsurface conditions for a normal building founded on soil-bearing spread footings. For purposes of this discussion a normal building shall be defined as a building of normal (moderate) sensitivity, located at a site where subsurface conditions are known to be, or are believed to be, reasonably uniform.

With this value as a basis, modifications for cases where normal conditions do not prevail can be readily established.

b. Buildings of High Sensitivity

Perhaps the best approach, simply is to make one exploration under each support.

c. Buildings of Low Sensitivity

Reduce requirement to two-thirds of that for "normal" conditions.

d. Where Unusually Erratic or Unusually Uniform Subsurface Conditions Occur

If subsurface conditions are unusually erratic, naturally, more careful exploration is required. Conversely, if conditions are unusually uniform, fewer borings are needed.

No attempt will be made to define when conditions are uniform on the one hand or unusually erratic on the other, but opening the spacing of borings to, say, 100 ft is not considered excessive for uniform conditions, nor need the spacing be reduced to less than 25-ft centers for erratic sites [1, 4].

e. Exceptions

See Sec. 2B.4.

2B.2 Where the Building Is to Be Supported on Spread Footings Bearing on Rock

Where this condition prevails, the bearing area will be uncovered during excavation and can be visually inspected. This inspection permits a judgment to be made whether the exposed surface is bedrock or a boulder and permits determination of the degree of soundness of the rock. The larger the area exposed, the more reliable this judgment can be. Suggested criteria for required numbers of borings are as follows.

a. Where Rock Is Exposed over a Large Area of the Building Prior to Construction or Will Be So Exposed after Excavation

Under such circumstance, visual examination of the rock surface provides a very reliable estimate of the type and quality of the rock. Accordingly, the requirement for subsurface investigations can be omitted entirely except in circumstances such as are indicated in Figs. 2-10*a* and *b* when the size or presence of defects in the rock, or the inclination of bedding planes, would affect the stability of the foundation. If the rock surface is not exposed prior to construction, some check borings should be made to determine that rock occurs and at what approximate level.

It should be noted that there is an element of risk if all of the rock borings are omitted, even if visual examination is possible (see Fig. 2-10*c*, for example). However, the risk factor is not materially increased because (1) a condition of the type illustrated in Fig. 2-10*c* probably would go undetected even if borings were made, unless they were made under every footing, and (2) the condition is unlikely to escape detection by visual examination of the rock surface if a relatively large area is uncovered.

b. Where the Rock Is Exposed in Pits, Only

For the case where preliminary investigation shows the rock surface to be relatively sound and uniform and where the exposed area is limited to pit excavations for individual footings, reduction in number of borings to one-half the requirement for normal conditions is applicable.

c. Intermediate Cases

For intermediate cases, an intermediate reduction in number of borings is applicable. Just how much should be implemented depends on the

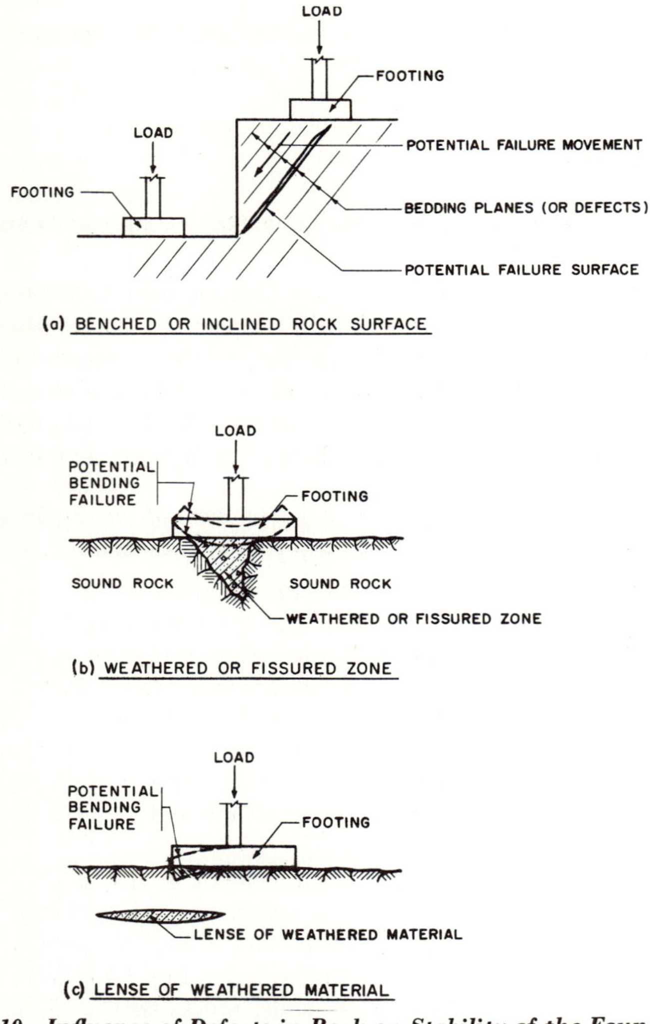

FIG. 2-10 Influence of Defects in Rock on Stability of the Foundation.

soundness of the rock surface and the size of the area which is exposed. The proper interpolation is left to the judgment of the design engineer.

d. Exceptions

See Sec. 2B.4.

2B.3 Where the Building Is to Be Supported on Piling (for exceptions, see Sec. 2B.4)

Where a building is to be supported on spread footings, the overlying soil is excavated to subgrade level, and the bearing area may be visually inspected to verify that the supporting material (at least that portion directly underlying the footing) is of adequate strength and limited compressibility as required to support the proposed building.

For pile-supported foundations this is not possible, except for the limited number of cases where open-end pipe or open-end shell casings are used. Even in such cases, however, the inspection is of little practical value because the bearing area is small, remote, likely to be disturbed, and often submerged.

As a result, the unknown (or calculated risk) factor in the design of a pile foundation is greater than that for a comparable spread footing, pier, or mat design. The use of pile-driving formulas represents an effort to compensate for this risk factor by utilizing penetration resistance to measure the density of the soil, in place, at the location of each pile. This is in lieu of the visual check and evaluation which can be made in the case of a footing. Needless to say, the substitution is not of comparable merit, and so safety factors assumed in the derivation of pile formulas conventionally are considerably higher than those used in the design of spread footings.

Both of these measures (approximation of the density of the soil by means of penetration resistance and use of higher safety factors) are logical and widely accepted. Still, the approach is an indirect one. Both beg the basic question: What is the class and density of the soil actually supporting the pile?

The direct approach would be to take a boring alongside of, or through, each pile. Of course, this is a matter of expense and, to the extent that the boring disturbs the supporting soil, would be undesirable. Still, where the foundation consists of high-capacity piers or caissons bearing on rock, this is precisely what should be done, and the provisions of the building code of the city of Baltimore (see Table 2-2) appear to have been drafted with this intent. (In the case of a caisson pile the data are provided automatically by the logs and screenings obtained during drilling of the socket.)

At the other extreme, for low-capacity piles or for friction piles of higher capacity but which are deeply embedded in a satisfactory bearing stratum, the load is spread out over a relatively large area; the performance of the foundation becomes dependent on average rather than local conditions in the soil, and other things being equal, the spacing of soil explorations for such a design should be about the same as for one employing spread footings.

For conditions intermediate between these extremes, intermediate measures are applicable. Just what these measures might be would be a matter of endless debate. It is the authors' opinion that the risk factor and sensitivity to local variations in soil properties are important considerations for all end-bearing piles and for friction piles of more than, say, 30 to 40 tons capacity and that, where the foundation is to be supported on such piling, the number of explorations should be increased to perhaps one for every 1,600 to 2,000 sq ft of building area from the norm of one for each 2,500 sq ft.

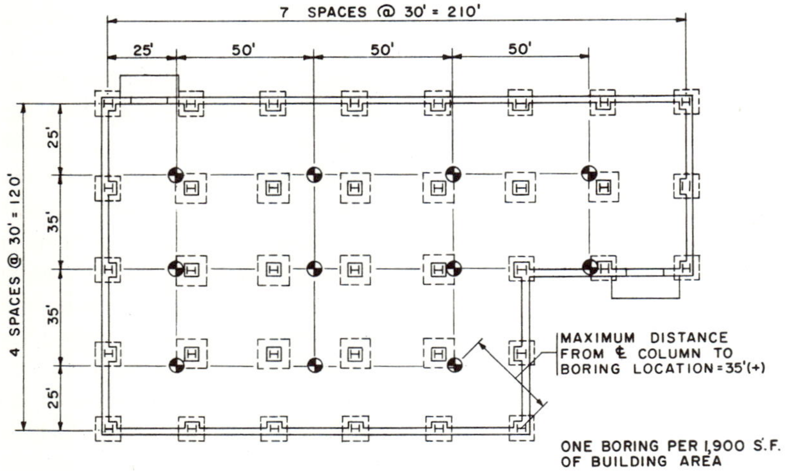

(a) BORING LOCATION PLAN-INTERMEDIATE SPAN FRAMING

(COLUMN LOCATIONS NOT PREDETERMINED – 11 BORINGS REQUIRED)

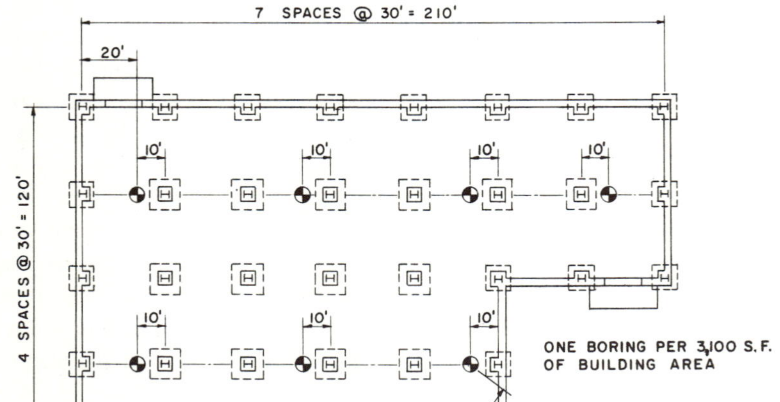

(b) BORING LOCATION PLAN-INTERMEDIATE SPAN FRAMING

(COLUMN LOCATIONS PREDETERMINED – 7 BORINGS REQUIRED)

FIG. 2-11

2B.4 Exceptions to Secs. 2B.1 to 2B.3

a. Cases Where the Column Locations Are Established before Implementing the Boring Program

Cases of this type often occur (for example, a field house, enclosed skating rink, hangar, or similar long-span construction) and, if recognized, offer an opportunity for economy in the boring program. The reason is that, with the column locations known, the relation between boring location and column location also is established, and the boring plan can be laid out using fewer borings without having any column more distant from a point of exploration than in the conventional case. This is illustrated in Fig. 2-11.

As shown in Fig. 2-11*a*, for an undetermined column pattern and with borings located on a 50-ft grid, the maximum possible distance from a column to a point of exploration is 25 ft × 1.414 = 35 (+) ft, and 11 borings are required to explore the foundation.

For a predetermined column pattern and keeping the same or a lesser maximum distance between the columns and a point of exploration as shown in Fig. 2-11*b*, only seven borings are required.

For long-span framing, a situation, such as is shown in Fig. 2-11*c*, evolves wherein only five borings are required to obtain a boring within 35 ft of each column location.

This criterion, namely, of locating a boring within such-and-such a distance from any given column rather than the more usual criterion of relating borings to area of the building, would apply only if soil conditions

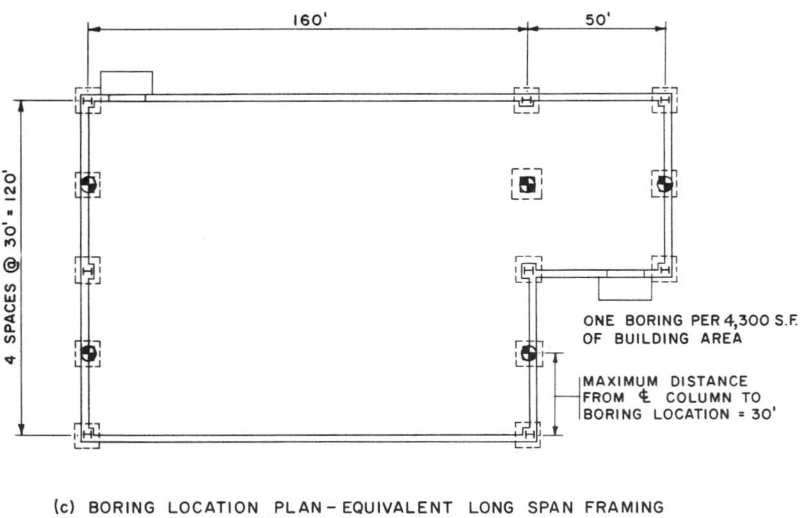

(c) BORING LOCATION PLAN – EQUIVALENT LONG SPAN FRAMING
(COLUMN LOCATIONS PREDETERMINED – 5 BORINGS REQUIRED)

FIG. 2-11 *(Continued)*

were found to be reasonably uniform. If found to be erratic, a pattern of borings, straddling and covering the entire loaded area of the soil, would be more appropriate. The reduction in number of borings would be less.

b. Small Buildings

The matter of required numbers of soil explorations for minor buildings such as one- and two-family residences, retail establishments, one- and two-car garages, sheds, etc., always is a matter of active debate. This is so because the cost of such explorations can become a significant part of the cost of the final construction and becomes burdensome, particularly where a single isolated building is to be erected. Where a development consisting of several buildings is contemplated, the cost of exploration is spread over the several buildings and becomes a less significant part of the total cost.

Soil explorations are required, even for small buildings. Burdensome or not, they should be made because the penalty, if some latent condition goes undetected, is infinitely more expensive than the cost of exploration. Such exploration need not necessarily consist of borings, however. Test pits, auger probings, etc., carried to the depths indicated in Sec. 2C are adequate and often more economical.

c. Buildings Supported on Fills

Two types of fills will be differentiated in this text, controlled and uncontrolled. A controlled fill is one placed and compacted under supervised conditions, using selected materials (see Sec. 5E, Chap. Note 3, for suggested specification for controlled fill). An uncontrolled fill is the result of unsupervised dumping of heterogeneous materials and includes all fills whose history and composition are not known and fills which have been placed without supervision.

A controlled fill is a good bearing material and the number of borings required to explore such a deposit is the same as that required to explore an equivalent area of natural ground.*

Uncontrolled fills are an entirely different condition. Because of their extreme variability and the possibility of inclusions of garbage, debris, voids, etc., they are an exceedingly treacherous material on which to found a building and must be explored with great care. Where a building is to be supported on an uncontrolled fill, it is essential to determine if there are any inclusions of mud, garbage, organic materials, cans, debris, voids or other unsatisfactory deposits or lenses. The use of test pits or larger-diameter bore holes (4-in. minimum) is required for this purpose,

* In some cases the field compaction records will be adequate to determine the bearing capacity of the material, and no borings are required at all, except to explore the soil underlying the fill.

and such pits or borings should be spaced approximately as required for the usual explorations, i.e., about 50 ft apart (1 per 2,500 sq ft of building area). Continuous samples should be recovered from the bore holes as the conventional technique of recovering samples at 5-ft intervals is not adequate to assure reasonable detection of unsatisfactory inclusions.

These borings or test pits should be considered as preliminary and should be supplemented by making a boring under each column (using standard techniques). If the preliminary borings or test pits were located under columns, then the supplementary borings may be omitted in the locations already explored. The depth of the borings should be as described in Sec. 2C.*

If either the preliminary borings or the succeeding explorations under each column show that the fill contains extensive inclusions of the type indicated, the situation is very treacherous indeed. Either the building should be carried on piles driven to bearing in strata below the fill or, if it must bear in the fill, the building should be designed (1) with sufficient articulation and/or flexibility to conform to probable differential settlements without being damaged or (2) as a stiff box capable of bridging any voids or inclusions of the maximum size of probable occurrence.

If the fill appears to be free of unsatisfactory inclusions, the situation is vastly improved, and the material may be treated as a natural deposit of equivalent classification. The magnitude of the probable settlements and the allowable bearing capacity would be determined as described in Chap. 5.

2B.5 Some Related Matters

a. The Problem of Precognition

It will be apparent from the preceding discussion and from the discussion in Sec. 2A relating to selecting appropriate types and methods of exploration that a more orderly and logical planning and a more efficient

* At first glance, so many test pits and borings are likely to appear to be a prohibitively expensive proposition. Actually, such is not the case. For example, assume a 20-ft-deep fill overlying a sand soil, a 22- by 22-ft column spacing, and a two-story industrial building. The column load would be about 100 tons, requiring, say, a three-pile group. The pile length might be 40 to 50 ft and for, say, cast-in-place piles, the three piles would cost about $700 to $900. The required boring for that column would be about 25 ft deep and would cost about $100. In addition, a test pit would be required for each fifth column. The test pit might be carried, say, 8 ft below the bottom of the footing, i.e., to about 12 ft depth and, if excavated with a backhoe might cost $250 to $300, or about $50 to $60 per column. Thus the increased cost per column is about $150, which is less than the cost of one pile, giving a potential saving of $500 to $700 per column or 50 cents per square foot of floor area in the building.

utilization of a program of soils exploration will result if the approximate subsurface conditions, the approximate type and size of the building, and the probable type of foundation are known before the program is started.

The type and size of the building usually are known before the site exploration is undertaken and, when the proposed construction is to be located on a developed site or in a developed area, the approximate subsurface conditions also can be estimated. Logs of previous or nearby borings provide the necessary information, if one has access to such data. Thus, for developed sites, efficient planning and utilization of the boring program is not a particular problem.

If the above described data are unavailable, or for undeveloped sites where the information is nonexistent, the use of stage borings is indicated, i.e., the boring program should be executed in two or more stages—a few widely scattered deep holes (perhaps auger holes), followed by a more detailed exploration. Unfortunately, the use of stage borings introduces a problem of delay and of multiple on-off charges. However, with proper planning, both or all stages can be accomplished under a single contract, without delay or extra cost, if a suitable unit price agreement is established, and intelligent supervision of the work is provided to interpret the requirements of the program as the information from the first holes is obtained. If the total number of borings required is few, then obviously only a one-stage program is practical, and this should be planned in expectation of the occurrence of unfavorable conditions.

Surprisingly, the importance of precognition of subsurface conditions is not widely recognized. Usual practice is to make borings the first order of business, as soon as the building is located. As indicated, unless the time schedule for preparation of the plans is so limited as to obviate other considerations, this is a wasteful practice and is not likely to provide the best type of information.

Related to the need for stage borings is the need for flexibility in the exploration program. In order to modify the scope and type of investigation as data from the early work are reported, it is essential to have a unit price contract, supervision by a person of competence and with the authority to order changes in the work should such appear advisable, and close liaison between the inspector (or the driller) and the supervisory engineer or architect.

b. Effect of the Height and Weight of the Building on the Required Number of Explorations

The reader's attention is directed to the fact that the height and weight of the proposed building are not mentioned in the previous discussion relating to number of explorations required and that the required number of such explorations, in fact, is independent of these two variables.

Section 2C REQUIRED DEPTH

2C.1 Depth of Soil Borings

The basic criterion which establishes the depth to which soil explorations for a building site should be penetrated may be stated as follows:

> The presence of any strata having characteristics of strength or compressibility which, if not considered in the design of the foundation, would adversely affect the performance of the proposed building, must be discovered.

Expressed in other terms, this requirement means that the explorations must penetrate to the depth at which the pressures imposed on the soil by the proposed construction are not of significant magnitude.

Just where the pressures cease to become significant is a matter of opinion. Rules of thumb, such as "at a depth twice the width of the largest footing," sometimes are used. The most rational of these rules of thumb is that suggested by Hvorslev [1] who, referring to a suggestion by De Beer, indicates that borings should penetrate to the depth at which the vertical stress caused by the proposed new construction is reduced to 10 percent, or less, of the original vertical stress at this depth.

Like all rules of thumb, this one is not precise nor does it cover all cases, but for the normal condition with which this text is intended to deal, it provides a very reasonable approach, is easy to use (see Fig. 2-12), and is recommended, subject to the following amplifications.

a. Computation of Vertical Stress Due to the New Construction

This may be done on the basis of Newmark's charts [Ref. 6 and Fig. 5-16], the 60° rule described in Chap. 5, or such other means as the designer may elect.

b. Minimum Penetration

For lightly loaded buildings, Hvorslev's 10 percent rule gives unrealistically low values for the required depth of exploration. Certain minimum values are applicable.

For very light buildings such as one- and two-family residences, retail establishments, sheds, one- and two-car garages, etc., unless hard material is encountered at shallower depth, the borings should penetrate at least 8 to 10 ft below the subgrade of the footings.

For heavier buildings, unless hard material is encountered at shallower depth, a minimum depth of exploration of 25 to 35 ft below the subgrade level of the footings is suggested.*

* This value corresponds to the 10-meter figure noted by Hvorslev [1] as appearing in the Manual (1937) of the German Committee on Foundation Investigations and is somewhat less conservative than the value of 50 ft prescribed by BOCA (and its adherents) for buildings 40 ft and more in height.

For pile-supported buildings, whatever the size, the investigations should be carried to sufficient depth to probe the soil underlying the pile tips.

c. Where Compressible Strata Are Encountered

For purposes of this discussion, compressible strata shall include deposits of fine-grained materials (silts and clays), strata of peat or other organic materials, and all uncontrolled fills.

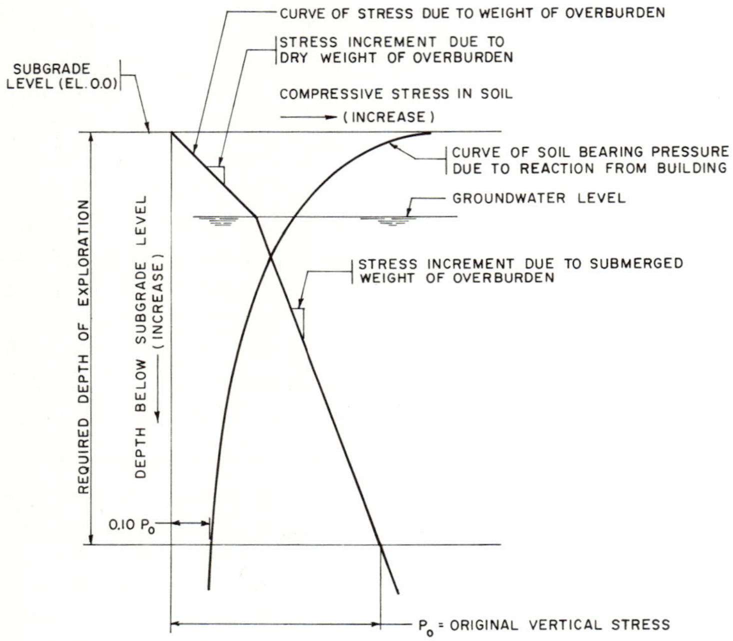

FIG. 2-12 Calculation of Required Depth of Soil Explorations.

When such materials are encountered, the first approach should be to see if the deposit can be completely penetrated by the explorations without recourse to unreasonable expense. To this end, one or two deep holes (100 ft or more in the case of heavy, or moderately heavy, buildings and 30 to 50 ft for light structures) should be sunk. If the deposit can be completely penetrated by explorations carried to this depth, then subsequent investigations also should fully penetrate the compressible strata.

If the deposit is so thick or deep that it cannot be fully penetrated by explorations carried to reasonable depth, consideration must be given to terminating the borings in the compressible strata. The required depth of penetration of the borings may be related to the compressibility of the soil strata, as follows:

1. When the strata encountered are of low compressibility (stiff or compact deposits), the depth of penetration should be increased to the point where the increased vertical stress is reduced to 7 percent, rather than 10 percent, of the original vertical stress at that level.*

2. When the strata encountered are of medium consistency, the same rule applies using a value of 5 percent rather than 7 percent.*

3. When the strata encountered are highly compressible (soft or loose consistency, organic deposits, or uncontrolled fills), the services of a soils specialist are advised.

CAUTION: The reader is reminded that if piles are to be used to support the foundation, soils explorations must penetrate below the tips of the piles.

2C.2 Depth of Coring

Where hard material is encountered, coring is required. Hard material is defined in the "Suggested Specification for Core Borings in Rock" (Sec. 2F, Chap. Note 3) as material showing a resistance of 120 blows or more per foot in the standard penetration test. Hard material is not necessarily rock. It may be a boulder or an obstruction. Accordingly, the first problem is to discover whether it is, in fact, *in situ* rock which has been encountered.

The determination is made by coring the material below the level of refusal in the bore hole, using a core barrel sampler. If concrete, bricks, wood, metal, or similar material is recovered, the material is not a natural soil and is cleared or punctured so that the boring may be carried deeper. If foreign material is not recovered, the material is either a boulder or a bed of rock (which, for purposes of this discussion shall include weathered as well as sound rock). To decide which, there are three alternative approaches, as follows:

1. The *geophysical methods* described in Sec. 2A.3 can be utilized. However, this involves the need for check borings and, since the rock level in the check borings must be ascertained, the problem still remains, albeit to a lesser degree.

2. *Local experience* can be used as a guide. In some locales, the top of bed rock is level and uniform, and boulders simply do not occur in the overlying soil formations.

3. *Interpretation of the core*, is the usual approach. In some cases, such interpretation is quite simple. For example, if the core shows a type of rock known from previous explorations in the immediate area to be geologically foreign to the area or shows an anomalous tilt or con-

* The derivation of these two rules of thumb is given in Sec. 2F, Chap. Note 5.

figuration of the rock strata, the anomaly would indicate that the material is not an *in situ* formation.

Usually, however, a clear anomaly does not appear, or the local rock formations are so heterogeneous that one cannot be sure. In such cases, the material must be cored to a sufficient depth to assure that, if the apparent rock material were actually a boulder embedded in a softer matrix, the foundation still would be safe. This determination can be made by a simple computation, as follows:

STEP 1: Assume or estimate (*a*) maximum column load, (*b*) maximum pile capacity, and (*c*) pile spacing.

STEP 2: Make a preliminary layout of the largest pile group (see Fig. 2-13*a*).

STEP 3: Determine spread of load and dimensions of equivalent bearing area below the assumed boulder (see Fig. 2-13*b*).

STEP 4: Check that bearing pressure does not exceed capacity of soil material encountered just above the assumed boulder.

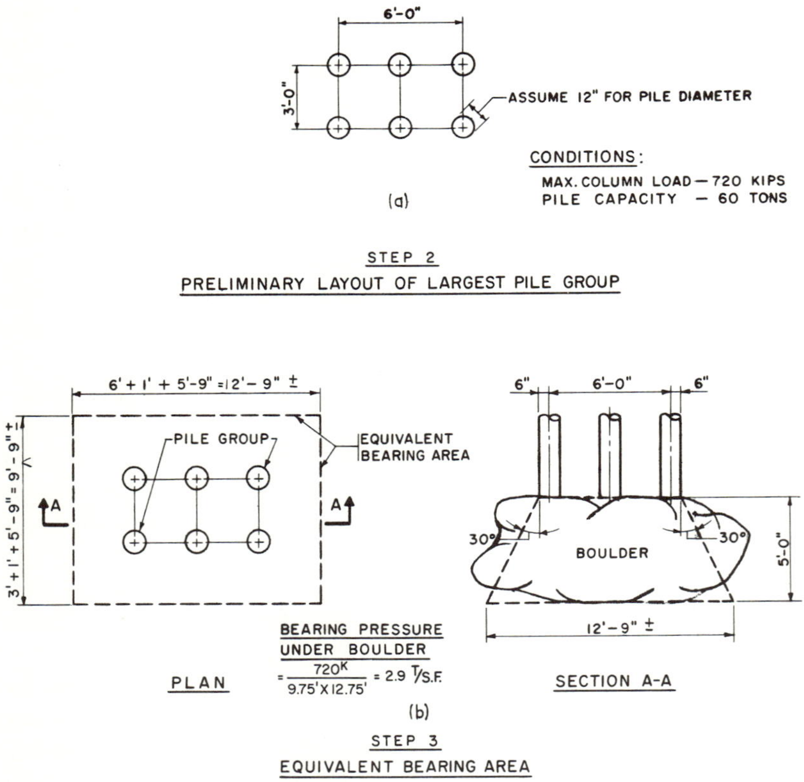

FIG. 2-13 Estimating Required Depth of Rock Core.

If the computed bearing pressure is greater than the capacity of the material encountered directly above the possible boulder, the depth of core should be increased until the requirement of Step 4 is satisfied.

This procedure is analagous to the 10 percent rule previously cited with regard to determining the required depth of penetration for soil borings. It assumes that the bearing capacity of the soil under the boulder (if it is a boulder) is as good as or better than that of the soil overlying the obstacle. Normally, this would be the case. Naturally, there are exceptions, and when in doubt, the depth of the core may be increased until the computed bearing pressure at the level of the bottom of the bore hole is any value which seems sufficiently conservative. However, 10 ft of coring should be adequate under any but very unusual circumstances.

This analytic procedure has the disadvantage of requiring some on-the-spot engineering by the boring inspector, supervisor, or liaison man so that for those who are willing to accept an increased risk factor, an alternative rule of thumb procedure described in Ref. 4, may be followed. This is to core 5 ft where the character and general location of the rock are known and to increase to 10 ft of core penetration (with one or two cores carried to 20 ft) where the character of the rock is not known or where boulders or irregularly weathered material are known to overlie the bedrock surface. The potential difficulty in applying this rule of thumb is in knowing where boulders or irregularly weathered material occur, as this often is precisely the question which is to be answered.

EXCEPTION: Where sound rock (i.e., unweathered or virtually unweathered material) is exposed at the surface prior to construction or where the surface of such rock will be exposed by stripping a large area of the building during excavation, borings may be omitted entirely, as discussed in Sec. 2B.4.

Section 2D LOCATION OF EXPLORATIONS

2D.1 General

Since the essential purpose of soil exploration for a building site is to ascertain the presence of subsurface conditions which would affect the performance of the proposed foundation, it is the conditions under the building in which the designer is interested. Conditions on adjacent lots seldom are of importance with regard to the design of the foundation per se, although they often are important with regard to other considerations, such as dewatering and stability of excavations and the settlement of adjacent structures. Therefore, except as discussed in Sec. 2D.2, insofar as the foundation design is concerned, the required explorations should be concentrated directly in the building area and should be spotted under the proposed columns, if these locations are known. If the locations of the columns are not known, the explorations should be distributed over the building area on an equal area basis, i.e., each boring should represent conditions in an equal area of the building. In addition, if there is to be a large open floor area, some borings should be made to probe the area under the ground slab.

If there are any known areas of poor subsurface conditions or if any such areas are discovered during the exploration program, then additional borings should be added as required to define their extent and to determine the pertinent soil characteristics.

2D.2 Use of "Sidewalk Borings"

Often the site of the building to be erected is already occupied and the proposed explorations must be carried out before the existing structures are demolished. In such case there are two approaches: (1) to take borings using a low-head rig inside of the existing building, or (2) to take borings in accessible areas around the perimeter of the building (so-called sidewalk borings).

Borings made inside the building are the better choice, if they can be made. Often, however, physical restrictions or legal problems make this impossible, and the use of sidewalk borings becomes the only recourse.

There is considerable disagreement as to the value of sidewalk borings. Many engineers are of the opinion that there are enough uncertainties in foundation design without making matters worse by having to extrapolate soil conditions under the building from borings made some distance away. They are correct, of course, but the degree of extrapolation required depends on just how far the borings are made from the limits of the building. There is some distance at which the required extrapolation is a reasonable one. As a rule of thumb, the authors suggest the acceptance of borings made not more than 25 ft beyond the limits of the building as representing soil conditions under the building. The 25-ft figure represents one-half the normal spacing required for soils explorations. As an additional precaution it is suggested that the number of sidewalk borings be limited to one-third of the total required number and that the borings be made in stages—those outside the building, first, by way of preliminary exploration, and those under the building later, perhaps after the site is cleared, as a verification and for greater detail.

Section 2E SOME MISCELLANEOUS MATTERS

2E.1 Disposition of Samples

It does not take long for an accumulation of soil samples and rock cores from successive jobs to become considerable and to create a problem of storage. At some point it becomes necessary to dispose of them. The question is, *When?* The answer involves two considerations, the legal requirement and the physical condition of the specimens.

From a legal point of view, if something were to happen to the building, the soil and rock samples can become evidence to verify or refute the propriety of the design assumptions. Therefore, there would be reason to keep the samples as long as they have potential value with regard to possible litigation. Unfortunately, there does not appear to be any statute of limitations on the designer's responsibility for the performance of the building. In fact, quite the contrary seems to be true. Accordingly, there is no cutoff limit with regard to retention of samples as a hedge against possible lawsuits. Therefore, the litigation problem provides no criterion by which to determine how long to keep soil samples unless it is desired to keep them in perpetuity. The controlling consideration is the condition of the samples themselves.

With regard to condition of the samples, bottle (or dry) samples, after the wax seal has been broken, dry out in a year or two and become of very limited value. Similarly, after a sample has been emptied from the bottle and handled for inspection three or four times, its value largely is lost because of disturbance. Thus, as a practical matter, after a year or two, bottle samples are not of much use. Rock cores, of course, remain a serviceable indicator of condition indefinitely.

It is concluded that on the basis of their physical condition, boring samples may reasonably be discarded two or three years after the building has been constructed and that, thereafter, reliance for information, should the need arise, must be placed on the boring logs or on new borings.

2E.2 Utilization of Old Borings

The matter of the extent to which one can properly utilize old borings to fulfill the needs of a current exploration program also frequently arises. There are three cases to consider:

1. If the old explorations were made using the desired techniques (standard penetration test, adequate depth of boring and of core, etc.) and the soil samples are available and are well preserved, then the old borings may be utilized as a part of the new program.

2. On the other hand, even if the explorations were made using the desired techniques, if the soil samples are lacking, dried out, or excessively disturbed, then the designer must depend on the boring logs alone for interpretation of the subsurface conditions, and a substantial part of the value of the explorations has been lost. The reason is that the designer's estimation of the probable behavior of a soil is largely dependent on his experience with the same or similar soils, and if he depends solely on data provided by the boring logs, his judgment on the probable behavior of the soil is dependent on someone else's opinion of the soil classification

and description, an opinion which may, or may not, coincide with his own. If the designer has full confidence in the person who classified the samples and in the classification system, or has had limited experience personally, then logs alone may be accepted in lieu of making new explorations. In the general case, however, the person responsible for the decision on type and capacity of the foundation should insist on viewing samples, and unless the proposed construction is so light and relatively insensitive that the precise soil classification does not matter, some new explorations should be made to recover sufficient samples to check the validity of the existing logs and to get a visual interpretation of what the log descriptions mean. A full-scale exploration program is not needed for this purpose; some check borings will suffice. A figure of one-half the number of explorations normally required is suggested where conditions are relatively erratic and is reduced to one-fourth the number normally required if the existing logs show conditions to be uniform. If the proposed new construction is light in weight and insensitive so that the subsoil conditions are not really critical, then, of course, the need for check borings is mitigated, and they may be omitted entirely.

3. In the third case, previous explorations were not made using the desired techniques. In such case, a set of new borings should be made.

Section 2F CHAPTER NOTES

1 Suggested Specification for Making of Wash Borings for Purposes of Recovering Dry Samples

a. Method of Making Borings

Borings shall be made by the continuous driving (and/or jetting) and cleaning out of a pipe casing and recovering samples as hereinafter specified. The pipe casing shall be the flush-joint type* and shall be (insert diameter: 2½-in. ID if dry samples are to be recovered, or 4-in. ID if undisturbed samples are desired) inch diameter. Borings shall penetrate all materials encountered to the required depth as specified on the boring location plan (for discussion of required depth see Sec. 2C).

Should hard material (which shall be defined as material exhibiting a resistance of 120 or more blows per foot when sampled in accordance with the procedure hereinafter described for the recovery of soil samples), boulders, timber, or other obstructions be encountered in driving the casing, the contractor shall penetrate the hard material or the obstruction by core drilling, as described in Sec. 2F, Chap. Note 3*a*. No extra pay-

* NOTE: the shouldered-joint casing tends to wedge in the ground due to the buildup of resistance at the shoulders, and in granular soils the apparent resistance to penetration for such casing is deceptively high.

ment will be made for penetrating hard material or obstructions, except that the depth of drilling required to penetrate such material will be included for payment at the contract unit price for core borings.

Soil samples shall be recovered with a split-barrel sampler having an outside diameter of 2 in., a wall thickness of 5⁄16 in. and a length, measured from the tip of the driving shoe to the check valve, of 24 in. A drive shoe of hardened steel shall be provided. The head of the sampler shall be provided with a ball check valve. The sampler, driving shoe, and head (or coupling) shall present a concentric flush surface, both interior and exterior. The sampler shall be of constant internal cross section, and the tip shall be kept sharp at all times. Use of dull, bent, or otherwise damaged sample spoons will not be permitted. The sampling tools shall be driven into the soil by means of a 140-lb weight having a free fall of 30 in. The impact of the weight shall be delivered to the sampler through a suitable driving head connected to the sampler by use of 1-in. extra-heavy pipe or by 1⅝-in.-OD drill rods. The number of blows required to drive the sampler for each 6 in. to the total penetration of 24 in. shall be recorded.

Soil samples shall be recovered from each boring at every change in soil formation and at intervals not to exceed 5 ft, except as ordered by the engineer (or architect). Before recovering the samples, the hole shall be cleaned to the bottom of the casing and any disturbed soil shall be removed. In saturated sands and silts, the drill bit or wash bit shall be withdrawn slowly to prevent loosening the soil around the hole. In no case shall a bottom-discharge fishtail bit be used for cleaning the hole nor shall the process of jetting through the sampler prior to sampling be permitted. The sampler shall be driven into soil which has not been disturbed by chopping, washing, hydrostatic imbalance, or other cause.

Immediately upon removal from the hole, a representative section of the sample recovered shall be removed and sealed in an approved, moistureproof, screw-top glass bottle or jar of not less than 16 oz capacity. Each sample container shall be clearly labeled to show the number of the hole and the depth below the surface from which the sample was recovered.

Where required for sample recovery, the sampling spoon shall be fitted with a flap valve. Should this procedure fail, an auger or a sand trap shall be used.

b. Samples and Records

The contractor shall deliver the soil samples packed in strong wooden partitioned boxes with hinged covers, hasps and padlocks to (insert location). Samples shall be packed so that they will not be damaged in shipment. The number of the hole from which the samples are representative, together with the box number, shall be painted on each box. A complete record of the borings shall be prepared, which shall show the location of each boring, the date taken, the character of the materials encountered, and the following data:

1. The elevation of the ground surface and the top and bottom elevation of the samples, referred to (insert datum)
2. The elevation of boundaries of soil strata
3. The size of the casing (if any)
4. The weight and drop of the hammer on both the casing and the sampler
5. The number of blows on the casing per foot of penetration
6. The number of blows required to penetrate the sampler for each 6-in. increment of penetration
7. The elevation of ground water together with any variations in this level noted during the progress of the boring work and a description of how and when the ground water level was observed
8. The location and total depth of each boring
9. Description of soil strata encountered
10. Any other information which may be helpful in identifying the sample and in recording the character of the subsurface conditions, such as cavities, obstacles, loss of wash water, use of special type sampler or traps, etc.
11. All abandoned or unsuccessful attempts at borings shall be reported

2 Suggested Specification for Recovery of Undisturbed Tube Samples

a. Recovery of Undisturbed Samples

1. The bore hole shall be cleaned to the level of recovery of the proposed sample by use of a jet clean-out auger or of a water jet having a nozzle which will direct the flow of water upwards. All disturbed materials or settled solids shall be removed.

2. The sample shall be recovered using a (insert diameter—2-, 3-, or 3½-in.-) ID seamless tube sampler formed from 16- or 18-gauge metal, having a length of 30 in., and fitted with a ball check valve at the upper end. The sampler shall be lowered to the bottom of the bore hole without impact, and pressed into the soil in a single continuous movement at a sufficiently slow rate to permit the check valve to pass the water in the tube without creating excessive back pressure. In firm materials, the sampler may be driven into the soil if so approved by the engineer (or architect).

3. After being penetrated, the sampler shall be freed from the soil by being rotated one full turn and shall be withdrawn. For soft rock, a waiting period of 3 to 5 min shall be observed before rotating the sampler.

4. The sample shall not be removed from the tube but shall be trimmed back from the ends of the tube, the space filled with hot paraffin, the tube capped with a metal or plastic cap and sealed with hot paraffin and adhesive tape.

b. Samples and Records

The provisions of Sec. 2F, Chap. Note 1*b* of the suggested specification for dry samples would apply and should be supplemented by requirements to the effect that the as-shipped weight of the tube plus sample are to be recorded and that undisturbed samples should be packed in wooden boxes using sawdust, excelsior, or similar material to prevent damage due to impact during shipment.

3 Suggested Specification for Core Borings in Rock

a. Penetration of Hard Material

Where hard material is encountered, such material shall be penetrated and sampled using a diamond-bit, double-tube, core barrel sampler providing not less than a 1⅜-in.-diam core. The sampler shall be advanced in increments of 5 ft until a core satisfactory to the engineer (or architect) is recovered.

The entire length of the cores recovered shall be packed in hinged, partitioned wooden boxes fitted with hasp and padlock and delivered to the engineer (or architect).

b. Samples and Records

The core-boring log shall state the percentage of core recovered as compared with the actual depth of drilling required to obtain same. Cores shall be labeled or marked to indicate the boring number, the elevation from which the core was recovered, the drill-run number, and the depth of the drill run. The alignment of the core from top to bottom shall be indicated and any unsuccessful attempts to recover core samples shall be noted.

4 Blow Count in Standard Penetration Test

Recording of blows in the standard penetration test for each 6-in. increment is becoming standard procedure. For a 24-in.-long sample spoon, there will be four such increments. Opinions differ on which two increments should be added to establish the N value, which is expressed in blows per foot.

The authors have indicated that the second and third increments are to be added. This same criterion has been adopted in ASTM standard D 1586-63T and is noted by Terzaghi and Peck [5].

Fletcher [3], on the other hand, suggests driving the sampler only 18 in. (three 6-in. increments) and taking as the N value the sum of the two lowest readings.

Still others (Schnabel [9], for example) suggest that the sampler be seated with a few light taps and that the N value be taken as the resistance for the first foot of penetration.

The differences in results using these several criteria are substantial. Other than the ASTM reference noted above, there is no accepted standard, and it is left to the reader's judgment to decide which criterion he prefers to adopt.

5 Derivation of Approximations for Depth of Exploration in Compressible Strata

As will be indicated in Sec. 5B.3, a tolerable magnitude of total settlement at working stress levels for a building of usual design, founded on or above soil strata such that the settlement develops over an extended period of time, is about 3 in.

The actual settlement of the building will consist of the known or calculated settlement due to consolidation of the strata encountered within the depth of exploration, plus an unknown amount of settlement due to consolidation of the soil strata below the depth of exploration. If the calculated settlement is to be limited to about 3 in., the required depth of soil exploration could be determined by assuming that the amount of the unknown settlement is of the same order of magnitude as the permissible overstress allowance, which for extreme conditions, such as are considered here, is 50 percent or about 1.5 in.

If it is further assumed that the depth of compressible material below the limit of exploration is some arbitrary, large value (say, 50 ft) and that the service life of the proposed building is 40 years, we may calculate the overload ratio to which explorations must be carried, as follows:

The total compression of a soil stratum of fine-grained-material under loads of reasonably wide extent is determined from the expression:

$$\Delta H = H\left(\frac{C_c}{1 + e_0}\right)\log\left(\frac{P_0 + \Delta P}{P_0}\right)$$

where ΔH = total compression of soil stratum, in.

H = initial thickness of soil stratum, in.

$\dfrac{C_c}{1 + e_0}$ = compression ratio (dimensionless)

$\dfrac{P_0 + \Delta P}{P_0}$ = overload ratio (see following text) (dimensionless) and the percent of consolidation at a given time interval (vertical drainage only) by:

$$U = f(T_v) \qquad \text{where } T_v = \frac{tc_v}{H^2}$$

where U = percent consolidation
T_v = time factor (dimensionless)
c_v = coefficient of consolidation (length²/time)
t = time
H = length of drainage path

The compression ratio $C_c/(1 + e_0)$ varies in value from about 0.1 to 0.4. For stiff to compact deposits, the range is from about 0.10 to 0.20 (0.12 will be assumed) and for medium deposits from about 0.20 to 0.25 (0.22 will be assumed). The corresponding approximate value of c_v is about 6×10^{-3} cm²/sec for stiff deposits, and 2×10^{-3} cm²/sec for medium deposits. The corresponding percentages of consolidation U at the end of 40 years as determined from the second equation above would be about 95 percent and 90 percent respectively.

The value of $(P_0 + \Delta P)/P_0$ is the overload ratio at the center of the stratum undergoing consolidation, and, considering only the settlement contributed by the material below the lower limit of the borings, the first equation above would indicate that it would be limited to 1.05 (log = 0.021) for stiff deposits and about 1.02 (log = 0.011) for medium strata. The permissible overload ratio at the lower limit of the borings would be somewhat larger, say 1.07 for stiff deposits and 1.05 for medium strata. Hence, the required depth of soil exploration would be to the level where $(P_0 + \Delta P)/P_0 = 1.07$ for stiff deposits, and 1.05 for medium strata.

REFERENCES

1. Hvorslev, M. J.: "Subsurface Exploration and Sampling of Soils for Civil Engineering Purposes," Waterways Experiment Station, Vicksburg, Miss., November, 1949.
2. "The Unified Soil Classification System," Technical Memorandum 3-357, prepared for the Office of the Chief of Engineers by the Waterways Experiment Station, Vicksburg, Miss., vol. I, March, 1953.
3. Fletcher, Gordon F. A.: Standard Penetration Test: Its Uses and Abuses, *Journal of the Soil Mechanics and Foundations Division, Proceedings of the American Society of Civil Engineers,* vol. 91, no. SM4, pt. 1, July, 1965.
4. Design Manual DM7 "Foundations," U.S. Navy, Bureau of Yards and Docks, Washington, D.C.
5. Terzaghi, Karl, and Ralph B. Peck: "Soil Mechanics in Engineering Practice," 2d ed., John Wiley & Sons, Inc., New York, 1967.

6. Newmark, Nathan M.: Influence Charts for the Computation of Stresses in Elastic Foundations, *University of Illinois Engineering Experiment Station Bulletin* 338, 1942.
7. Burmister, Donald M.: Principles and Techniques of Soil Identification, *Proceedings, Twenty-ninth Annual Meeting of the Highway Research Board,* December, 1949.
8. Stubbings, John E.: discussion of "Standard Penetration Test: Its Uses and Abuses," *Journal of the Soil Mechanics and Foundations Division, Proceedings of the American Society of Civil Engineers,* vol. 92, no. SM2, March, 1966.
9. Schnabel, James J.: discussion of "Standard Penetration Test: Its Uses and Abuses," *Journal of the Soil Mechanics and Foundations Division, Proceedings of the American Society of Civil Engineers,* vol. 92, no. SM2, March, 1966.
10. Yang, N. C.: Condition of Large Caissons during Construction, National Research Council, Highway Research Board, *Highway Research Abstracts,* no. 75, 1965.

CHAPTER 3

Depth of the Foundation

There are three basic criteria which establish the depth to which a building foundation must be carried. These are:

1. The construction must bear on, or in, a bearing stratum of adequate capacity and stiffness to support the applied loads without failure of the soil mass and without excessive settlement.

2. The strata underlying the bearing stratum also must be of adequate capacity and stiffness to resist the bearing pressures imposed at each level.

3. The foundation must bear in stable ground below the influences of erosion, frost heave, or volumetric changes in the soil due to changes in moisture content or chemical reaction.

Criteria 1 and 2 relate to stress and strain, i.e., possible overloading or undue deformation of the soil. The depth of embedment required to prevent such overload or undue deformation is determined by analysis. The matter is discussed in Chap. 5.

Criterion 3 relates to the effects of external influences which are a function of the site and of the type of soil at the site but which are independent of the building and the load. Determination of the embedment required to offset their effects is predicated more on judgment and experience than on analysis. Criterion 3 is discussed in the following sections.

The depth of embedment to be used in the final design is the largest of the values dictated by the three criteria.

Section 3A FROST HEAVE

The foundation for a building should be set below the maximum depth of frost penetration in order to prevent heave of the structure due to freezing of the underlying soil. This heave is caused, under natural or artificial freezing conditions, by the formation and growth of ice lenses in the soil and is a common phenomenon which occurs in all but a few coarse-grained soil materials. Foundations placed in or above the zone of frost penetration tend slowly to lift as the underlying soil freezes and then suddenly to subside when the mass thaws. In fine-grained soils, there is an additional problem in that the shear strength of the soil material is decreased when it thaws after being frozen. This loss of strength occurs because the thaw liberates moisture soaked up into the soil during freezing so that the moisture content of the soil is greatly increased as compared to the condition existent before freezing. As an example, tests by Broms and Yao [1] showed a loss of strength of a silty clay, subject to a single cycle of freezing and thawing, in excess of 90 percent of its original strength. A foundation bearing on soil, subject to such a loss of strength, probably would fail.

There are two basic ways to safeguard a foundation against frost heave. One is to set the bearing surface deep enough so that the ground provides insulation against the cold. The second is to set the footings at shallower depth and to provide artificial insulation to the same effect.* Another device is to set the foundation on a nonsusceptible soil material (Sec. 3A.6).

3A.1 Depth of Frost Penetration

Whichever approach is followed, it is first necessary to know the approximate maximum depth of frost penetration at the proposed site. For this purpose a chart, drawn from data supplied by the U.S. Weather Bureau, is given in Fig. 3-1. The values indicated in the chart necessarily are approximate only. They should be corrected to conform to the dictates of local experience, as discussed in Sec. 3A.1, if such experience is available. Correction for other factors, such as type of soil and location of the footings, also should be made, as discussed in Secs. 3A.2 to 3A.8.

Measurement of the depth of frost penetration is made from the lowest level of the adjoining ground surface or adjoining pavement surface, which is exposed to frost, to the subgrade of the footing (see Fig. 3-2). If the footings are offset from the surface exposed to frost, then the measurement is not made according to depth but according to the dis-

* As a corollary, where neither approach is practical, such as in permafrost regions, insulation may be provided against the penetration of thawing rather than the penetration of freezing.

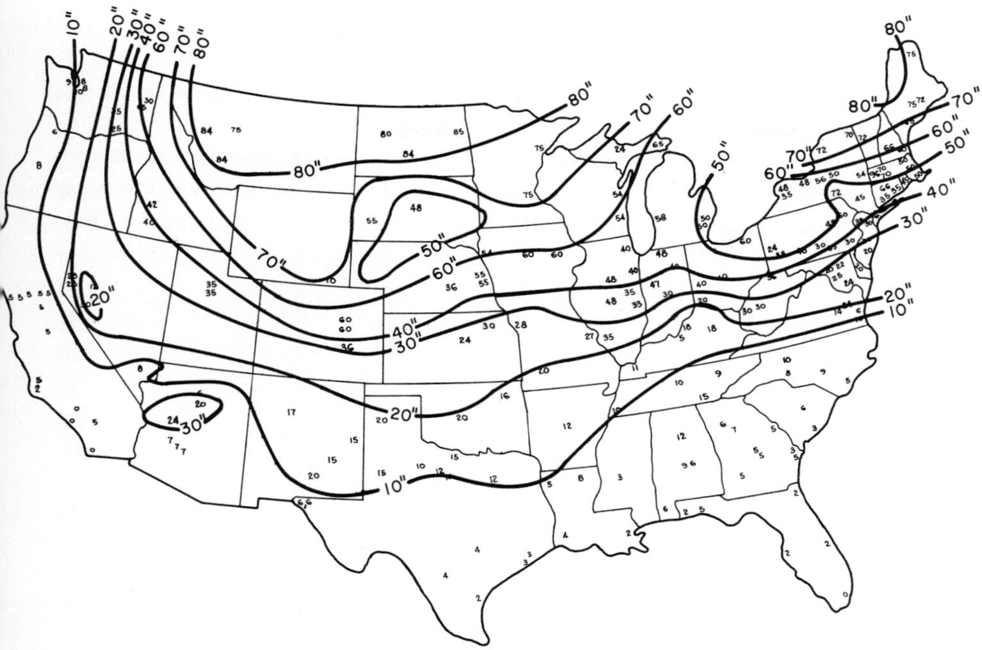

FIG. 3-1 ***Frost Penetration Map.*** Isobars represent recorded maximum frost penetrations as determined from data provided by the U.S. Weather Bureau. Small figures indicate the actual measurements. Isobars are necessarily approximate and should be modified to reflect local variations in climate or exposure.

tance from the edge of the bearing area to the nearest point of frost exposure (Fig. 3-2, Case III).

3A.2 Correction for Local Experience

Frost depth varies locally depending on exposure, depth of snow cover, altitude, proximity to bodies of water, and, in highly urban areas, the presence of adjacent heated premises. The values indicated in the frost-depth chart (Fig. 3-1) should be corrected for the effects of these local variations, if there are data to do so. Unfortunately, accumulation of the required data requires an extended period of time, and so is beyond the scope of general practice. If local building laws exist and contain criteria which are applicable to the local situation, they are the best guide. If there are no local regulations, the required depth can be estimated from the performance of adjacent buildings. Otherwise, the designer must extrapolate from Fig. 3-1 in accordance with his own judgment.

3A.3 Reduction in Interior of the Building

It is common practice to reduce the embedment of footings in the interior of a closed and heated building to a value less than the depth of the potential frost penetration at exterior ground surfaces. The practice usually is defended on the basis that the interior of the building is heated, and the interior footings, therefore, are not subject to frost

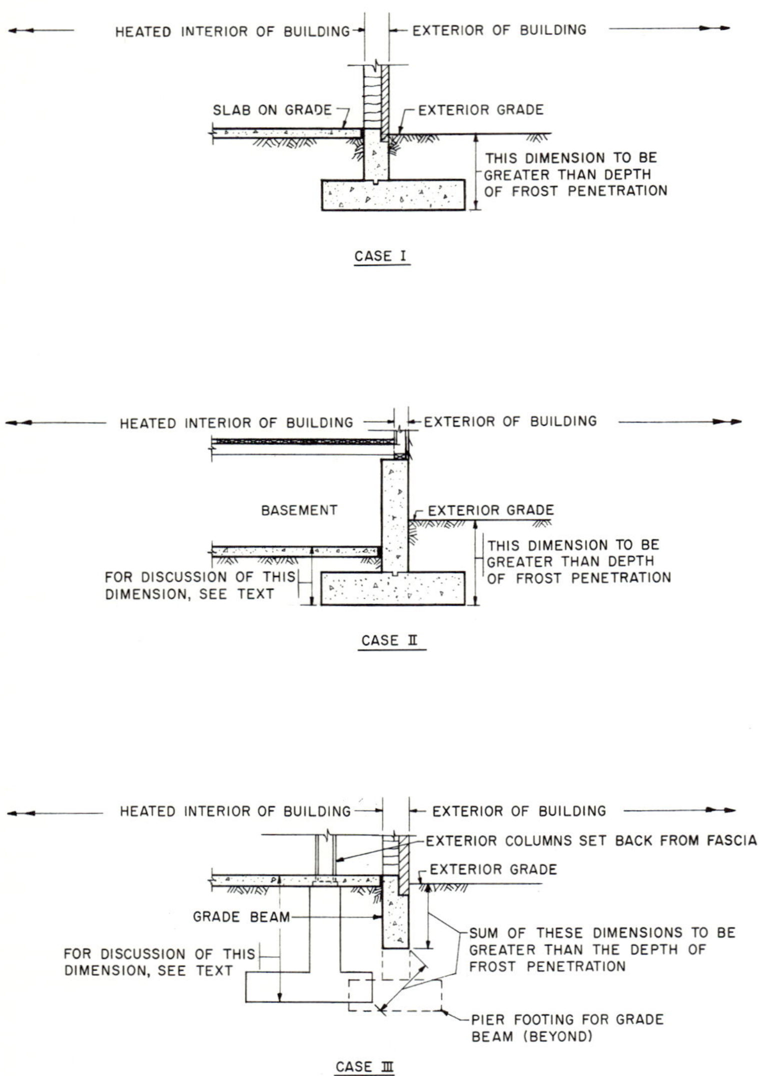

FIG. 3-2 Measurement of Depth of Frost Penetration.

action. In the event that the heating system should fail or be turned off, it is argued that the building itself offers a degree of insulation against the intrusion of extreme temperature.

The stated reasons seem to be valid, and the authors are not aware that the practice has produced any particularly unsatisfactory installations. A reduction of embedment to one-third or one-half of the nominal frost depth is common, apparently successful, and results in a saving in the costs of concrete and excavation.

3A.4 Penetration under Pile Caps

The question sometimes arises whether the requirement for embedment below the depth of frost penetration applies to pile caps. Except where the column loads are sufficiently heavy to offset the heave pressure of the frost* or where the cap is attached to piles of sufficient pullout capacity to resist the heave, the answer is, *Yes!*

Another exception would be where the piles project above grade, and there is a free space between the underside of the pile cap and the ground. This would preclude any potential heave action.

3A.5 Penetration under Cold Storage and Refrigerator Rooms

A serious frost-penetration problem may occur beneath cold storage, refrigerator, and similar spaces within a building. The depth of freezing in such cases can be surprisingly large, often making it impractical to carry the foundations down to adequate depth to get below the frost level. Tschebotarioff [2] cites an example of a refrigeration plant in a semitropical climate, operating at $-10°F$ for a period of approximately 15 years. The plant suffered severe damage because the underlying ground had frozen to a depth of 28 ft.

In cases of this type, the only sure protection is the installation of warm air ducts or of pipes with a circulating heated fluid to prevent loss of heat from the underlying soil.

3A.6 Reduction in Penetration for Non-frost-heaving Soil

In highway design, the usual approach to the problem of frost penetration is to place the pavement on a base course of a soil material which is not susceptible to frost heave and to drain such soil so that it cannot become inundated.

* As an extreme, the heave pressure would be the pressure required to prevent the formation of ice. This is about 300 to 400 psi, or 20 to 30 tsf. This is a large force, but for high-capacity piles in the 200-ton range or more, the column load (dead load only) usually is sufficient to offset the heave pressure. Where such is the case, some saving in the cost of the foundation can be effected by reducing the embedment of the pile caps, raising the soffit into the zone of frost penetration.

The Building Code of the city of Boston has a provision recognizing the potential of a corresponding practice in the design of the foundation for a building, as follows:

> Foundations of detached garages or similar accessory structures not exceeding eight hundred square feet in area and not over one story high, and grade beams of all structures, need not be carried more than one foot below an adjoining surface exposed to natural freezing if the underlying soil to a depth of at least four feet beneath the surface, and extending at least four feet outside the building, is sand, gravel, cinders, or other granular materials containing not more than five percent (by weight) passing a No. 200 mesh sieve.

The practice is a valid one. The problem is to determine what types of soil are not susceptible to frost heave. Opinions vary somewhat. For example:

a. Reference 4 in Chap. 2 describes two types of soil as non-frost-susceptible:

I. Types GW, GP, SW, SP (Unified Classification System), clean and with 1 percent, or less, of material smaller than 0.02 mm (silt and clay fraction)

II. Uniform sands with 10 percent, or less (by weight), of material smaller than 0.02 mm

b. The provisions of the Boston code are cited above. The soil material described (granular material containing not more than 5 percent, by weight, of particles passing a no. 200 mesh sieve) corresponds to GW, GP, SW, and SP in the Unified Classification System and is similar to group (I) noted in Ref. 4 in Chap. 2.

c. Beskow [7] indicates that a soil having less than 15 percent passing the 0.062-mm sieve and less than 22 percent passing the 0.125-mm sieve is non-frost susceptible.

The above descriptions of soils which are not susceptible to frost heave differ from each other somewhat with regard to the maximum allowable percentage of fines, but all describe an essentially granular material (gravel and sand) having little or no fine particles. Such materials are found in nature, and where the footings bear thereon, the depth of embedment can be reduced to a value in consonance with the requirements for interior locations. If the material does not occur naturally at the proposed site, some saving may be effected by excavating the existing frost-susceptible material to frost depth and replacing it with nonsusceptible material. This involves some cost for the replacement material, but saves on the concrete in the piers. The economy, if any, depends on the relative quantities and prices of the two materials.

In theory, clay soils also should be insensitive to frost heave as their permeability is low, thereby limiting the amount of moisture which can be drawn up into the soil to form the ice lenses which cause the heave. Clay soils are not proof against heave, however, given enough time to offset the low permeability, and the authors are not aware of any instances where reliance has been placed on the presence of clay soils to reduce frost penetration.

3A.7 Penetration under Foundations Bearing on Rock

It is usually accepted that if the foundation bears on sound rock, it may be founded at levels above the limit of frost penetration. This is done on the basis that there are no pores in the rock to contain water which might freeze. Of course, it is necessary that the rock be sound

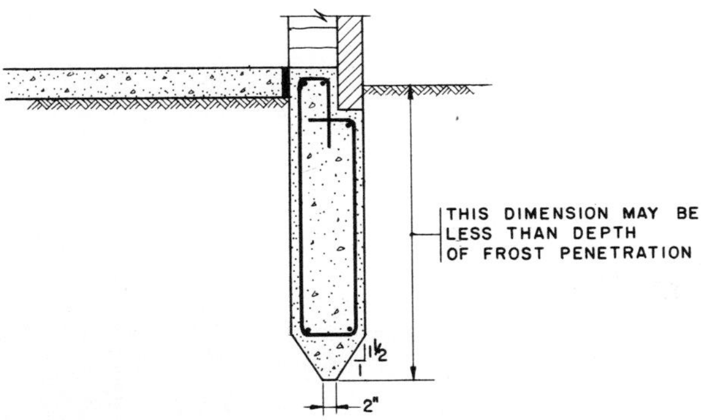

FIG. 3-3 Sharpening of Underside of Grade Beam.

(and massive) because any large or extensive seams, cracks, or areas of disintegration in the rock would act as reservoirs for water and become a source of difficulty. To this end, when the excavation is made and the proposed bearing surface is uncovered, the area should be inspected, any large seams or cracks flushed out and grouted. Any areas of disintegration should be cut out and filled with concrete.

3A.8 Penetration under Grade Beams

Another question which often arises is whether grade beams should be embedded in accordance with the considerations applicable to the embedment of footings and pile caps.

There are two divergent schools of thought in this matter. Many engineers and architects embed grade beams the same depth as the footings on the basis that the beams are just as important a structural element as the footings and just as susceptible (or perhaps more susceptible) to frost heave. Others follow a practice of embedding grade beams less than the footings, sometimes sharpening the bottom of the grade beam so that it will "knife" through the soil in the event of any tendency of the soil to heave (see Fig. 3-3).

It is difficult to rationalize why heave should be any different a problem with a load-bearing element like a grade beam than with an even more heavily loaded element like a footing or a pile cap. But the practice of reduced embedment for grade beams (often on a rule-of-thumb basis of one-half the assumed depth of frost penetration) is common and, apparently, very often successful. The true reason, probably, is that the frost never does get under the beam, partly because of the heat from the building and partly because the usual assumptions for values of frost penetration are conservative.

FIG. 3-4 Erosion of Ground around a Footing. A case of erosion due to the action of wind. The total depth of erosion has been 18 to 24 in.

In any event, it is good practice to provide top and bottom reinforcement in the beam just in case there is some heave, and if there is a slab on grade behind the beam, the sum of the depth plus width of the beam should equal the frost depth so that the slab will not be heaved. The efficacy of sharpening the bottom of the beam is questionable.

Section 3B EROSION

In warm climates, there is no frost and no need to get below frost depth. Theoretically, quite shallow footings are feasible. However, some embedment still is required to prevent the footing being undermined by erosion due to wind or by the runoff of rainwater from the downspouts or the sidewalls of the building (see Fig. 3-4).*

* The additional problem of volume changes in the soil due to changes in moisture content is discussed in Sec. 3C.

Normally, erosion is not a serious problem. Adequate embedment is provided simply by burying the exterior footings out of sight. However, if very shallow footings (less than 18-in. embedment) are being considered, the erosion problem also must be evaluated.

Section 3C VOLUME CHANGES IN THE SOIL

3C.1 Due to Changes in Moisture Content

Partially saturated clay soils which undergo a change in moisture content undergo a corresponding change in volume due to the resulting variations in capillary stresses. An increase of moisture content tends to cause swelling, and a decrease (desiccation) tends to produce shrinkage.

This phenomenon gives rise to the relatively common problem of movement of buildings which are founded in partially saturated, highly preconsolidated clay soils of high plasticity in regions where the strata are subject to large seasonal variations in moisture content. For example, buildings in regions where such conditions exist tend to settle or heave depending on the season in which they are built. Buildings constructed during the wet season experience settlement of the exterior footings as the soil desiccates in the dry season, and those constructed during the dry season may heave several inches as the ground swells during the rainy period. Even worse than this, since the changes in moisture content are greater around the perimeter footings than near the interior supports, the building is repeatedly racked, back and forth, with resulting damage.

The problem also is encountered in arid climates where the surface evaporation is several times larger than the rainfall. In such case, the ground under a building is shielded against evaporation, whereas the ground outside is not protected. The result is a migration of moisture to the soil beneath the building, an increase in soil moisture content, and a resulting continual heave. Moisture from leaking pipes will produce the same effect.

The opposite effect occurs where all or a portion of the building is heated (boiler room, for example), and the ground slab is not adequately insulated or is not provided with a proper vapor barrier. In such case, desiccation of the underlying soil and consequent settlement may result. Similarly, in humid regions, settlements due to desiccation are produced by periods of drought or by transpiration of moisture drawn from the root system of nearby vegetation [8].

Problems of movement due to changes in moisture content are most pronounced with light buildings such as one- and two-family residences, light industrial structures, and retail establishments. For buildings which impose a heavy load on the soil, the desiccation problem remains, but the swell phenomenon becomes of lesser consequence as the downward pressure of the weight of the building offsets the swell pressure (see Sec. 4E).

The problem of preventing or minimizing volume changes in soil due to changes in moisture content is solved simply by preventing or minimizing the change in moisture content. This can be done by (1) carrying the footings down deep enough to get below the active zone in the soil, (2) using piers or piles to the same effect, or (3) removing the clay soil to the depth of the active zone and replacing it with a coarse-grained, inert fill.

In any case, peripheral areas around the building should be paved, insofar as possible, and the site should be graded to provide positive drainage away from the building. Downspouts from the roof should discharge well away from the footings, and substantial eaves should be provided.

Sewage pipes and water lines should not pass under or adjacent to the building unless specially designed to prevent leaks. Trees should not be planted closer to the building than a distance equal to their mature height. Shrubbery adjacent to the building should be minimized.

If the above described measures are impractical, one may resort to the alternates of constructing a stiff building capable of riding up and down with the heave and settlement without damage, of making provision for jacking and leveling as described in Sec. 3F, or (where swell, not desiccation, is a problem) of proportioning the foundation so that the bearing pressure exceeds the swell pressure.

These solutions are relatively simple and easy to apply. Initially, however, it must be established (1) if the problem really exists (i.e., if the soil at the site is a type which is subject to volume change) and (2) what the depth of the active zone (zone in which volume change takes place) in the soil may be. Approximate answers to these questions may be determined, as follows:

a. Types of Soil Which Are Susceptible to Volume Change Due to Changes in Moisture Content

No specific test has been developed to determine the susceptibility of a given soil to volume changes due to changes in moisture content. However, Seed, Woodward, and Lundgren [4] have tentatively related the swell potential to the plasticity index.* The relationship is shown in Fig. 3-5. A low swell potential corresponds to a plasticity index of about 30, or less; medium swell potential corresponds to a plasticity index between 30 and 40; and where the plasticity index exceeds 40, the problem

* Plasticity index (PI) = liquid limit − plastic limit. For description of tests to determine liquid limit and plastic limit, see "Atterberg Limits" in Ref. 5 or 17 of Chap. 5.

of potential swell is likely to be great. Sowers [9] reports a similar relation.

b. Determination of Depth of Active Zone

In areas where the problem is well recognized, local experience is the best guide to this determination.

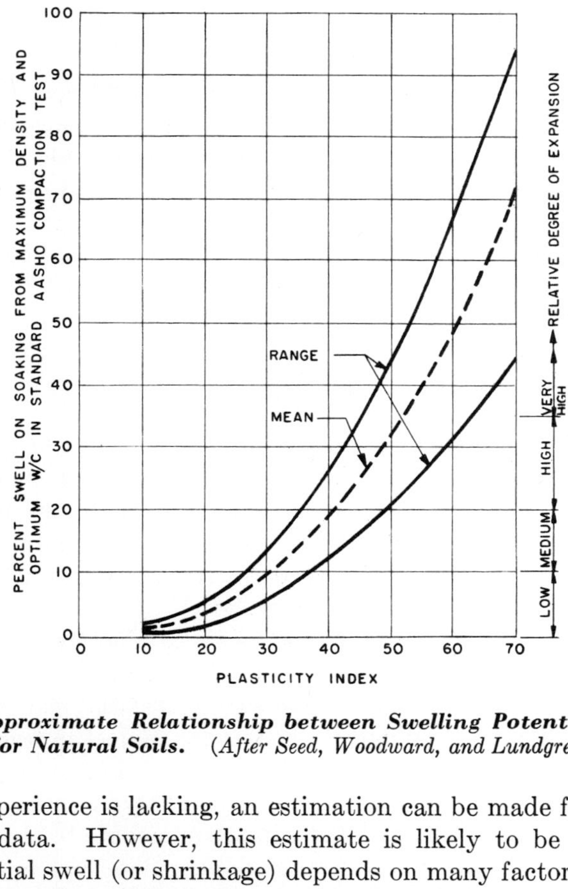

FIG. 3-5 Approximate Relationship between Swelling Potential and Plasticity Index for Natural Soils. *(After Seed, Woodward, and Lundgren, Ref. 4.)*

If such experience is lacking, an estimation can be made from consolidation test data. However, this estimate is likely to be problematic as the potential swell (or shrinkage) depends on many factors other than the properties of the soil, including the initial moisture content, the availability of water, the types of dissolved ions in the water, the confining pressure, and the time and temperature [3].

As a rough approximation, Ref. 4 of Chap. 2 notes a depth of 4 to 6 ft as the active zone in soils subject to swelling and a depth of 8 to 12 ft as the zone where potential desiccation of the soil is a problem. Sowers [9] suggests a depth of 5 to 10 ft as a common limit.

These depths correspond to the usual height of basement construction, so that the provision of a basement usually is sufficient to place the footings for the building in stable ground. This is not always true, however, and caution is required. A particular exception is where the problem is that of potential desiccation due to the provision of artificial heat. Here, much greater depths of activity frequently occur, and usually it is impractical to found the construction deep enough to get below the active zone. For such cases the provision of insulation plus a proper vapor barrier is required.

3C.2 Due to Chemical Change

Shales and clay shales containing certain minerals, principally pyrites and anhydrite, tend to oxidize or hydrate upon exposure to air or water and to swell. This phenomenon is not often serious with building foundations of a routine nature. However, if such a condition is suspected, preventive action requires that the excavation not be opened until all is in readiness to place the concrete and that the concrete be placed as quickly as possible, using a seal coat of shotcrete or bitumen if any delay in operations is envisioned.

Section 3D PERMAFROST PROBLEMS

As with other problems of potential volume change of the supporting soil (Secs. 3A and 3C), establishing a foundation in stable ground in a permafrost region requires that the construction be carried below the level of the zone of seasonal freezing and thawing. In addition, however, care must be taken that the proposed construction does not of itself create a change in the limits of the frozen material.

The problem is not a difficult one. It is handled exactly as described in Sec. 3A, i.e., by setting the level of the foundation below the limits of the active zone, by insulating the soil against thermal changes, or by founding the construction in soil which is not subject to frost heave. An alternate means, which is seldom employed in temperate regions because of its expense, would be to "float" the building on the active material, either using a rigid structure that can accommodate to changes in grade level, or using a blanket of riprap or similar material to provide a rigid distribution mat in lieu of providing rigidity in the structure.

What is difficult, however, is to determine the probable limits and depths to which the proposed new construction will affect the permafrost zone. This involves a relatively complex and, because of variable environmental factors, largely indefinite analysis of heat balance. Accordingly, the usual solution is to avoid the issue by bearing the foundation in the permafrost and then to insulate and ventilate between the building and the ground so that the presence of the building does not change the heat regimen in the ground. Many details have been evolved

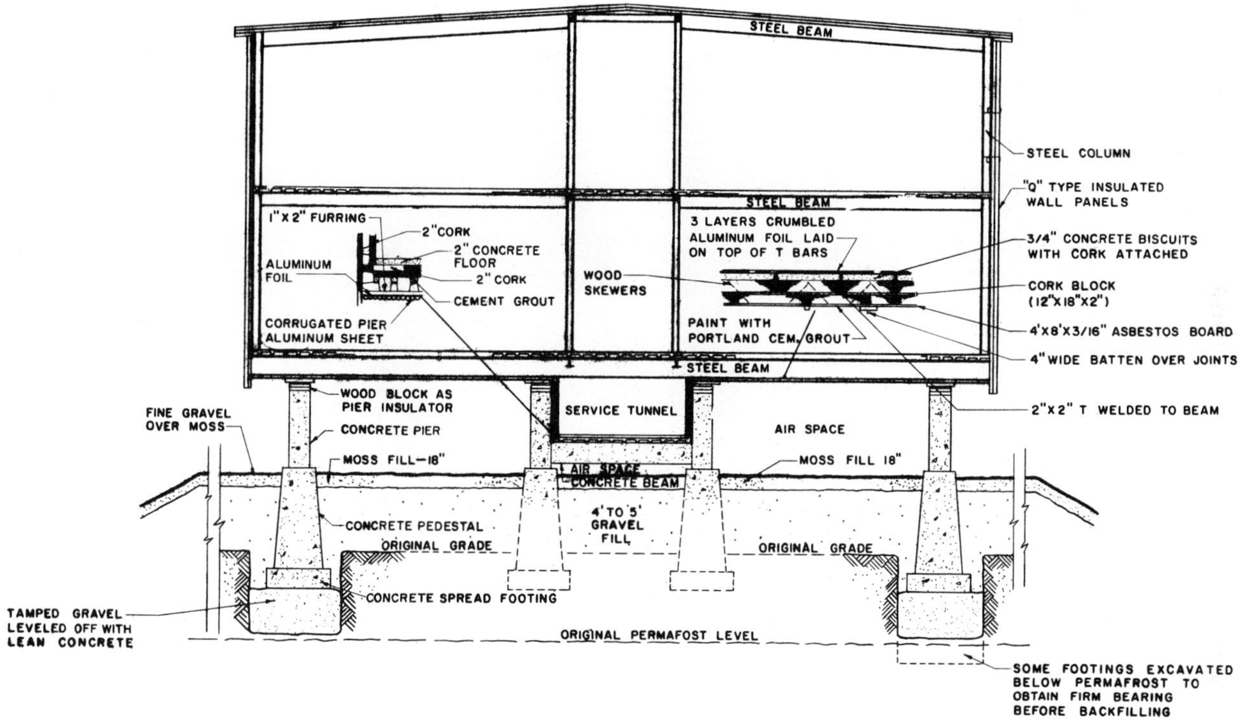

FIG. 3-6 *(After Dickens and Gray, Ref. 6.)*

for this purpose, a typical example of which is shown in Fig. 3-6. For other details, the reader is referred to Refs. 5 and 6.

As an alternative solution, the soil in the active zone may be excavated and replaced with materials which are not susceptible to frost heave, as discussed in Sec. 3A.6. The foundation then is set to bear on the nonsusceptible materials.

Finally, the foundations may be permitted to bear in the potential active zone if a source of artificial refrigeration is provided to keep the soil under the footings permanently frozen. A buried circulatory system is the usual approach. This solution is seldom used, however, because of its high cost.

Section 3E ROOT GROWTH

Figure 3-7 shows a concrete sidewalk slab heaved by the root growth of the adjacent tree. The same thing will happen to the ground floor

FIG. 3-7 Heave Due to Root Growth. The sidewalk slab has been heaved about 2 in. because of the growth of roots of the adjacent tree. A similar problem is encountered with very shallow foundations and, sometimes, with basement walls.

slab in a building if a grade beam or perimeter wall is not provided to a sufficient depth (perhaps 2 ft) to intercept the roots and prevent their penetration under the building. Shallow foundations are similarly affected. Bulging of block walls in the basement also may occur.

Section 3F MISCELLANEOUS MATTERS

3F.1 Temporary and Portable Structures

Light small temporary buildings like tool sheds, or house or construction trailers are in frequent use and conventionally are set on sills (or grillages) laid directly on the surface of the ground or pavement. Such construction does not conform to the requirements and recommendations of the previous sections. Nevertheless, in most cases it performs quite satisfactorily. That it does so is due partly to the small size and

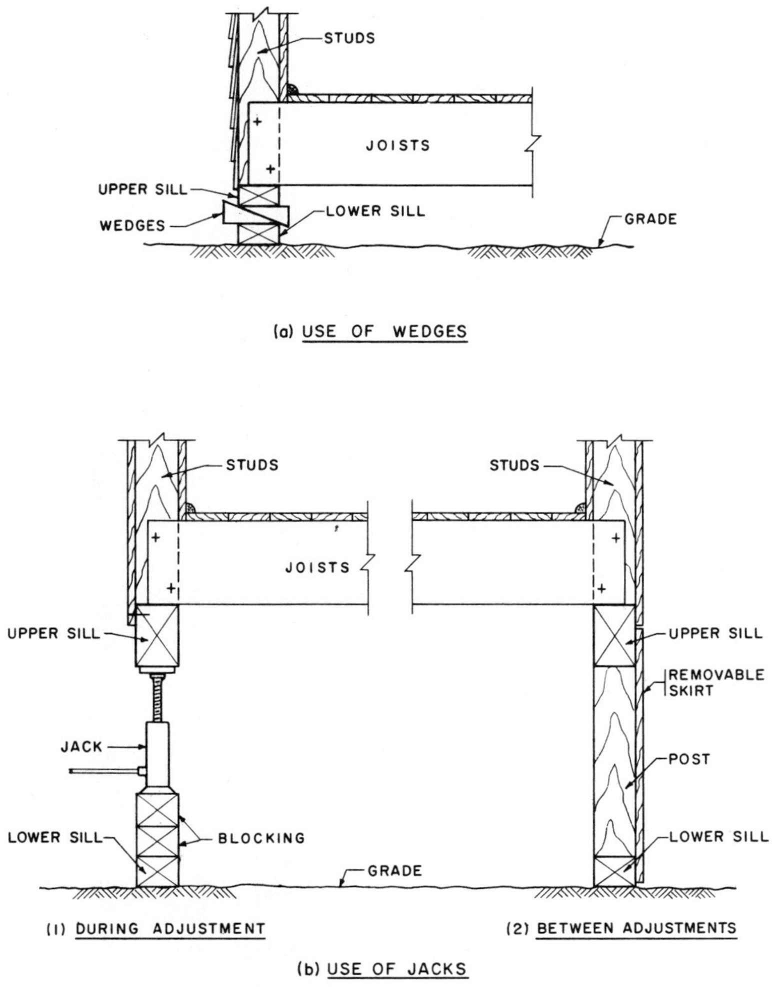

FIG. 3-8 Support of Temporary or Portable Buildings.

relative rigidity of the superstructure which enables it to bridge the differential displacement of the supports, and partly because what little damage or distortion does occur is overlooked because of its minor nature, the minimal hazard engendered thereby, and the generally utilitarian purpose of the construction.

There is an increasing tendency, however, toward the use of larger and more complex temporary or portable buildings (portable classrooms, for example) in which the relative rigidity of the superstructure is not pronounced and which, therefore, would be more susceptible to damage in the event of differential displacement of the supports. Moreover, the occupancy is such that the potential danger due to relative displacement of the supports cannot be treated in a cavalier manner.

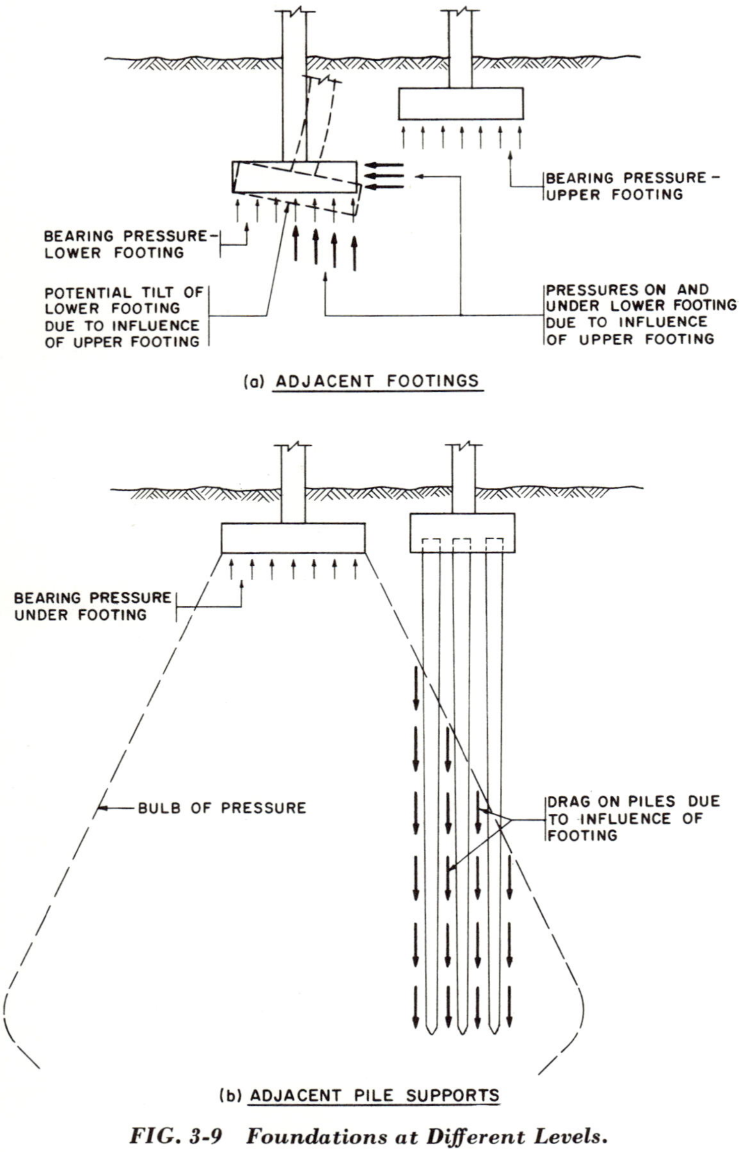

FIG. 3-9 Foundations at Different Levels.

Such buildings can be built without conventional foundations and can be satisfactorily utilized, but maintenance effort is required if an acceptable level of risk is to be assured.

Provision for anticipated maintenance effort commonly consists of providing jacking points, wedges, or comparable devices (see Fig. 3-8) between the posts supporting the building and the sills on which the

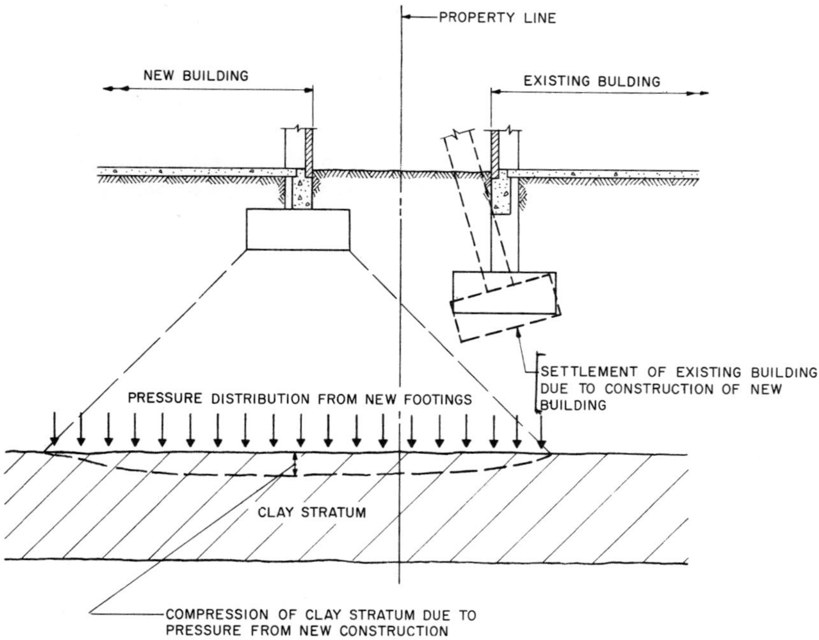

FIG. 3-10 Influence of Construction of New Foundation on Performance of Existing Construction.

posts bear. The purpose is to bring the posts to bearing and the building back to level in the event of displacement of the sills (whether due to settlement, frost heave, or other cause). Naturally, a competent and reliable periodic inspection and maintenance effort must accompany the provision of such devices because they are of no use unless the required adjustments are made.

Inspection and adjustment should be made at least once a year, preferably following the spring thaw and again just before the winter season.

3F.2 Foundations at Different Levels (see also Chap. 5)

Not only is it important to establish the level of the foundation at adequate depth, but the relative elevations of the neighboring elements (footings or pile caps) also must be considered.

FIG. 3-11 Subsidence of Ground. The bottom course of brick was embedded below grade when the building was constructed. The ground, which consists of several feet of fill, subsided about 8 in. Subsequent erosion revealed the condition of the electrical conduit shown. The conduit, fortunately, is flexible and is undamaged. Water and sewer pipes in the area have not been so fortunate. There have been several breaks.

FIG. 3-12 Subsidence of Ground. The fence post is embedded in concrete to about 24 in. below grade. The ground, which consists of miscellaneous granular fill, including ash and brick, has subsided. The differential subsidence in 24 in. has been 5 in., as shown.

For example, the pressures under the higher footings (or pile caps) will tend to load those at a lower level (see Fig. 3-9), and the footings must be adequately separated to prevent this.

A rule of thumb in common use for this purpose is to limit the difference in level between two adjacent footings to one-half the clear distance between footings and to neglect the influence of the adjacent footings on each other. The basis for this seems to be an assumed 30° distribution

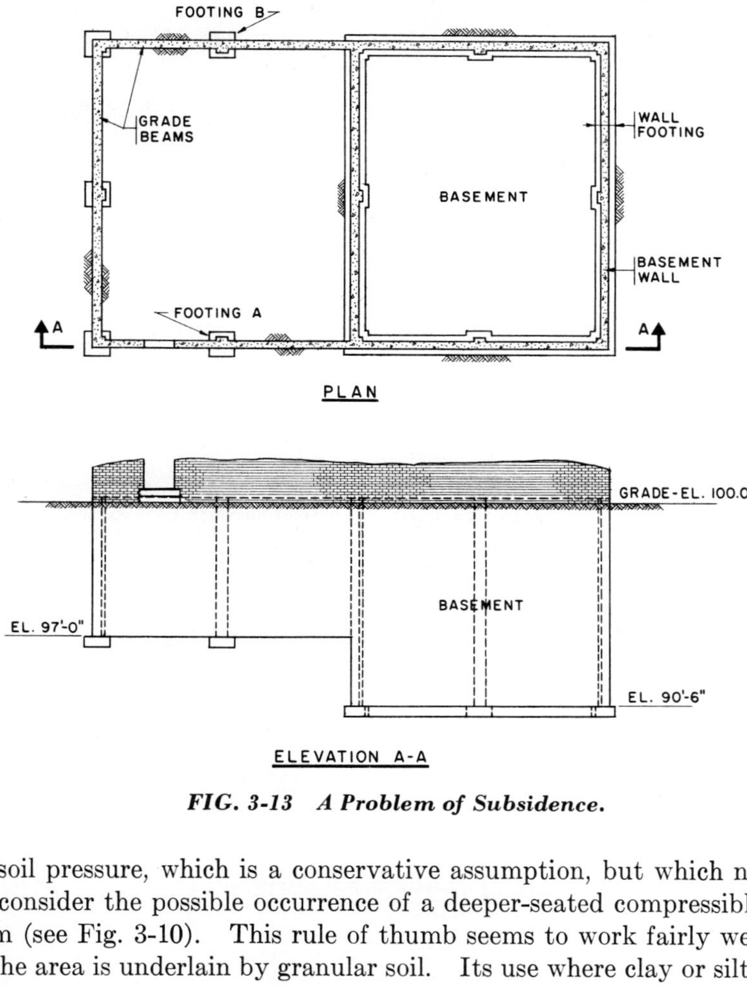

FIG. 3-13 A Problem of Subsidence.

of soil pressure, which is a conservative assumption, but which neglects to consider the possible occurrence of a deeper-seated compressible stratum (see Fig. 3-10). This rule of thumb seems to work fairly well only if the area is underlain by granular soil. Its use where clay or silt strata occur at the site is not advised. Where such is the case, the effects of adjacent construction should be analyzed by the methods described in Chap. 5.

A second consideration in founding the footings for a building at different levels is that of potential subsidence. Figure 3-11 shows an example of subsidence of the ground from around the basement wall of a building. The total is 8 in. Figure 3-12 shows how the ground has subsided about 4 in. around fence posts in the same area. The posts are embedded about 24 in. In both cases, the ground is granular fill with some ash and with fragments of brick.

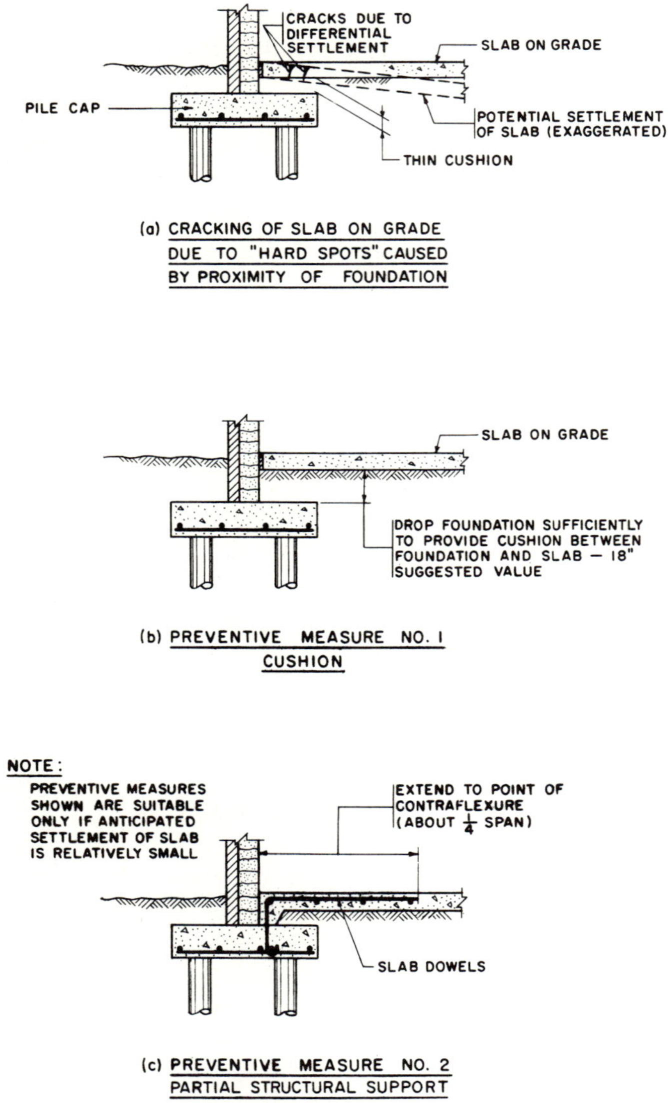

FIG. 3-14 Influence of Depth of the Foundation on the Performance of a Slab-on-grade.

These photographs show the amount of compaction that can occur in a limited depth of ground (4 in. in a depth of 24 in. in the case of the fence posts and 8 in. in a depth of about 60 in. in the case of the basement wall). To illustrate the potential effect on a building, assume that the proposed design contemplated a condition such as that shown in Fig. 3-13. If the compaction with time of the ground were anything like the order of magnitude shown in the photographs of Figs. 3-11 and 3-12, it would not be long before footings *A* and *B* were hanging from, rather than holding up, the intersecting grade beams.

3F.3 Slab on Grade

Another consideration controlling the embedment of shallow foundations is the presence of a slab on grade over such foundation. The matter is illustrated in Fig. 3-14.

The depth of the foundation should be established so that a cushion is provided between the slab and the foundation (Fig. 3-14*b*). Alternatively, the slab may be given a whole or partial structural support system (Fig. 3-14*c*).

3F.4 Construction Problem

The final consideration which will influence the designer's determination of the depth of embedment of the foundation is the problem of construction.

The reader is reminded that it is particularly important to keep the subgrade of the foundation above the groundwater level, if possible, and at or above the level of the foundations of adjacent buildings. The latter requirement relates to minimizing the potential need for underpinning.

REFERENCES

1. Broms, Bengt B., and Leslie Y. C. Yao: Shear Strength of Soil after Freezing and Thawing, *Journal of the Soil Mechanics and Foundations Division, Proceedings of the American Society of Civil Engineers,* vol. 90, no. SM4, pt. 1, July, 1964.
2. Tschebotarioff, Gregory P.: "Commentary on Foundations Provisions of Proposed Building Code for New York City" (unpublished).
3. Parcher, James V., and Ping-Chuan Liu: Some Swelling Characteristics of Compacted Clays, *Journal of the Soil Mechanics and Foundations Division, Proceedings of the American Society of Civil Engineers,* vol. 91, no. SM3, pt. 1, May, 1965.
4. Seed, H. Bolton, Richard J. Woodward, and Raymond Lundgren: Prediction of Swelling Potential for Compacted Clays, *Journal of the Soil Mechanics and Foundations Division, Proceedings of the American Society of Civil Engineers,* vol. 88, no. SM3, pt. 1, June, 1962.
5. Technical Publications TP-Pw-11, "Arctic Engineering," U.S. Navy, Bureau of Yards and Docks, Washington, D.C., Mar., 1955.

6. Dickens, H. B., and C. M. Gray: Experience with a Pier-supported Building over Permafrost, *Journal of the Soil Mechanics and Foundations Division, Proceedings of the American Society of Civil Engineers,* vol. 86, no. SM5, pt. 1, October, 1960.
7. "Comprehensive Report: Investigation of Construction in Arctic and Sub-arctic Regions, 1945–1948," U.S. Army, Corps of Engineers, 1950.
8. Cooling, L. F., and W. H. Ward: Some Examples of Foundation Movements Due to Other than Structural Loads, *Proceedings, Second International Conference on Soil Mechanics and Foundation Engineering,* Rotterdam, 1948.
9. Sowers, G. F.: Shallow Foundations, in G. A. Leonards (ed.), "Foundation Engineering," McGraw-Hill Book Company, New York, 1962.

CHAPTER 4

Design Loads and Stability Requirements

One of the first requirements in the design of a foundation (or of any structural member) is to determine the applied loads. This would seem to be a simple matter. In practice, it is not. In fact, it is a major phase of the design procedure.

There are a number of points involved in determining the design loads for foundation elements wherein opinions differ as to proper procedures to be used. These points are discussed in Secs. 4A through 4G.

Opinions also differ about required safety factors with regard to stability (overturning and sliding). These are discussed in Sec. 4H.

A checklist summary of the loads to be considered in the design of foundations is given in Table 4-1.

Section 4A DEAD LOAD

The design load for foundation components includes the dead-load reactions from the superstructure. It also includes the dead load of the foundation (weight of concrete and of soil). In practice, however, the design computations often are simplified by neglecting the weight of the foundation and of the overlying soil, as discussed below.

4A.1 Weight of the Foundation

For convenience of computation, many designers omit the weight of the footings, pile caps, or piers in calculating the subgrade reactions or the pile loads. They include, instead, the weight of an equivalent volume of soil. This considerably simplifies the computations, and the error involved usually is small, being limited to the difference in weight of the concrete (150 pcf) versus the weight of the soil (say, 100 to 120 pcf). The practice is acceptable where the foundations bear on rock or granular materials or for other cases where the allowable bearing pressure is, say, 2 tsf or more, since in such a case the weight which is neglected is only a

TABLE 4-1 ***Summary of Loads to Be Considered in the Design of Foundation Elements***

Vertical loads:
1. Dead load
    a. Column reactions
    b. Weight of walls, footings, piers, and pile caps
    c. Weight of overlying soil
    d. Weight of slabs on grade
    e. Weight of piling
2. Live load
    a. Loads due to occupancy
    b. Surcharge (including live load on ground floor slabs)
3. Buoyancy
4. Drag (negative skin friction)
    a. Due to remolding of soil
    b. Due to general settlement of soil strata at the site
5. Heave
    a. Due to freezing of ground
    b. Other causes

Lateral loads:
1. Wind forces
2. Earthquake forces
3. Lateral earth pressure
4. Fluid pressures

Impact:
1. Impact equivalents related to moving and dynamic loads

small percent of the bearing capacity. However, for lesser bearing capacities and for compressible soils where a difference of a couple of hundred pounds per square foot would be significant with regard to settlement, the error may be large, and the actual weight of the footings, piers, and pile caps should be used.

4A.2 Weight of Overlying Soil

The exact portion of the weight of soil overlying the footings or pile caps which should be included in the load to be supported by the footings or pile caps depends on the relative displacement of the foundation as compared with the subsidence of the surrounding and overlying soil (see Fig. 4-1). It is conventional to make the computation on the basis

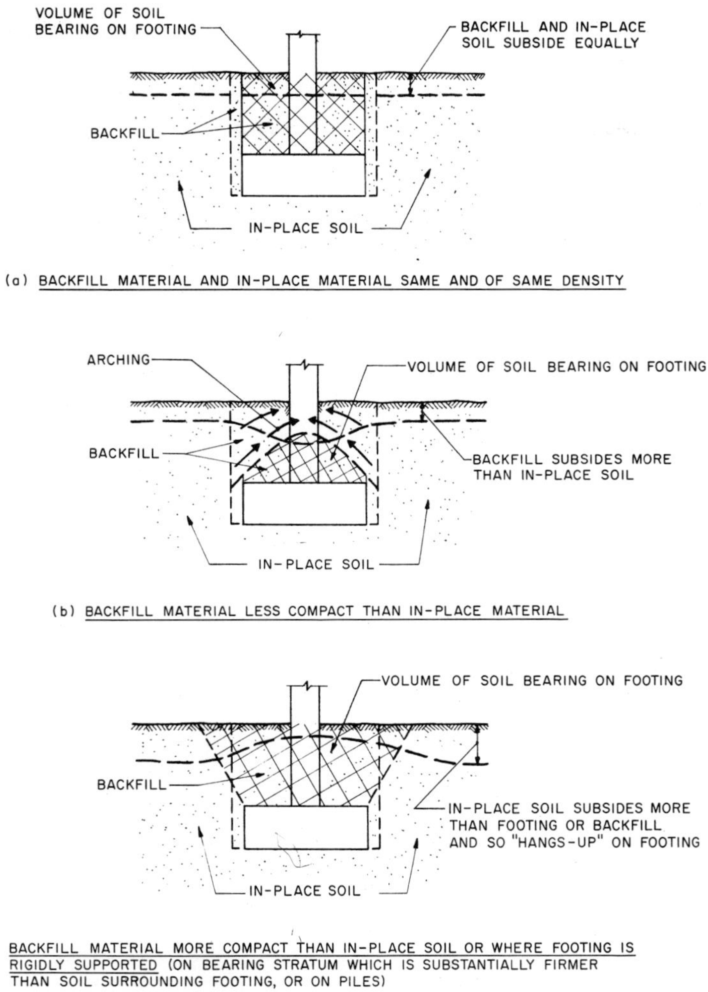

FIG. 4-1 Effect of Relative Rigidity of Foundation and Soil on Weight of Soil Which Reacts on Buried Footing.

of including all load contained within vertical planes projected upward from the limits of the pile cap or footing (see Fig. 4-2).

In connection with this matter of the weight of the overlying soil, another frequent practice in design is to omit such weight from the calculation of load to be supported by the footing. In fact, this practice is expressly written into the Building Code of the City of Boston (1962), which states:

> The loads to be used in computing the pressure upon bearing materials directly underlying foundations shall be the live and dead loads of the

structure, . . . including the weight of the foundations and of any immediately overlying material, but deducting from the resulting pressure per square foot the total weight of a one-square-foot column of soil, including the water in its voids, which extends from the lowest immediately adjacent surface of the soil to the bottom of the footing, pier or mat.

Since allowable soil pressure (the determination of which is discussed in Chap. 5) is the allowable increase in "net" pressure, i.e., the allowable value over and above the stabilized weight of the overburden existing at

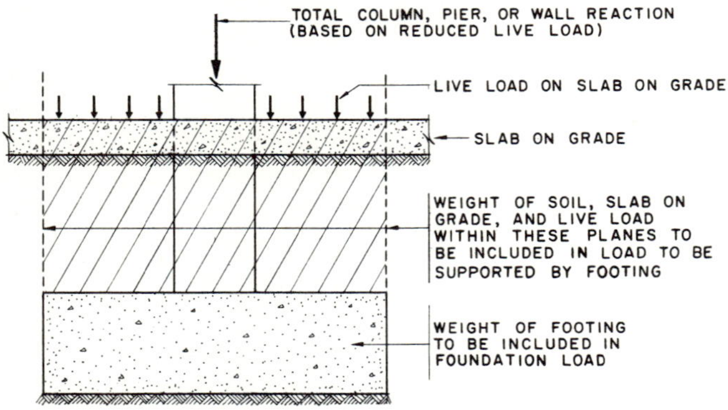

FIG. 4-2 Basic Foundation Loads.

the time that the borings were made, this device of omitting the weight of the overlying soil from the computation of load usually gives results which are numerically correct and has the advantage of providing a shortcut in the computation.

However, for reasons which are discussed in Sec. 5D.1, the device does not always give correct answers, and a safer approach is to follow the physical reality, i.e., to include the weight of the overlying soil as a load and then to increase the allowable bearing pressure to allow for the confining effects of the overburden, to the extent that such increase is legitimate.

4A.3 Pile Reactions

a. Weight of Overlying Soil

In calculating pile reactions the weight of the overlying soil may not be neglected, even as a computational device. The reason is that described in Sec. 4A.2; namely, omission of the weight of the overlying soil from the computation of load on a footing is justified only on the basis that allowable bearing pressures, as conventionally determined, represent allowable increases in "net" pressure. Allowable pile loads, however, are governed by several considerations besides soil bearing pressure, such as stresses in the material and driving formulae, and while omission of the weight of overlying soil might give results which are numerically correct with regard to considerations of soil pressure, such omission would not give correct values for use in the computation of stress in the piles, etc.

b. Soil under the Pile Cap

The soil under and around the pile cap, conventionally, is assumed *not* to provide any vertical support for the cap (although it does provide lateral restraint). The reason is that if conditions are such as to require the use of piling, the soil in which the upper portions of the piles are embedded may be presumed to be relatively loose or compressible. The piles, if properly driven, therefore, would be stiff as compared to the supporting soil under the cap and so would pick up most of the load on the cap. This does not mean that the underlying soil actually carries zero load. It does not, and there are some cases (such as the effect of a new construction on an adjacent construction as in Fig. 3-9) in which the load on the soil under the cap must be considered. In the general case, however, where only vertical loads are being considered, the piles should be proportioned to carry 100 percent of the load. Whatever support the soil under the cap can provide is then merely an additional margin of safety.

c. Weight of the Pile

Whether or not to include the weight of the pile itself in the computation of total load on the pile is a matter of debate. Usually the load is neglected, largely on the premise that it is small and that the accuracy with which we can determine allowable pile loads is so gross that the refinement is not justified. Except for a possible circumstance in which the use of some monster pile were being considered, this convention seems reasonable.

Section 4B LIVE LOAD

The design load for foundation components also will include the live-load reactions from the superstructure. However, these reactions are *not* those due to the full live load but should be reduced for the probability factor relating to contributory area, as specified in most building codes (see Sec. 4I, Chap. Note 1).

Another important point is that the live load on the ground or basement slab should be included in the load to be supported. The importance of this matter is illustrated in Fig. 4-3, which shows a design condition encountered in a steel supply warehouse, taken from an actual case in which the design floor load was 4,500 psf. The effect of the partial floor loading condition on the equilibrium of forces on the potential failure plane and, consequently, on the stability of the footing was the controlling design condition.*

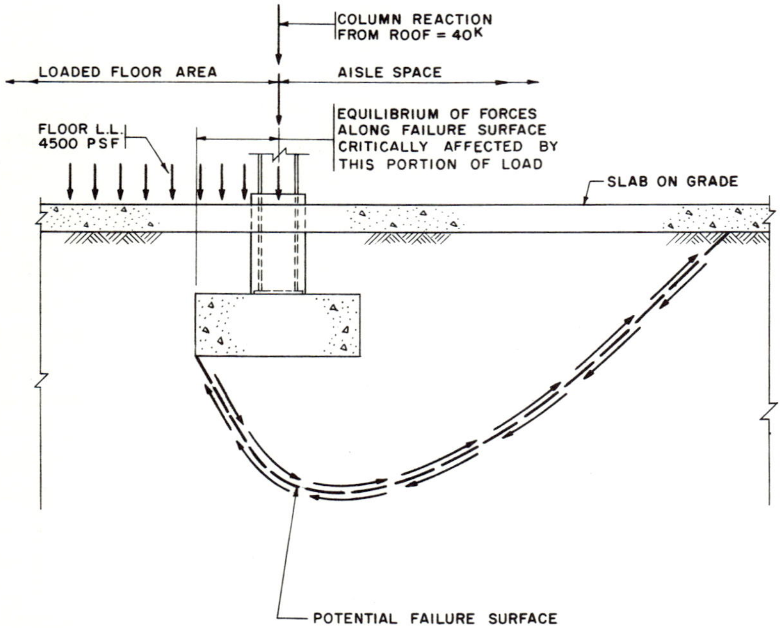

FIG. 4-3 Influence of Surface Live Load on Stability of Foundation.

Section 4C BUOYANCY

The reactions, as calculated in Secs. 4A and 4B, should be reduced for the buoyant effect of the groundwater. The weight of the footings or pile caps and of the overlying soil will be reduced by this effect.

More important, the hydrostatic uplift due to groundwater will tend to raise light constructions such as basement slabs or buried tanks. Cases of "floating" swimming pools and cofferdams have occurred. Sowers [15] notes a case where a balanced floating boxlike pumping plant rose 2 ft when the site became inundated by an unprecedented flood.

Calculation of the buoyant forces of groundwater on the foundation does not involve any special considerations.

* It should be noted that the condition of full live load over the entire floor was not critical, only the partial loading. This may account for the fact that this load is so often neglected in design.

Section 4D DRAG (NEGATIVE SKIN FRICTION)

Where a pile is driven through compressible strata (such as fill, peat, clay, or silt) to resistance in relatively firm material, there is a tendency for the compressible strata to consolidate (or compress) by an amount in excess of the downward displacement of the pile under the applied loads. The reason is that the pile passes through the compressible material and bears in firmer strata (see Fig. 4-4). Under such circumstance, the soil around the pile tends to "hang up" on the pile, exerting a downward drag which can be of considerable magnitude, particularly if the compressible strata are located at some depth so that the overlying strata also hang up on the pile.

This phenomenon is quite important. Van Weele [9], for example, reports that in portions of Holland, where the piles are driven through deposits of soft clay and peat into bearing in sand, the drag force amounts to 20 percent of the total pile load.

In the United States, a number of failures of pile foundations have been traced to the occurrence of drag conditions [1].

In general, the effects of negative skin friction and the conditions under which drag forces are likely to develop are well recognized. However, there are no standard or accepted procedures whereby one can estimate the amount of drag to be anticipated. The following summarizes the procedures which have been used, plus those suggested by the authors.

Two cases occur. In the first, the compressible strata are in a state of equilibrium before introduction of the piling, equilibrium which is upset by the remolding incident to installation of the piles (Fig. 4-4*a*). In the second case, in addition to the effects of remolding, either the compressible strata themselves are in a state of nonequilibrium (active consolidation), as shown in Fig. 4-4*b*, or the condition of equilibrium is upset by the addition of new load (surcharge), as shown in Fig. 4-4*c*.

CASE 1 *Drag Due to Effect of Remolding:* Consider first, the case in which drag is induced by the effect of remolding alone. The amount of drag depends (among several lesser factors) on the amount of consolidation which takes place in the remolded layer, on the width of the remolded zone, and on the stress-strain relation in shear for the soil or soils in which the piles are embedded.

As an extreme, the total drag force cannot exceed the lesser of the following values (see Fig. 4-5):

1. The total force due to the shear stresses q_{d1} developed at the interface between the remolded zone and parent soil (this force will not exceed some fraction of the total weight of the soil within the remolded zone).

2. The adhesion Q_{d2} which can be developed between the remolded soil and the piles (ultimate values of unit adhesion q_{d2} are given in Table 4-2).

In general, the first limitation (a fraction of the total weight of the soil within the remolded zone) is the lesser value and therefore governs. The potential drag force due to remolding may therefore be computed if

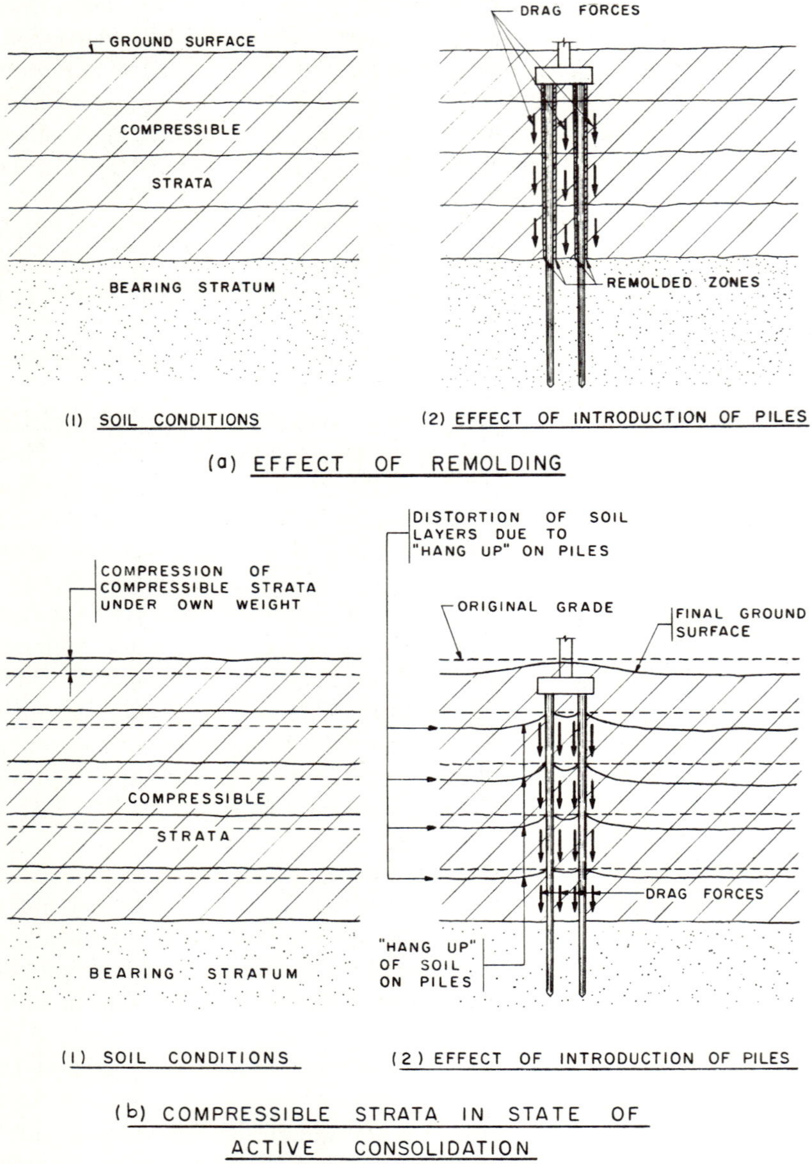

FIG. 4-4 Drag on Piles.

numerical values are assigned for (1) the extent of the remolded zone and (2) the fraction of the total weight of the remolded soil which must be supported by the piles.

The extent of the remolded zone may be estimated from the sensitivity* of the soil. For example, Ref. 2 indicates that for relatively insensitive clays (degree of sensitivity less than 2 or 3), the remolded zone is small and may be neglected. For highly sensitive clays (sensitivity equal to 4 or more), the remolded zone extends 12 to 18 in. around each pile.†

The fraction of the weight of the remolded soil which is supported by the piles is computed by statics.

A sample computation of total drag force due to remolding is illustrated in Fig. 4-6. For this illustration it has been assumed that, as an upper limit, the remolded zone in the soil extends a distance around each pile equal to half the pile spacing. It also has been assumed that the potential adhesion between the soil and the piles exceeds the weight of the remolded material. As indicated above, this is the usual case. If desired, however, the value of the adhesive force can be computed, and

* Degree of sensitivity is defined as

$$S_t = \frac{\text{unit cohesion, undisturbed sample}}{\text{unit cohesion, remolded sample}}$$

† Other sources suggest that the width of the remolded zone extends to a distance of 1½-pile diam [10], or that it has a negligible effect [11].

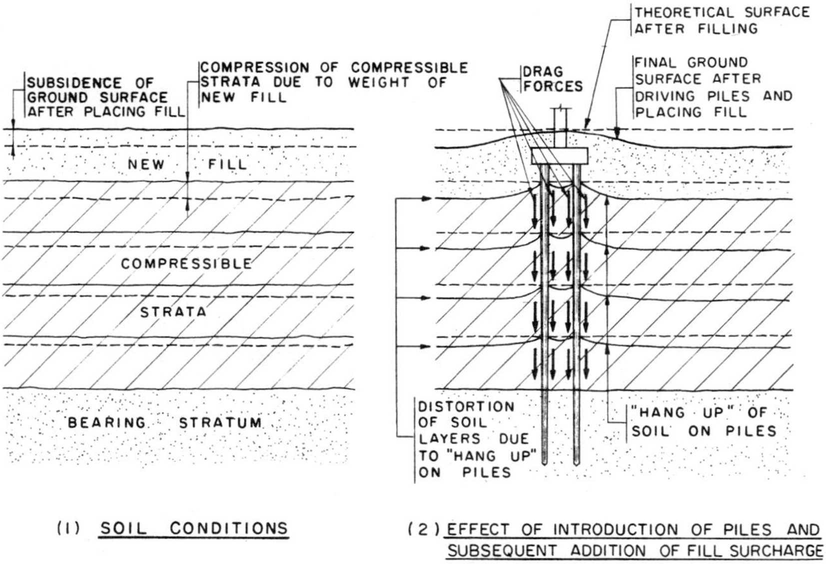

(c) EFFECT OF ADDED WEIGHT

FIG. 4-4 *(Continued)*

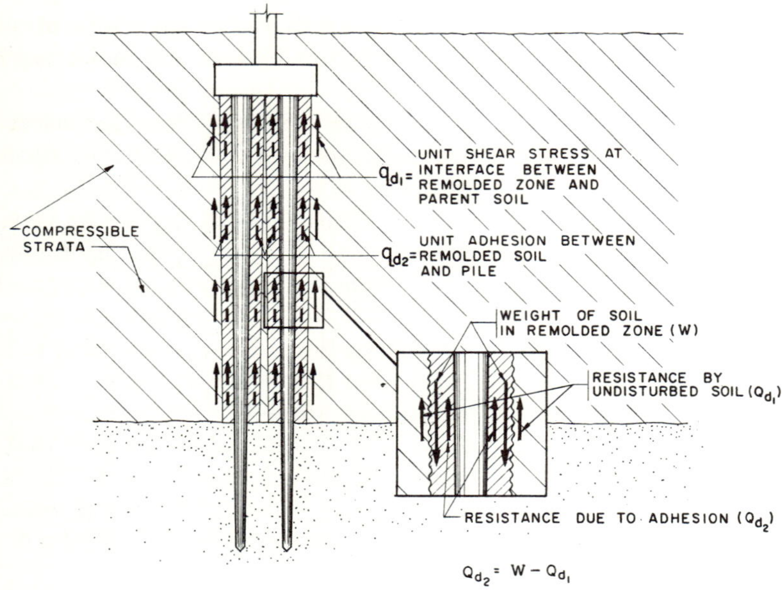

FIG. 4-5 Drag on Piles Due to Remolding of Compressible Soil.

the relative values of adhesion force and weight of remolded soil can be checked as indicated in Fig. 4-7. It is emphasized, however, that the drag cannot exceed the *lesser* of the two forces (adhesion or weight of remolded material).

The procedure illustrated in Fig. 4-6 is not intended to represent an exact solution. For example, it neglects the fact that the total movement in the lower limits of the soil strata is small and so does not impose

TABLE 4-2 ***Values of Adhesion between Soil and Pile (Clay Soils)****

Pile type	Consistency of soil	Adhesion, psf
Timber and concrete	Very soft	0–250
	Soft	250–480
	Medium stiff	480–750
	Stiff	750–950
	Very stiff	950–1,300
Steel	Very soft	0–250
	Soft	250–460
	Medium stiff	460–700
	Stiff	700–720
	Very stiff	720–750

* *After Tomlinson, M. J.* [3], *by permission.*

the same drag as does a similar unit of soil at the top of the stratum. Rather, the procedure represents an upper limit to the potential drag force.

Evaluation of drag due to remolding by the above described method is sufficiently accurate for most cases. The reason simply is that the mag-

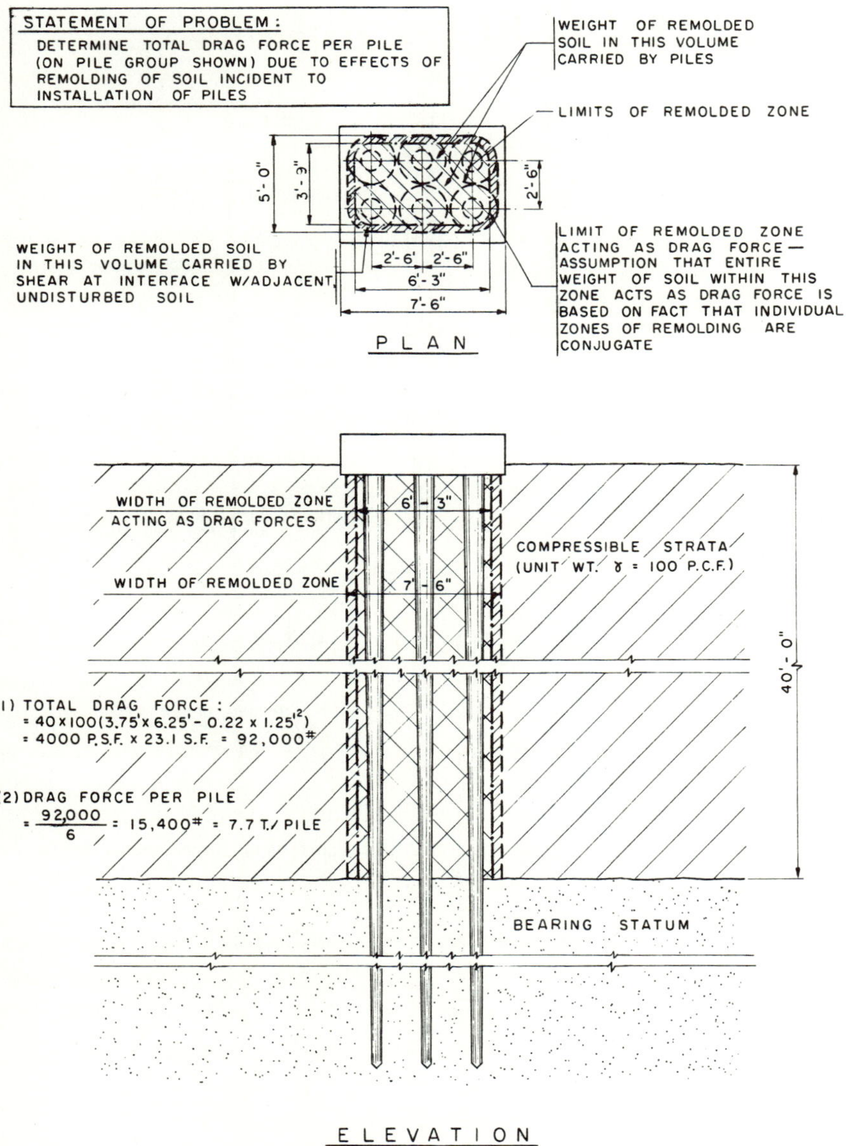

FIG. 4-6 Calculation of Drag Force Due to Weight of Remolded Soil.

nitude of the drag force so computed is relatively small compared to the design capacity of the piling and so can be absorbed in the usual allowable overload capacity of the piling, or can be supported by providing an extra pile or two. This is illustrated by the example of Fig. 4-6. The figure represents a group of six timber piles, embedded 40 ft in a compressible

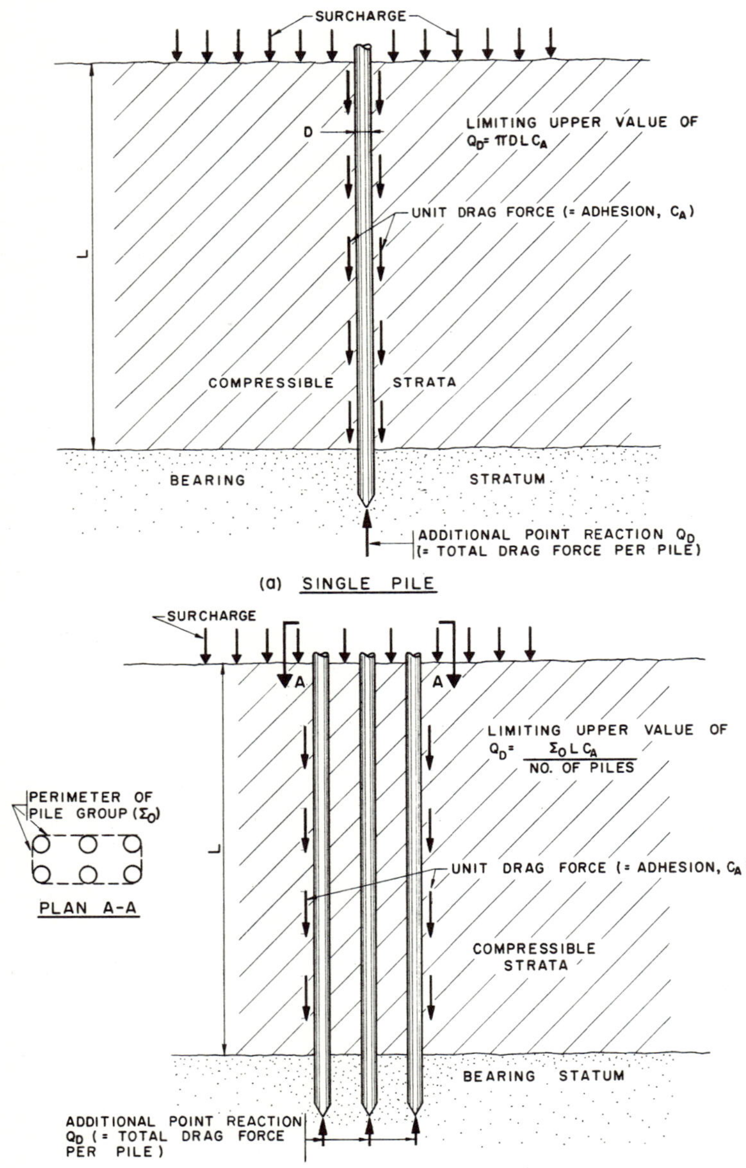

FIG. 4-7 Computation of Adhesion on Piles.

stratum and spaced at 2 ft 6 in. center to center. This is a fairly extreme case. If the compressible stratum is highly sensitive (again an extreme condition), the computation would give an estimated upper limit of the potential drag force of 7.7 tons per pile.

For timber piles of 25 ton capacity, and assuming an allowable overload due to drag of 15 percent (as recommended, for example, in Ref. 2), the drag force would cause a reduction in the apparent pile capacity of only 4 tons per pile which could be absorbed either by driving the piles to a slightly higher capacity (29 tons) or by adding one extra pile to the group. Neither solution imposes a severe penalty.

This example also illustrates why the principal troubles with drag are encountered with low-capacity members (say 10- to 15-ton timber piles). The total drag for any given condition is the same regardless of the capacity of the pile. Thus, whereas the overload factor in the indicated example is only about 30 percent, it would be 50 percent for a 15-ton pile and 75 percent for one of 10-ton capacity. Similarly, it would be only 25 percent for a 30-ton pile and 13 percent for a 60-ton pile.

CASE 2 *Drag Due to General Subsidence:* In the second case, as previously noted, the effects of continuing consolidation of the strata penetrated by the pile, either under the dead weight of the strata or because of the imposition of surcharge, are to be added to the effects of remolding. Again upper limits can be placed on the amount of drag which can be developed. Several approaches are available.

METHOD 1: The drag force cannot exceed the unit adhesion (Table 4-2) which can be developed between pile and soil, multiplied by the length and perimeter of the pile or (as shown in Fig. 4-7) of the pile group, whichever is the lesser value. The reason is that if the adhesion capacity were to be exceeded, relative movement between the pile and the soil would take place, tending to dissipate the drag forces. For the case illustrated in Fig. 4-6, assuming a medium-stiff clay as the compressible stratum (for which the unit adhesion is about 600 psf), the drag force per pile would be about

$$\frac{19 \text{ ft} \times 40 \text{ ft} \times 600 \text{ psf}}{6 \text{ piles}} = 76{,}000 \text{ lb or } 38 \text{ tons/pile}$$

This is an exceedingly high value, and computations based on this method would preclude the use of low- to medium-capacity piles for support of the building. Therefore, the validity of the computation should be checked by other methods, as follows:

METHOD 2: The drag force also cannot exceed the ultimate shear strength of the soil multiplied by the length and perimeter of the pile group, since if such strength were exceeded, the soil again would move and dissipate the drag force. The shear strength to be used in the computation should be the drained strength, as measured in the triaxial test. This computation usually gives a higher upper limit than the computation based on adhesion (Method 1), however, and so is not generally a controlling consideration.

METHOD 3: The drag force is assumed to be equal to the pullout capacity of the pile. The basis for this assumption is that drag results from the tendency for the soil to move downward with respect to the pile, and therefore a similar effect would be realized if the pile tended to move upward with respect to the soil (action equals reaction). This premise is likely to give results which are quite conservative, but again, it is an upper limit (and likely to be a more tenable one than the perimeter-adhesion or perimeter-shear results). Its use requires an evaluation of the pullout capacity. A series of pullout tests is one way to do this. This is a good procedure if the pile is permitted to set long enough for the adhesion of the surrounding soil to develop before the tests are conducted and if some sort of stub pile, representing the section embedded in the compressible strata, is used. The approach has the disadvantage of cost of the tests, of delay, and of the impracticality of evaluating group action of the piles, but it has the important advantage that the effect of the shape of the pile (taper, smooth versus rough wall, etc.) on the magnitude of the drag force is reflected in the test results.

METHOD 4: Johannessen and Bjerrum [8] suggest that negative skin friction can be computed from the confining pressure of the surrounding soil and suggest the equation

$$\tau = \bar{\sigma}_v K \tan \bar{\phi}_a \tag{4-1}$$

where τ = unit shear
$\bar{\sigma}_v$ = effective vertical stress
K = ratio of vertical to lateral stress
$\bar{\phi}_a$ = effective friction angle between pile and soil

This equation may be recognized as an expression of the fact that friction is proportional to normal pressure. Evaluation of the variables $\bar{\sigma}_v$, K, and $\bar{\phi}_a$ requires extensive laboratory testing, plus judgment.

METHOD 5: A fifth approach would be to assume that the drag force on the pile (which occurs because the presence of the pile prevents consolidation of the compressible strata) is equal to the force distributed in the soil which would produce the same amount of consolidation (again, action equals reaction). The approach is illustrated in Fig. 4-8.

For purposes of numerical evaluation, it is necessary to make an assumption as to the distribution of the drag force along the length of the pile. Its intensity probably is greater at the top of the pile than at the bottom since the tendency toward relative movement of the soil with respect to the pile is greater at the top. Therefore, an assumption of a triangular distribution is suggested, as indicated in the figure. A curvilinear distribution may be assumed, if desired. More conservative results will result from assumptions in which the greater part of the drag force is assumed to act at the top of the pile.

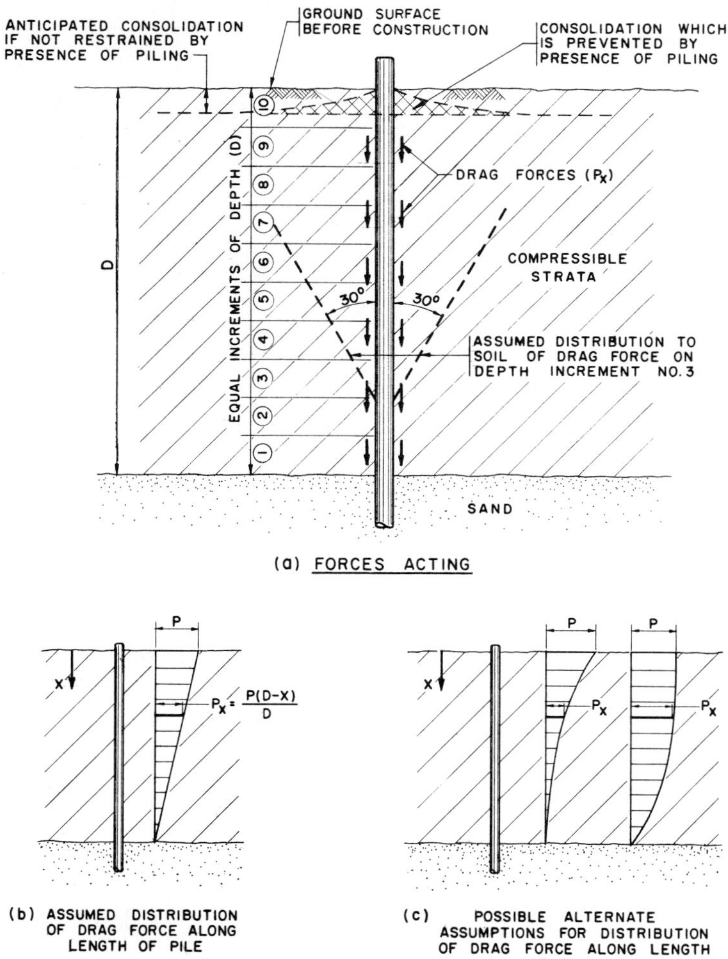

FIG. 4-8 Evaluation of Drag Force on Pile—Method 5.

The distribution of the drag force within the soil mass may be approximated on the basis of the conventional 30° distribution indicated in Fig. 4-8*a* (and discussed in Chap. 5), or may be evaluated from theoretical considerations as discussed in Refs. 4 and 5.

An example of a computation worked out following Method 5 is contained in Sample Computation 4-1.

SAMPLE COMPUTATION 4-1

Drag Force on Piles

STATEMENT OF PROBLEM

Determine total drag force on single pile, as indicated, due to consolidation of clay soil resulting from placement of 6-ft depth of sand fill after or immediately prior to installation of pile.

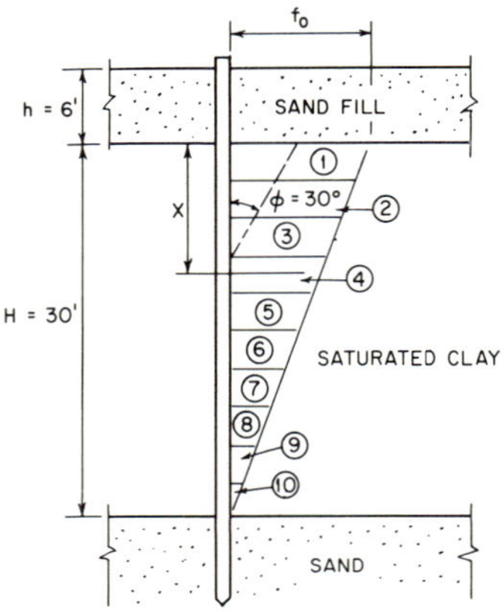

GIVEN

1. Unit weights:
   Sand fill: $\gamma_s = 120$ pcf
   Clay: $\gamma_c = 95$ pcf
2. Clay is fully consolidated under its own weight.
3. $C' = \dfrac{C_c}{1 + e_0} = 0.2$

PROCEDURE

STEP 1: Compute amount of consolidation which would occur in clay stratum under weight of sand fill, if there were no piling.

Settlement due to surcharge:

$$S = HC' \log\left(\frac{P_0 + \Delta P}{P_0}\right)$$

$$P_0 = \frac{30}{2} \times 95 = 1{,}425 \text{ psf}$$

$$\Delta P = 120 \times 6 = 720 \text{ psf}$$

$$S = 30 \times 0.2 \log\left(\frac{1{,}425 + 720}{1{,}425}\right) = 1.065 \text{ ft}$$

STEP 2: Compute shear (drag) forces acting between perimeter of pile and abutting soil which, when applied to the soil, would produce the same total consolidation of the clay stratum as that produced by the weight of the sand fill (step 1).

Drag on pile per unit of length at distance x from top of clay layer $= f_x$.

Assume that the distribution of the shear (drag) force varies linearly with depth, as shown in Fig. 4-8*b*, i.e., $f_x = f_0\left(1 - \dfrac{x}{H}\right)$, where f_0 is the drag per unit of length at the top of the clay layer.

Divide H into $n = 10$ segments: $\Delta x = \dfrac{H}{n} = \dfrac{30}{10} = 3$ ft

Drag force in one segment: $F_x = f_x(\Delta x) = f_0\left(1 - \frac{x}{H}\right)\Delta x$

$$= f_0\left(1 - \frac{x}{n\,\Delta x}\right)\Delta x$$

Assume total drag force on each segment (F_x) is distributed to the soil at an angle $\phi = 30°$, as indicated, then

Let ΔPF_x be the average change of unit pressure in the depth x of clay soil overlying the point x caused by the force F_x.

$$\Delta P_{F_x} = \frac{F_x}{\pi\left(\frac{x}{2}\tan\phi\right)^2} = \frac{f_0\left(1 - \frac{x}{n\,\Delta x}\right)\Delta x}{\pi\left(\frac{x}{2}\tan\phi\right)^2}$$

$$\text{But } x = i\,\Delta x - \frac{\Delta x}{2} = \frac{\Delta x}{2}(2i - 1)$$

Where i is the number of the segment being considered, starting from the top.

$$\Delta P_{F_x} = \frac{f_0\left(1 - \frac{\Delta x(2i-1)}{2n\,\Delta x}\right)\Delta x}{\frac{\pi}{4}(\tan^2\phi)(2i-1)^2\frac{(\Delta x)^2}{4}}$$

$$= \frac{8f_0(2n - 2i + 1)}{n\pi(\tan^2\phi)(2i-1)^2\,\Delta x} = \frac{8f_0(21 - 2i)}{10 \times 3.14(0.58)^2(2i-1)^2(3)}$$

$$= \frac{(21 - 2i)}{(2i-1)^2} f_0 \times 0.252$$

P_0 (original average stress in clay stratum above the level x due to weight of clay)

$$= \tfrac{1}{2}(\gamma_c)\left(i\,\Delta x - \frac{\Delta x}{2}\right) = \frac{\gamma_c}{4}\Delta x(2i-1) = \tfrac{95}{4} \times (3)(2i-1) = 71.2(2i-1)$$

$S_x = xC' \log\left(\frac{P_0 + \Delta P}{P_0}\right)$, where S_x is the consolidation of the depth x of the clay stratum due to the force F_x acting on the segment at depth x.

Using: $x = \left(i\,\Delta x - \frac{\Delta x}{2}\right)$, $C' = 0.2$, and $\Delta x = 3$ ft

$S_i = \left(i\,\Delta x - \frac{\Delta x}{2}\right) 0.2 \log\left(\frac{P_0 + \Delta P}{P_0}\right)$, where S_i is the total consolidation of the clay stratum due to the drag force on segment i.

$$S_i = 0.3(2i - 1)\log\left(\frac{P_0 + \Delta P}{P_0}\right)$$

Substituting the values of P_0 and ΔP previously derived:

$$S_i = 0.3(2i - 1) \log \left[\frac{(21 - 2i)f_0 \times 0.252}{(2i - 1)^3 71.2} + 1 \right]$$

For $i = 1$:

$$S_1 = 0.3 \log \left[\frac{4.79 f_0}{71.2} + 1 \right]$$

Similarly, settlement S_i is obtained for $i = 2$ through 10, and the result is tabulated below for different values of f_0.

S \ f_0	1,500	1,200	1,000	800	600	500
S_1	0.602	0.573	0.550	0.521	0.485	0.461
S_2	0.575	0.509	0.458	0.401	0.332	0.294
S_3	0.321	0.269	0.231	0.191	0.148	0.126
S_4	0.168	0.137	0.115	0.093	0.070	0.059
S_5	0.091	0.073	0.061	0.049	0.037	0.031
S_6	0.051	0.040	0.034	0.026	0.020	0.017
S_7	0.028	0.026	0.019	0.014	0.011	0.009
S_8	0.014	0.012	0.011	0.008	0.005	0.005
S_9	Negl.	Negl.	Negl.	Negl.	Negl.	Negl.
S_{10}	Negl.	Negl.	Negl.	Negl.	Negl.	Negl.
ΣS	1.850	1.639	1.479	1.303	1.108	1.002

Total $S = 1.065$ ft (from Step 1)
By interpolation,

$$f_0 \approx 560 \text{ lb per ft}$$

Total drag:

$$F = 560 \times {}^{30}\!/_{2} = 8{,}400 \text{ lb per pile}$$

The advantages of Method 5, as compared to nos. 1 through 4, are that the computation reflects the effects of surcharge loading and of preconsolidation (if any) of the compressible strata. These effects are reflected in the computation by reason of their influence on the calculation of the amount of anticipated consolidation which would occur if the compressible strata were not restrained. Pile group action also is considered.

The computation assumes, of course, that the adhesion between pile and soil is not exceeded, i.e., that there is no relative movement between the surface of the pile and the abutting soil.

In practice, each of the above methods will give different results for the drag force developed in a particular situation. The value to be used for design would be the *least* of the values so obtained.

The reader's attention is directed to the basic difference in approach between Case 1 (drag due to remolding) and Case 2 (drag due to subsidence). In Case 1 it was determined (1) that the drag could not exceed a certain amount, (2) that this amount was negligible (or small), and, therefore, (3) that the drag could be neglected or compensated for by adding a few extra piles, by driving to a little higher capacity, or by accepting some relatively innocuous penalty.

In contrast, where the problem is one of subsidence, the drag forces are likely to be large, so large in fact that for low-capacity piles often they cannot be handled at all, necessitating the use of fewer piles of higher capacity.

Accordingly, where a serious problem of drag is anticipated, such as where the piles penetrate strata of peat, organic clays or silts, soft clay deposits, or recent fills, particular caution should be exercised in the design.

Section 4E HEAVE FORCES

Heave forces are those due to frost and to swell of the ground. They include two components: the heave pressure on the underside of the foundation and the grip of the soil along the sides.

The heave pressure on the underside of the foundation usually is eliminated by founding the structure below the active zone (Chap. 3).

The adhesion of the soil to the perimeter of the foundation also can be eliminated. The usual method consists of isolating the foundation from the soil by a sleeve or blanket of weak spongy material which will not rot out, such as mineral wool, vermiculite, or an oil-wax mixture. In most cases, however, this is not necessary. Experience indicates that perimeter grip is not a serious problem except in unusually severe cases, e.g., a very light structure supported by a foundation which has a large perimeter in contact with a highly active soil.

In cases where, for whatever reason, it is found necessary to estimate the forces due to heave, the heave pressure can be determined by simulating field conditions in the laboratory. The usual procedure is to place a sample of the foundation soil material in the consolidation ring apparatus and to measure the pressure necessary to maintain the sample thickness as (1) moisture is added or (2) as the sample is frozen.

At the present time (1968) the only reliable means for estimating the grip forces also are laboratory tests which seek to simulate field conditions.

Section 4F LATERAL FORCES

4F.1 Wind and Earthquake

The lateral loads on the superstructure due to the forces of earthquake and wind will produce vertical and horizontal reactions on the foundations. The vertical reactions, which include uplift, are due to the overturning moment. The lateral reactions resist the tendency of the foundation to slide. The computation of such forces is made in accordance with conventional well-known procedures and will not be considered here. However, the application of wind and earthquake forces to design involves two matters which bear special discussion: (1) allowable overstress factors and (2) uplift resistance.

a. Allowable Overstress Factors

The allowable overstress factor for wind and earthquake load is an expression of the improbability of ever realizing a juxtaposition of maximum wind or earthquake force simultaneously with maximum vertical live load. In its usual form it is stated:

> For stresses due to wind (or earthquake) acting alone or in combination with other loads, the design stresses may be increased one-third over basic allowable values, provided the section thus furnished is not less than that required for the normal combination of dead and live load.

In foundation design, however, the engineer usually deals with allowable loads rather than allowable stresses. In addition, where allowable stresses are used, the values vary depending on embedment and other factors. Lastly, allowable loads may be predicated on ultimate capacity rather than on stress. For all of these cases, an increase in allowable stress is meaningless and an alternate approach to the allowance for probability is required. This alternate is that instead of an increase in allowable stress, the designer incorporates a decrease in load. The principle may thus be restated as follows:

> One-fourth of each of the loads acting on the element being designed may be neglected where the element is subject to wind (or earthquake) acting alone or in combination with other loads, provided the section thus provided is not less than that required for the normal combination of dead and live load.

EXCEPTION: An exception to the above statements must be made when the loading condition being considered already includes an allowable overstress (or probability) factor. An example would be when a temporary structure was being designed for wind load. The design merits an allowable overload factor reflecting the improbability of simultaneous occurrence of maximum wind and maximum vertical load and another overload factor reflecting the temporary nature of the construction. Designs proportioned for thermal stresses plus wind (or earthquake) forces are in a similar category.

For such multiple combinations, the allowable overload factors are not additive, i.e., the allowable overload factor for thermal stress (1.33) plus wind load (1.33) is not $1.33 \times 1.33 = 1.77$, nor is it $1.33 + 1.33 = 2.66$. In fact, the exact value probably could be debated endlessly. As a practical matter, however, an upper limit value of 1.50 (corresponding to a one-third decrease in the applied load) is used, this value being established in several design specifications (AASHO, etc.) as the largest overload factor to be permitted under any combination of loading conditions. This limit provides for a minimum margin of safety of 10 percent* under the possible (albeit unlikely) concurrence of all maximum loading conditions.

In this connection, it is important to note that the live-load reduction factor (Sec. 4I, Chap. Note 1) conventionally considered in calculating the reactions from the superstructure also is an overload factor reflecting the improbability of the simultaneous occurrence of loads (in this case full live load on all areas of the building contributing load to the member in question). As such, it comes under the same category as the thermal stresses or considerations of temporary construction described above. Structural elements of the foundation, proportioned for the combination of wind (or earthquake) load plus reduced live load at a one-third increase in allowable stress (or its equivalent, a one-quarter reduction in total load), should be capable of supporting the combination of wind (or earthquake) plus *full* live load without exceeding allowable stress or allowable load values by more than 50 percent.

b. Uplift Resistance

Only the dead-load forces, including the weight of the foundation discounted for buoyancy, should be assumed to resist uplift forces; i.e., it should be assumed that the building is unoccupied when the earthquake or hurricane hits.

There is some debate whether the weight of the soil overlying the footings should be included in the computation of dead load.† The

* Conventional minimum safety factor = 1.65 (steel in tension)

$$\frac{1.65}{1.50} = 1.10 \text{ or 10 percent margin of safety against yield}$$

† References on this question are rare. The Uniform Building Code is the only one of the standard codes reviewed by the authors to make specific reference to the problem and indicates that the weight of the overlying soil shall be included in the calculation of the dead-load moment assumed to resist the overturning moment due to lateral loads.

issue relates to the possibility that such soil might be inadvertently excavated and that its stabilizing effect would not be available for a period of time. In general, the weight of such soil is not of important magnitude and can be neglected without serious consequences. Also, the probability of such excavation being open at the same time that disaster strikes is extremely low. Accordingly, many designers overlook these factors and include the weight of the overlying soil in the computation of the forces resisting uplift.

NOTE: For purposes of design, the forces of wind and earthquake are not considered to act simultaneously.

4F.2 Earth Pressure

Foundation walls, including retaining walls, must be designed to resist the lateral earth pressures acting thereon plus the effects of surcharge. Lateral earth pressure (including hydrostatic pressure due to ground water conditions) is not considered a condition subject to an allowable overload factor, and the stresses and forces resulting therefrom should be added directly to the effects of dead, live, and other loading conditions.

If the wall penetrates below groundwater level, any differential hydrostatic pressure which might occur should be added to the lateral earth pressure, with the lateral earth pressure due to the weight of submerged soil being taken as that due to its buoyant, not dry, weight.

Computation of lateral earth pressure may be made as described in any of several standard references [4, 5].

Section 4G IMPACT

With certain exceptions, impact forces may be neglected in the design of the foundations for buildings because they are small in comparison to the total of the static components of the load and are largely dissipated in the mass of the foundation. The exceptions are:

1. Where the foundation bears on loose (N less than 10) granular soils. For such soils, the compaction due to impact vibration can cause serious settlements, and foundations subject to dynamic loads of frequent occurrence should not be founded on such materials.

2. Where the ratio of the live load causing impact reaction on the foundation to the total static load (dead plus live) acting thereon is large (say, in excess of one-third*); the effect of the impact is not small and

* The value of ⅓ is derived, as follows:

$$\text{Amount of overload due to impact} = \frac{K(\text{LL causing impact})}{\text{total load (DL + LL)}}$$

where K is impact factor.

(Footnote continued on page 97.)

must be considered by adding the appropriate impact percentage to the impacting live load. The impact percentage to be added will generally be given in the design specification for the project (AASHO for truck loading, AREA for railway, or AISC for cranes). In the absence of such data, the following approximate values may be used for the design of buildings:

1. Truck loading—10 percent
2. Passenger car loading—negligible
3. Crane runways—25 percent of lifted load or 15 percent of wheel loads, whichever is greater
4. Monorails—10 percent for hand-operated and 25 percent for electrically operated trolleys
5. Machinery supports—reciprocating or heavy power-driven units: 50 percent; light shaft or motor-driven units: 25 percent
6. Structures subject to blast loading—see Sec. 4I, Chap. Note 2

Section 4H STABILITY

The foundation under the action of the lateral loads described in this section must be stable, i.e., it must not overturn or slide.

The factor of safety against overturning should be at least 1.5 (see Table 4-3) calculated as the sum of the righting moments divided by the sum of the overturning moments. As described in Sec. 4F, the calculation of the righting moments should include a deduction for the effects of buoyancy and should include a positive moment reflecting the weight of the soil overlying the footings. Any live load in the building should be neglected in the computation.

The factor of safety against sliding also should be at least 1.5 (see Table 4-3). Again the calculation of the resisting forces should consider any effects of buoyancy and should consider the building free from any live load. There is some debate, however, whether to include passive earth pressure in the calculation of resistance.

Amount of overload which might be considered negligible, say 5 percent.

Assumed average value of $K = 0.15$.

Therefore the ratio

$$\frac{\text{LL causing impact for which impact can be neglected}}{\text{DL} + \text{LL}} = \frac{0.05}{0.15} = \frac{1}{3}$$

If the value for K were 0.25, then the ratio of

$$\frac{\text{LL for which impact could be neglected}}{\text{DL} + \text{LL}} = \frac{0.05}{0.25} = \frac{1}{5}$$

Some authorities (Ref. 7, for example) recommend that passive pressure be disregarded. The reasons generally given are (1) that such resistance might be removed inadvertently by excavation and (2) that if such resistance is provided by soil subject to freezing action, the strength of the soil could be drastically reduced upon thawing. Possible dissipation of soil strength due to root growth also has been cited [7].

TABLE 4-3 ***Recommended Factor of Safety Against Stability Failure (From Various Sources)***

Source	Against overturning	Against sliding	Remarks
Terzaghi and Peck [7]	(See remarks)	1.5	Resultant within middle third
Henry [13]	2.0	1.5 minimum 2.0 preferred	
Hough [6]	On rock, 1.5 On soil (see remarks)	...	Resultant within middle third
BuDocks—DM7 [2]	1.5	1.5	Not considering passive pressure
		2.0	Considering passive pressure
Dunham [14]	1.75	1.5–2.0	
Uniform Building Code	1.5		
Southern Standard Building Code	1.5		
BOCA Code	1.33		
City of Baltimore	1.5		
Los Angeles	1.5	...	Resultant within middle half
Cleveland	1.5		
Pittsburgh	1.5		
St. Louis	1.33		
New York City (new)	1.5	1.5	

Other sources [2] recommend a compromise solution wherein the passive pressure is neglected unless the sliding surface is below the frost depth, in which case the passive resistance on that portion of the foundation below frost depth is included in the stability computation (a factor of safety of 2.0 is required for such case, however, in lieu of the more conventional value of 1.5).

Still other designers include the full passive pressure in their calculations.

The reasons for disregarding passive pressure in the stability computation, as cited above, are valid. However, they do not always apply.

In a warm climate, for example, freezing of the ground does not occur, and if the grade around the foundation is paved, dissipation of resistance from root growth, seepage, or inadvertent excavation of the soil around the foundation is an unlikely occurrence.

A problem which always exists, however, is that the development of passive earth pressure requires movement of the foundation. Worse still, the amount of movement is likely to be erratic due to variations in the soil and vagaries in the imposed load. In general, such movements are not acceptable. Foundations are not supposed to move. Therefore, except where the foundation bucks up against a substantial slab or other construction which can resist the lateral forces by friction or is unusually deep, it is recommended that the passive pressure be disregarded in the stability computation.

In addition to passive pressure, the forces resisting sliding include the frictional resistance between the foundation and the underlying soil and the lateral load capacity of any piling.

The lateral load capacity of piling is discussed in Chap. 6.

The frictional resistance between the foundation and the soil is equal to the normal pressure times the applicable coefficient of friction. For medium- to coarse-grained granular soils and for foundations bearing on rock, a value of 0.5 is suggested as a conservative value for the coefficient of friction. For fine-grained granular materials and for clayey or silty medium sands, this value should be reduced to about 0.40 to 0.45, with the lower value applicable for deposits containing a relatively higher percentage of fine material.

For silts and clays, the shear strength of the soil may be less than the frictional resistance between soil and foundation so that the failure plane may pass partly or wholly through the soil beneath the foundation. This is not the case for granular materials since the coefficients 0.4 to 0.5 correspond to a friction angle ϕ of 21.8° to 26.5°, which would be unrealistically low values for a granular soil. Therefore, for silts and clay soils, two conditions must be checked. One represents a failure plane along the interface between soil and foundation for which a somewhat reduced friction coefficient of 0.35 to 0.40 is recommended, and the second represents possible failure planes (of which several configurations may have to be checked) in the soil beneath the foundation for which the conventional analysis of translational failure (Fig. 4-9) applies. For a discussion of the analysis of translational failure see chap. 7 of Ref. 2 or chap. 8 of Ref. 6.

In the case of pile-supported structures, the frictional resistance between the foundation and the underlying soil should be entirely discounted because of the aforementioned tendency for the underlying soil to settle away from the underside of the pile cap, thereby dissipating such resistance. For piled structures, resistance to lateral load must be provided by the piles themselves, or by some form of deadman anchorage.

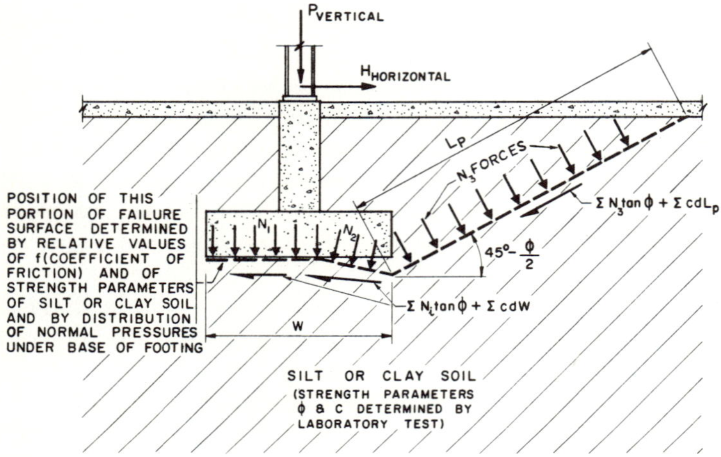

REQUIRED FOR STABILITY:

$$H \leq \Sigma N \tan\phi + \Sigma\, cdW + (\Sigma N \tan\phi + \Sigma\, cdL_p)\cos\left(45^\circ - \frac{\phi}{2}\right)$$

FIG. 4-9 Analysis of Translational Failure.

Section 4I CHAPTER NOTES

1 Live-load Reductions

Design live loads consist of assumed concentrations (representing elements of machinery, equipment, or moving wheel loads) and assumed equivalent distributed loads (representing the weight of the occupants, furnishings, partitions, and stored materials or goods).

In building design in general, and in the design of the foundations in particular, the equivalent distributed loads are by far the more important consideration.

These equivalent distributed loads represent assumed maximum accumulations. The probability that such accumulations will occur simultaneously over large areas of the building is less than the probability of their occurrence over limited areas. Most building codes recognize this fact and provide that as the area of the building which is supported by each column or pier increases, the column or pier reactions may be calculated on the basis of a reduced magnitude of the assumed distributed load. Some typical provisions are:

a. *New York City Building Code* (1938)

In structures intended for storage purposes all columns, piers or walls and foundations may be designed for eighty-five per cent of the full assumed live load. In structures intended for other uses the assumed live load used in designing all columns, piers or walls and foundations may be as follows:

one hundred per cent of the live load on the roof,
eighty-five per cent of the live load on the top floor,
eighty per cent of the live load on the next floor,
seventy-five per cent of the live load on the floor next below.

On each successive lower floor, there shall be a corresponding decrease in the percentage, provided that in all cases at least fifty per cent of the live load shall be assumed.

b. *USAS A58.1* (1962)

Reduction in Live Loads

Roof Live Loads. No reduction shall be applied to the roof live load.

Live Loads 100 Pounds per Square Foot or Less. For live loads of 100 pounds or less per square foot, the design live load on any member supporting 150 square feet or more may be reduced at the rate of 0.08 percent per square foot of area supported by the member, except that no reduction shall exceed either R as determined by the following formula, or 60 percent:

$$R = 100 \times \frac{D + L}{4.33L}$$

in which R = reduction, in percent
D = dead load per square foot of area supported by the member
L = design live load per square foot of area supported by the member.

Live Loads Exceeding 100 Pounds per Square Foot. For live loads exceeding 100 pounds per square foot, no reduction shall be made, except that the design live loads on columns may be reduced 20 percent.

2 Blast Loading

For special structures designed to resist unusually severe transient loadings (such as nuclear blast–resistant buildings and underground structures), a more complex mathematical analysis which takes account of the interaction of the soil and structure must be resorted to, and is not treated here. Reference is made to standard texts such as Ref. 12.

REFERENCES

1. Discussions: Meetings of Code Committee, Metropolitan Section, ASCE, 1964–1965.

2. Design Manual DM7, "Foundations," U.S. Navy, Bureau of Yards and Docks, Washington, D.C.
3. Tomlinson, M. J.: The Adhesion of Piles Driven in Clay Soils, *Proceedings, Fourth International Conference on Soil Mechanics and Foundation Engineering,* 1957.
4. Terzaghi, Karl: "Theoretical Soil Mechanics," John Wiley & Sons, Inc., New York, 1943.
5. Tschebotarioff, G. P.: "Soil Mechanics, Foundations, and Earth Structures," McGraw-Hill Book Company, New York, 1951.
6. Hough, B. K.: "Basic Soils Engineering," The Ronald Press Company, New York, 1957.
7. Terzaghi, Karl, and R. B. Peck: "Soil Mechanics in Engineering Practice," 2d ed., John Wiley & Sons, Inc., New York, 1967.
8. Johannessen, I. J., and L. Bjerrum: Measurement of the Compression of a Steel Pile to Rock Due to Settlement of the Surrounding Clay, *Proceedings, Sixth International Conference on Soil Mechanics and Foundation Engineering,* vol. 2, 1965.
9. Van Weele, A. F.: Negative Skin-friction on Pile Foundation in Holland, *Proceedings, Symposium on Bearing Capacity of Piles,* Central Building Research Institute, Roorkee, India, 1964.
10. Casagrande, A.: The Structure of Clay and Its Importance in Foundation Engineering, *Proceedings of the Boston Society of Civil Engineers,* vol. 19, no. 4, April, 1932.
11. Cummings, A. E., G. O. Kerkhoff, and R. B. Peck: Effect of Driving Piles into Soft Clay, *Transactions of the American Society of Civil Engineers,* vol. 115, 1950.
12. Norris, C. H., R. J. Hansen, M. J. Holley, Jr., J. M. Biggs, S. Namyet, and J. K. Minami: "Structural Design for Dynamic Loads," McGraw-Hill Book Company, New York, 1959.
13. Henry, F. D. C.: "The Design and Construction of Engineering Foundations," McGraw-Hill Book Company, New York, 1956.
14. Dunham, C. W.: "Foundations of Structures," 2d ed., McGraw-Hill Book Company, New York, 1962.
15. Sowers, G. F.: Shallow Foundations, in G. A. Leonards (ed.), "Foundation Engineering," McGraw-Hill Book Company, New York, 1962.

CHAPTER 5
Allowable Soil Bearing Pressure

Allowable soil bearing pressure is limited by two considerations: (1) the foundation must be safe against shear failure of the supporting soil, and (2) postconstruction settlement must not be excessive. The bearing pressure to be used for design is the lesser of the values dictated by the two requirements.

In the general case, these considerations are somewhat involved. For a specific case (a given type of foundation bearing on, or in, a given class of soil), however, many of the aspects of the general treatment are superfluous, and the problem is much simpler than might at first appear. Table 5-1 summarizes the procedures to be followed in determining the capacity of a loaded area bearing on different classes of soil and should be referred to before tackling any specific problem.

Section 5A SAFETY AGAINST SHEAR FAILURE

The safety of a proposed foundation against shear failure of the soil is gaged by comparing actual bearing pressure on the loaded area with the ultimate bearing capacity of said area, reduced by a suitable safety factor. Methods for estimating the ultimate bearing capacity are given in Sec. 5A.1. Recommendations as to applicable safety factors are contained in Sec. 5A.2. Methods for computing actual bearing pressures are given in Sec. 5C.

5A.1 Determining Ultimate Bearing Capacity

Much attention [1–4] has been devoted in the development of soil mechanics to the derivation of equations expressing the ultimate bearing

TABLE 5-1 ***Procedures for Estimating Allowable Soil Bearing Capacity***

Type of soil	Procedure
Rock, hardpan, gravel soils	1. Determine actual bearing pressures as described in Sec. 5C. 2. Determine allowable bearing pressure from empirical values indicated in Sec. 5D.2, modified for correction factors as described in Secs. 5D.3 to 5D.6. 3. Actual bearing capacity to be less than or equal to allowable capacity, as modified.
Sand soils	1. Determine actual bearing pressures as described in Sec. 5C. 2. Estimate allowable total settlement of structure (Sec. 5B.3). 3. Determine allowable bearing pressure from empirical values indicated in Sec. 5D.2, modified for relation between estimated allowable total settlement and assumed total settlement of 1 in. used in establishing the indicated empirical relation. 4. Modify determination of allowable bearing pressure for factors described in Secs. 5D.3 to 5D.6. 5. Actual bearing capacity to be less than or equal to allowable capacity, as modified.
Clay and silt soils	1. Determine soil properties, ϕ, c, and C'_c. 2. Determine actual bearing pressures as described in Sec. 5C. 3. Estimate allowable settlement (total or differential) of structure (Sec. 5B.3). 4. Determine ultimate capacity of each bearing area (Sec. 5A). 5. Determine probable actual settlement of structure (total or differential) as described in Sec. 5B.1. 6. Determine allowable bearing pressure as the lesser of the following: *a.* Ultimate bearing capacity divided by factor of safety. *b.* Comparison of allowable settlement with probable actual settlement.
Miscellaneous soil materials	See Sec. 5D.2*g*.

capacity of foundations. Three cases commonly are differentiated: (1) concentrically loaded shallow bearing area, (2) concentrically loaded deep bearing area, and (3) eccentrically loaded bearing areas. The results generally are expressed as follows:

a. Concentrically Loaded Bearing Area at Shallow Depth (Footings and Mats)*†‡

I. Continuous strip footing:

$$q_{\text{ult}} = \overbrace{cN_c}^{\text{cohesion}} + \overbrace{\Sigma\gamma_d D_f N_q}^{\text{surcharge}} + \overbrace{\frac{\gamma_d B N_\gamma}{2}}^{\text{friction}} \tag{5-1}$$

II. Square or rectangular footing:

$$q_{\text{ult}} = cN_c\left(1 + 0.3\frac{B}{L}\right) + \Sigma\gamma_d D_f N_q + 0.4\gamma_d B N_\gamma \tag{5-2}$$

III. Circular footing:

$$q_{\text{ult}} = 1.3cN_c + \Sigma\gamma_d D_f N_q + 0.6\gamma_d r N_\gamma \tag{5-3}$$

where q_{ult} = ultimate bearing capacity (per unit of area).

c = unit cohesive strength, usually taken as one-half the unconfined compressive strength (undrained specimen – unconfined compression test). May be taken as intercept on cohesion axis of Mohr envelope derived from triaxial-test data. If triaxial test is utilized, base Mohr envelope on total stresses. Triaxial test must be conducted so that drainage conditions will resemble those in field. This is a difficult and complicated task [see Ref. 21] for which the services of a specialist are advised. The data in Table 5-2 are useful for rough order-of-magnitude estimations.

$\Sigma\gamma_d D_f$ = confining pressure of soil surrounding base of footing (equal to weight of a column of dry soil measured from base of footing to surface of adjacent grade).

γ_d = dry unit weight of the soil. Evaluate by laboratory test or, for purposes of estimation, use 100 to 120 pcf for sand and gravel soils and 80 to 100 pcf for clays and silts.

* A shallow bearing area is defined as one in which the depth of the bearing area below the level of the adjacent ground is less than the width of the bearing area. For such a depth, friction and adhesion on the sides of the foundation conventionally are neglected.

† After Terzaghi and Peck [5].

‡ Assumptions:

1. Soil is uniform (ϕ, c, γ) to a depth below bearing area greater than the width B of the bearing area.

2. Water level is at a depth below bearing area greater than the width B of the bearing area. Where water level is above the level of the bearing area, substitute γ_s (submerged unit weight of soil) for γ_d in Eqs. (5-1) to (5-3). Where water level is below bearing area, but within a depth B below such area, substitute $\gamma_s + F(\gamma_d - \gamma_s)$ for γ_d in Eqs. (5-1) to (5-3). For values of F, see Fig. 5-2.

B = width of bearing area.
L = length of bearing area.
N_c, N_q, N_γ = bearing capacity factors related to the friction angle ϕ of the soil (see Fig. 5-1).* For cohesionless materials, friction angle may be obtained from direct shear test. For mixed soils, determine friction angle from slope of Mohr envelope derived from triaxial-test data. Use Mohr envelope based on total stress. Drainage conditions in test should simulate those in field. As with determination to cohesive strength, this is a difficult task for which the services of a specialist are advised. See Table 5-2 for approximate values of ϕ for different kinds of soil.

In Eqs. (5-1) through (5-3) the first term represents the effect of cohesion, the second the effect of surcharge, and the third the effects of friction.

CAUTION: Equations 5-1 to 5-3 do not apply to footings on *steeply sloping ground.*

b. Concentrically Loaded Deep Bearing Area† *(Piers and Piles)*

The expression for ultimate bearing capacity of deep foundations takes the form [7]:

$$q_{\text{ult}} = \overbrace{[\text{Eqs. (5-1) to (5-3)}]}^{\text{end bearing}} + \overbrace{\frac{\Sigma(C_A + P_H \tan \delta)\, dA}{\text{area of base}}}^{\text{friction and adhesion}} \tag{5-4}$$

where C_A = adhesion force on element dA of shaft or base
P_H = lateral pressure of confining soil on element dA of shaft or base
δ = angle of friction between shaft or base and soil
dA = elemental area (see Fig. 5-3)

Equation (5-4) expresses the ultimate-bearing capacity as the sum of its two components: (1) end bearing and (2) the resistance offered by friction and adhesion along the sides of the foundation.

* An exception to the use of the data given in Fig. 5-1 would be where the bearing material consists of loose (N equal to 10, or less) granular soils, or highly sensitive clay (sensitivity equal to 10, or more). For these soils, failure of the soil mass can occur progressively, shearing first the soil near the edges of the load and progressing outward, or may occur as a large compression of the soil under the bearing area without general shear failure. For such cases, Terzaghi [1] has suggested an empirical correction to the bearing capacity factors for general failure. The corrected factors are shown as dashed lines in Fig. 5-1 and are based on reducing the strength factors ϕ and c by one-third.

† A deep bearing area is defined as one in which the depth of the bearing area below the level of the adjacent ground is greater than the width of the bearing area.

The amount of end bearing is evaluated from Eqs. (5-1) through (5-3). The resistance along the sides is evaluated from the adhesion of the abutting soil and from the lateral confining pressure of the soil multiplied by the tangent of the angle of friction between the pile and the surrounding soil.

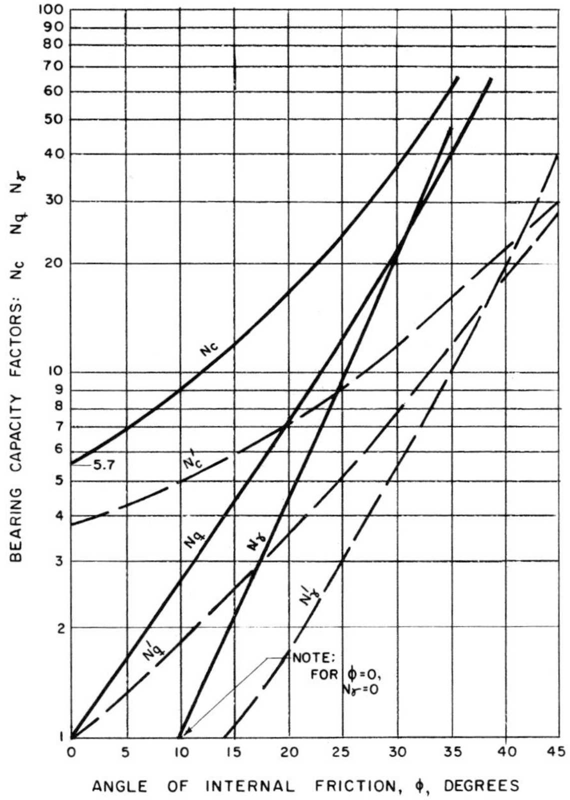

FIG. 5-1 Bearing-capacity Factors. *(From Terzaghi, Ref. 1.)*

Neither the adhesion of the abutting soil nor the friction angle is precisely known. Some approximate values for friction angle are given in Tables 5-3 and 5-4. Approximate values of adhesion between clay soils and various materials are given in Table 4-2. For sand or gravel soils, the adhesion conventionally is assumed to be zero. For silts or for mixed soils intermediate values would apply.

The confining pressure of the soil (which is the "at rest" pressure, K_n) also is not precisely known. To date (1968), no practical method for calculating at-rest pressure has been developed, and approximate values are used. A value of $K_n = 0.50$ commonly is used for granular soils.

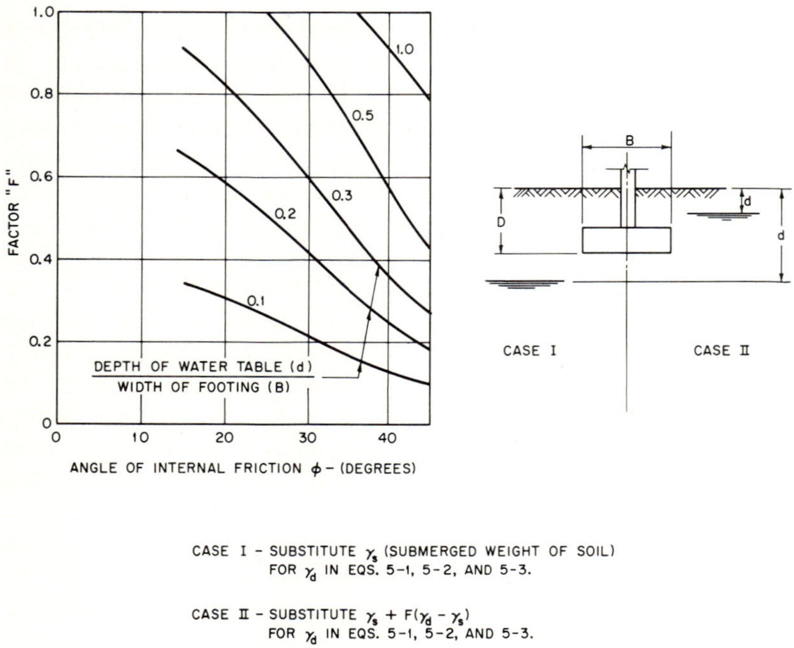

FIG. 5-2 Correction of Bearing Capacity for Effects of Shallow Groundwater.
(From Meyerhof, Ref. 6.)

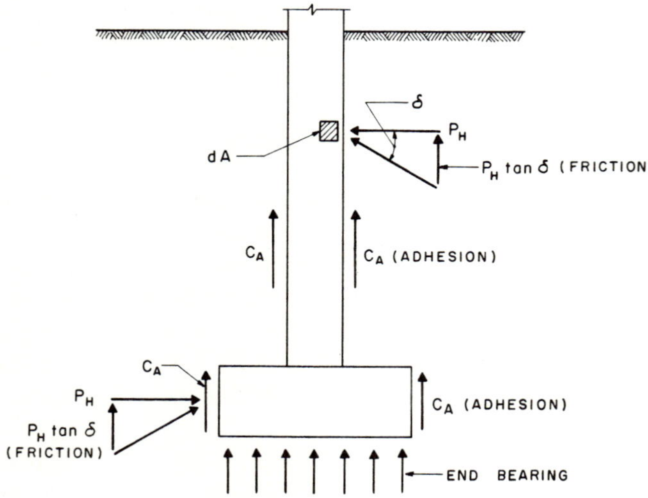

FIG. 5-3 Deep Bearing Area.

For very soft clay soils, the value may be increased to 1.0; i.e., a fluid condition is assumed.*

The problem is further complicated by the fact that the values of adhesion and friction, as indicated above, are ultimate values and are not developed until there has been relative movement of the foundation with respect to the confining soil. The amount of movement required for development of such resistance varies with the stiffness (stress-strain relation) of the soil in which the foundation is embedded: the stiffer the soil, the greater the resistance per unit of movement.

Where the sides of the foundation are embedded in relatively loose or compressible soils, relatively little resistance from adhesion and friction will be developed within the limits of movement which (1) can be tolerated in the design or (2) can take place before a base failure under the bearing area occurs. On the other hand, where the sides of the foundation are embedded in stiff (or compact) soils, the adhesive and frictional resistance may be fully developed within these limits.

Because of all these approximations and uncertainties, calculation of resistance along the sides of a foundation is likely to be an educated guess at best, and precise determination of the amount of load-bearing capacity of a foundation contributed by such frictional and adhesive resistance seldom is attempted. Instead, the approach is to establish certain extreme conditions in which said resistance is assumed to be either negligible or fully developed, to attempt to categorize each individual case into one of these extremes, and to estimate the resistance as follows:

I. *Where Side Resistance Normally Is Neglected.* Two cases occur where adhesive and frictional resistance on the sides of the foundation patently will be a small part of the total resistance. These are

1. A shaft embedded in loose or compressible strata with base on firm material (Fig. 5-4*a*)
2. A construction in which the ratio of side surface to bearing surface is small (Fig. 5-4*b*), i.e., a shallow footing

For such cases, the adhesive and frictional resistances on the sides usually are neglected.

II. *Where Side Resistance Is Fully Developed.* There also are two cases where the side resistance is considered to be fully developed. These are

1. a "friction" pile (Fig. 5-5*a*)
2. a concrete-filled rock socket (Fig. 5-5*b*)

* For further discussion of K_n values see Ref. 17, article 10-8, et seq.

In the case of the friction pile, the point resistance is assumed to be negligible, and all of the load is assumed to be supported by friction and adhesion along the sides. At usual working loads, most long piles embedded in uniform soil deposits would conform to this criterion. The frictional resistance is evaluated from penetration resistance or static load tests as described in Chap. 6.

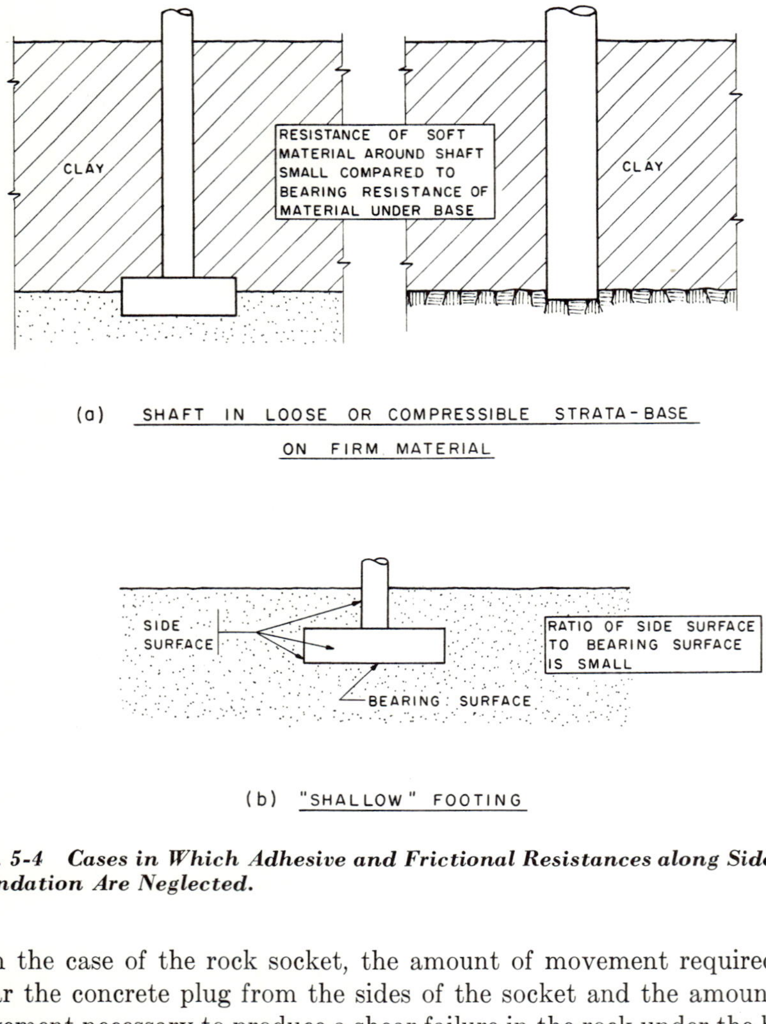

FIG. 5-4 Cases in Which Adhesive and Frictional Resistances along Sides of Foundation Are Neglected.

In the case of the rock socket, the amount of movement required to shear the concrete plug from the sides of the socket and the amount of movement necessary to produce a shear failure in the rock under the base are of the same order of magnitude—very small. Therefore, the two resistances are physically compatible and may develop simultaneously. The frictional resistance is determined as described in Sec. 8F.3. The end-bearing resistance is determined as described in Sec. 5D.2.

III. *Intermediate Cases.* Between the extremes described in Secs. I and II there are a number of different types of foundations which can be expected to develop some sensible load carrying capacity by reason of adhesion or friction along their sides. Of these, only for three types (piles, drilled piers, and sunk or driven constructions such as caissons

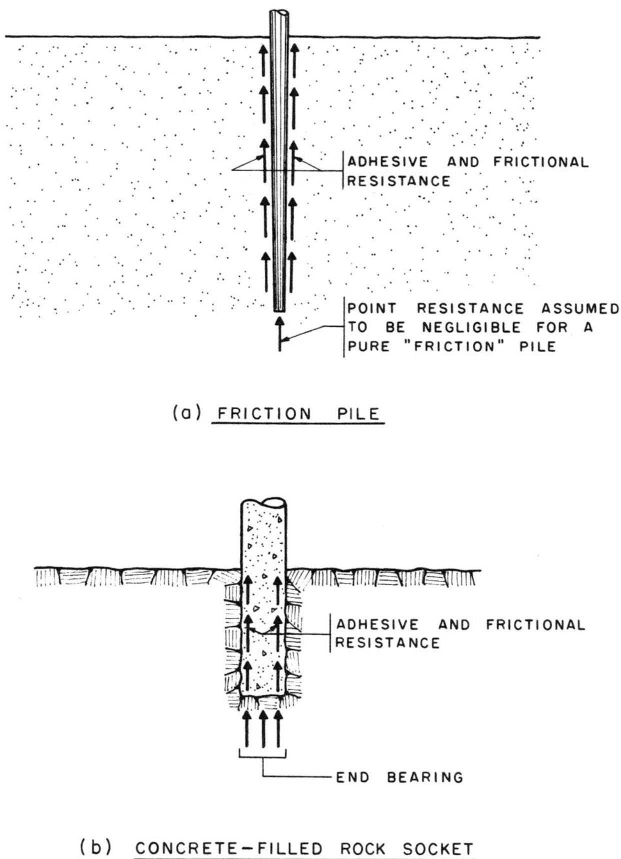

FIG. 5-5 Cases in Which Adhesive and Frictional Resistances along Sides of Foundation Are Assumed to Be Fully Developed.

and sheet pile structures) has any real attempt been made to evaluate such capacity.

For piles, the frictional and adhesive resistance can be differentiated from the end-bearing resistance by use of instrumented or cyclical load tests as discussed in Chap. 9. Normally, however, it is only the total capacity of the pile which is of interest, and it does not much matter how that resistance is divided between end bearing and friction.

A discussion of the capacity of drilled piers [which consist of a concrete shaft, with or without an enlarged base, cast into an open hole drilled into the soil (Fig. 5-26)] is contained in Sec. 5E, Chap. Note 1.

Caissons and sheet pile structures are unusual in building work, and their use and capacity are not considered in this text. They are relatively specialized forms of construction and are best handled by a specialist. For collateral reading matter on this subject, the reader is referred to Refs. 14–16.

For other types of construction (which include piers placed in open excavations and confined by backfill, deep footings, and box foundations) the resistance offered by friction and adhesion along the sidewalls usually is neglected. The rationale is that such constructions are confined not by undisturbed soil but by backfill, and there is a common prejudice against considering backfill as a load bearing material. The validity of this prejudice may be debated since a properly placed backfill is as good a bearing material as the parent soil. Nevertheless, the convention remains and provides a conservative approach.

TABLE 5-2 ***Approximate Soil Strength Parameters***

Type of soil	Strength parameters	
	ϕ	c
1. *Gravel soils:*		
a. Clean, dense, well-graded	38–40°	Zero
b. Clean, poorly graded	36–38°	Zero
c. Clayey or silty	32–35°	Zero
2. *Sand soils:*		
a. Clean, dense, well-graded ($N \geq 50$)	40° or more	Zero
b. Dense (N = 30–50)	36–40°	Zero
c. Medium (N = 10–30)	32–36°	Zero
d. Clayey or silty	30–34°	Zero
e. Loose (N less than 10)	Determine by test	Zero
3. *Silt soils:*		
a. Inorganic	30–32°	Determine by test
b. Organic	Determine by test	Determine by test
4. *Clay soils:**		
a. Stiff (N = 8–16)	(Usually assumed to be zero)	1,000–2,000 psf
b. Medium (N = 4–8)	(Usually assumed to be zero)	500–1,000 psf
c. Soft (N = 2–4)	(Usually assumed to be zero)	250–500 psf
d. Very soft (N less than 2)	(Usually assumed to be zero)	Determine by test

* After Terzaghi and Peck [5].

TABLE 5-3 ***Recommended Coefficients of Friction f for Concrete or Masonry on Soil (From Various Sources)***

Type of soil	Recommended value of f	Source
Granular soils:		
General	0.60	Ref. 9
Rock	0.70	Ref. 7
	0.70–1.00	Ref. 8
Gravel	0.60	Ref. 10, 12
	0.55–0.60	Ref. 7
Sandy gravel	0.70–1.00	Ref. 8
Sand-gravel mixtures	0.55–0.60	Ref. 7
Sand:		
General	0.40	Ref. 10, 12
By grain size:		
Clean	0.55	Ref. 5
Coarse	0.55–0.60	Ref. 7
Fine-medium	0.45–0.55	Ref. 7
Fine	0.35–0.45	Ref. 7
Silty	0.45	Ref. 5
Silty or clayey (medium-coarse)	0.45–0.55	Ref. 7
Silty or clayey (fine-medium)	0.35–0.45	Ref. 7
By density:		
Compact	0.70–0.80	Ref. 8
Loose	0.60–0.70	Ref. 8
Cohesive soils:		
Dry clay	0.50	Ref. 10, 12
Very stiff clay	0.40–0.50	Ref. 7
Wet or moist clay	0.33	Ref. 10, 12
Silt	0.35	Ref. 11
Fine sandy silt	0.30–0.35	Ref. 7
Nonplastic silt	0.30–0.35	Ref. 7
Silty clay	0.30–0.35	Ref. 7
Stiff and medium stiff clay	0.30–0.35	Ref. 7

TABLE 5-4 ***Friction Factors (tan δ) for Soil on Various Construction Materials***

Construction material	Type of soil	$\tan \delta$
Steel	Granular	0.3–0.4
	Cohesive	0.2–0.3
Concrete	See Table 5-3	$\tan \delta = f$ from Table 5-3

c. Eccentrically Loaded Bearing Areas

Usual practice when dealing with an eccentric load on a footing or pile group is to determine the allowable bearing pressure on the basis of an assumed concentric loading and then to proportion the footing so that the maximum bearing pressure imposed by the loads, computed as $P/A + Mc/I$, does not exceed this allowable value (see Fig. 5-6).

This and alternate approaches are discussed in Sec. 5E, Chap. Note 2.

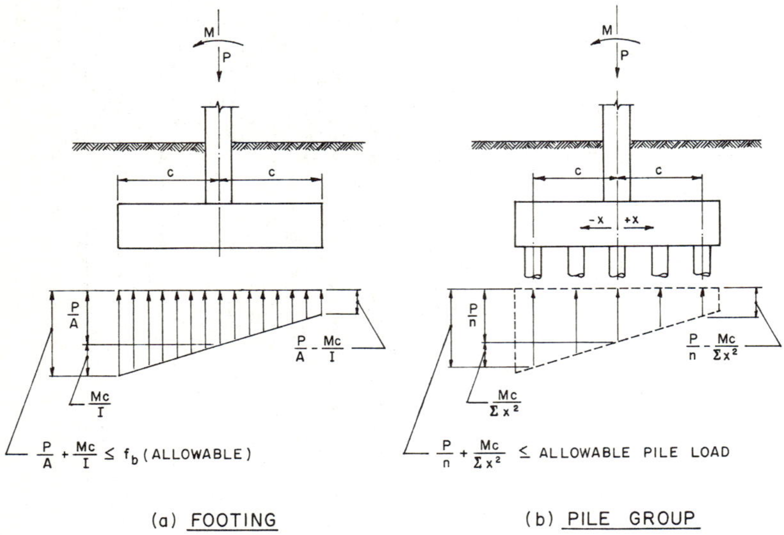

FIG. 5-6 Conventional Treatment of Eccentricity of Foundation Load.

d. Stratified Soil Deposits

Equations (5-1) through (5-3) are based on the assumption that the soil within the shear zone below the bearing area is homogeneous. In practice, most natural soils are, more or less, stratified and variable. Theoretical solutions to the problem of ultimate bearing capacity have been developed only for a very limited number of simple cases (Ref. 46, for example).

Sowers [48] suggests that where the variation of soil-strength properties within the depth of the shear zone does not exceed 50 percent, the computation of ultimate-bearing capacity can be made using a weighted average of such properties. Where the variation exceeds 50 percent, the bearing capacity is computed on the basis of the strength of the softest stratum but with the load intensity at the level of the soft stratum being computed on the basis of distribution of the load as described in Sec. 5C.

5A.2 Required Factor of Safety

Having determined the *ultimate* bearing capacity of the soil material underlying and confining the foundation, the next step is to determine the tentative *allowable* capacity as governed by considerations of stability (shear failure of the soil). This tentative allowable capacity is equal to the ultimate capacity divided by an appropriate factor of safety. The problem is to select the appropriate factor.

There probably are as many opinions in this matter as there are foundation designers. There is no simple solution. Some of the published expressions of opinion and practice are as follows:

1. Design Manual DM7, U.S. Navy [7]. Factor of safety between 2 and 3 is suggested for dead load plus normal live load depending on the

TABLE 5-5 ***The Selection of Factors of Safety on the Basis of the Average Sensitivity of a Clay to Remolding****

Degree of sensitivity	Suggested factors of safety (F)	
	Permanent structures	Temporary structures
High	3.0	2.5
Medium	2.7	2.0
Slight	2.5	1.8
Not sensitive	2.2	1.6

* After Tschebotarioff [17].

nature of the structure and the reliability with which subsoil conditions have been determined. Value reduced to 1.5 to 2 for dead loads plus maximum live load. No differentiation is made as to type of soil.

2. Terzaghi and Peck [5] suggest a value of 2.0 to 3.0 for *clay soils*, with the shear strength determined from unconfined compression tests carried to failure. The value of 2.0, presumedly, would be applicable for temporary structures and 3.0 for permanent construction.

3. Tschebotarioff [17] suggests for *clay soils* that the factor of safety should vary with sensitivity as well as the permanence of the structure and suggests values as shown in Table 5-5. The unconfined compression test is suggested as the basis for determination of the shear strength.

4. Krynine [18] suggests a value of 1.5 to 2.0. No differentiation is made as to the type of soil.

5. Henry [8] suggests a safety factor of 3.0, but notes that this value is not limited so much by the need for safety against shear failure as a requirement relating to limiting settlement of the bearing area.

6. Anderson [27] in an illustrated example uses a safety factor of 2.0.

7. The Israel Standard Code of Practice [53] indicates a factor of safety of 3.0, reduced to 2.0 when analysis is based on extensive site investigation and tests.

a. Discussion

Summarizing the above recommendations, the proposed factors of safety range from 1.5 to 3.0, with the lower values applicable for temporary structures or to conditions of infrequent combination of load and the upper values applicable for permanent construction of a sensitive nature and/or founded in sensitive soil. Thus, the range of applicable values is established. The problem is to place the situation under design within this range.

We may narrow the range somewhat by considering the nature of the overload factors for temporary construction and infrequent loading conditions. In Sec. 4C the problem of allowable overstress factors was discussed. It was noted that, because of the increasing complexities of design, the modern trend toward expression of allowable overstress is to take the overload provision directly out of the applied load. For example, the improbability of occurrence of full design live load on all areas of the building is already compensated for in the usual live-load reduction factors. The allowable overstresses for wind, earthquake, temporary construction, and other infrequent loading conditions are already compensated for by reducing the applied load by one-fourth or one-third, as discussed in Chap. 4. Therefore, these influences need not be, nor are they properly, included in the determination of the applicable safety factor for bearing capacity. The range of applicable values may be narrowed to about 2.0 to 3.0 and applied to loads which have been corrected as above for (1) live-load reductions due to area considerations and (2) infrequent loading conditions, such as wind, earthquake, temporary construction, etc.

As a further reflection on the problem, let us consider the load factors, predicated on ultimate design or plastic theory, which are used in modern specifications for the design of the superstructure of the building. These load factors are related to dead and live load, as follows:

I. *Concrete Design.* American Concrete Institute, Building Code (ACI 318-63).

$$\text{Required ultimate strength} = \frac{1}{\phi}(1.5 \text{ dead load} + 1.8 \text{ live load}) \tag{5-5a}$$

The ϕ value varies from 0.70 for tied columns to 0.90 for flexural members and reflects the sensitivity of the different types of members to vagaries of workmanship or other unknowns. The resulting load factor for an assumed ratio of dead to live load of 1.0, nominally, is about 1.8 to 2.4.

II. *Steel Design.* American Institute of Steel Construction, Specification for the Design, Fabrication, and Erection of Steel for Buildings (AISC, 1963).

Design load factor = 1.70 (dead + live load) for simple and continuous beams and 1.85 (dead + live load) for continuous frames.

The resulting load factor, based on the formation of plastic hinges, is 1.70 to 1.85. The factor on ultimate strength is somewhat higher.

III. *Cable-supported Structures.* The American Iron and Steel Institute has promulgated a specification for the design of cable-supported structures. It is required that the ultimate strength of bridge wire cables be computed on the basis of load factors, as follows:

1.5 dead load + 3.0 live load
2.5 (dead load + live load)
2.0 (dead load + live load + wind or earthquake effect)

The differentiation of safety factor as applied to dead versus live load, which appears in these design standards, is based on the fact that the dead-load component of the total load is more precisely known than is the live-load component. This is a logical conclusion which should be applicable to the design of foundations as well as to the design of the superstructure, and some foundation designers do consider it.

The ϕ factors of the concrete code, which reflect the sensitivity of the member to adverse variations in material strength, workmanship, etc., also would have a logical application in the design of foundations as a reflection of the sensitivity of the building and of the soil materials.

Accordingly, the authors suggest the following design factors for proportioning for safety against shear failure of the supporting soil.

$$\text{FS (factor of safety)} = \frac{1}{\phi}\,\frac{(1.5\text{ dead load} + 1.8\text{ live load})}{(\text{dead load} + \text{live load})} \qquad (5\text{-}5b)$$

The following ϕ factors are suggested:

1. For the foundations of buildings of low to medium sensitivity (see Sec. 2B, page 28, for suggested classification), founded on granular deposits, in built-up areas where the performance of the bearing stratum has been tested by previous construction—0.85.

2. For the foundations of buildings of low to medium sensitivity founded in fine-grained deposits of low to medium sensitivity or in undeveloped areas and for the foundations of buildings of medium to high sensitivity (continuous framing, thin ornamental masonry, arch,

shell, or vault construction) founded in granular deposits in built-up areas—0.75.

3. For the foundations of buildings of medium to high sensitivity and for the foundations of buildings of low to medium sensitivity, founded in fine-grained deposits of high sensitivity—0.60.

The resulting safety factors, for an assumed ratio of dead to live load of 1.0, will be 2.0 to 2.8, with an upper value of 3.0 when the live load is much heavier than the dead load. These values correspond to the range 2.0 to 3.0 previously indicated.

Section 5B SETTLEMENT

For any given structure there is a certain amount of settlement (in terms of either total or differential displacement) which can be tolerated without (1) creating an unsafe condition by overstressing the structure or (2) creating an unacceptable maintenance or aesthetic problem by causing walls and partitions to crack or by disrupting utility services. The foundation must be designed so that the anticipated settlements do not exceed this amount, and the tentative allowable bearing pressure determined in accordance with the text of Sec. 5A must be modified, if necessary, to meet this limitation.

The first step in the calculation of allowable bearing capacity as limited by settlement is to estimate the probable amount of settlement, assuming that the building is infinitely flexible. This estimate is then corrected for the fact that the structure has rigidity. The amount of settlement (in terms of both total and differential displacement) which can be tolerated is then estimated and compared to the estimate of probable displacements. If the tolerable values are greater than anticipated values, all is well. If the converse is true, the design must be modified or other corrective actions undertaken.

5B.1 Estimating Probable Amount of Settlement

The settlement of a building derives from two sources: (1) deep-seated consolidation of the soil due to time-dependent volume changes and (2) short-term settlement due to shear strain occurring at shallow depth beneath the individual loaded areas (footings). The latter is referred to as contact settlement. The total settlement is the sum of the two components, corrected for rigidity of the superstructure, and is approximated as follows:

a. Settlement Due to Long-term Consolidation

Three cases must be considered: (1) where the structure is founded on or above a stratum of saturated, fine-grained material, (2) where founded on or above a partially saturated stratum of fine-grained material, and (3) where founded on or in a coarse-grained (sand or gravel) material.

CASE 1 *Structure Founded on or above Saturated Fine-grained (Clay or Silt) Strata:* Terzaghi's theory of consolidation has become widely accepted as the standard for the computation of settlement due to consolidation and is used for the evaluation of settlements of foundations founded on, or overlying, saturated clay or silt strata. The solution for the case of one-dimensional compression generally is adopted, and since the anticipated service life of a building is several decades, in general, it is only the potential total settlement (and not the rate) which is of interest. For this limiting case, the settlement may be obtained from the pressure-void ratio curve, using the expression

$$\Delta H = \Sigma H \left(\frac{\Delta e}{1 + e_0}\right) \tag{5-6}$$

where ΔH = total settlement
H = original thickness of increment of compressible stratum being considered
Δe = change in voids ratio in increment of compressible stratum due to change of applied pressure
e_0 = original void ratio (before change of applied pressure)

Determination of the change of applied pressure is discussed in Sec. 5C.

Having made this determination, the calculation of total settlement is carried out by subdividing the total thickness of the compressible strata into increments H and summing the right-hand side of Eq. (5-6) for the total number of these increments.

The e_0 value is determined in the laboratory, and the applicable value of Δe is similarly evaluated from the laboratory pressure-void ratio diagram *e-p* as determined from the oedometer (consolidation) test. Several *e-p* curves may be required if the soil strata are variable.

For most clay and silt deposits it has been found, empirically, that the *e-p* curve, when plotted on semilog paper, takes the form shown in Fig. 5-7 in which the curve consists of two more or less straight lines. One line represents recompression from zero initial stress (and the corresponding void ratio e_0 remaining after removal of the specimen from the ground) up to the maximum preconsolidation stress under which the specimen previously had attained equilibrium, and the second line represents the stress-strain relation in the specimen for unit pressures exceeding the preconsolidation load. For such a diagram, Eq. (5-6) takes the familiar form:

1. For applied pressure less than p_c

$$\Delta H = H \frac{(C_r)}{1 + e_0} \log \left(\frac{p_0 + \Delta p}{p_0}\right) \tag{5-7a}$$

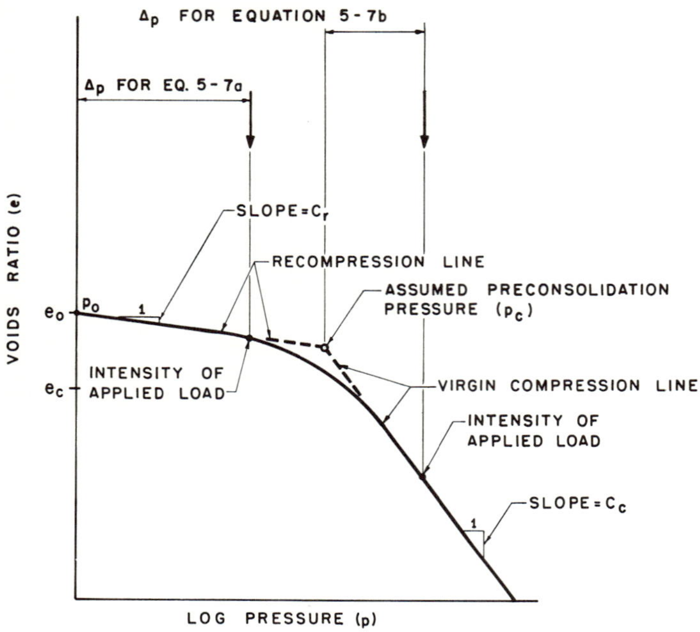

FIG. 5-7 Typical e-p Diagram.

where C_r = the recompression index and equals the slope of the recompression line

p_0 = average initial effective pressure

Δp = the average change in pressure in the stratum under consideration

p_c = average preconsolidation pressure in stratum

2. For applied pressure greater than p_c

$$\Delta H = H \frac{(C_c)}{1 + e_0} \log \left(\frac{p_c + \Delta p}{p_c} \right) \tag{5-7b}$$

where C_c equals the compression index and equals the slope of the virgin compression line.

If the range p_0 to the applied pressure encompasses both the recompression range and the virgin compression branch, two computations are required, one using C_c and the other using C_r. The total settlement is the sum of the two. Where the range p_0 to the applied pressure is encompassed within one branch of the curve only, a single computation will suffice. If the soil is variable, then Eq. (5-7) must be applied to layers of incremental thickness as is done with Eq. (5-6).

One advantage of Eq. (5-7), as compared with Eq. (5-6), is that the values of C_c and, more important, of $C_c/(1 + e_0)$ have been related empirically to the index properties and void ratio for the soil. For silt and clay soils, these relations are

$$C_c = 0.54(e_0 - 0.35)^* \qquad (5\text{-}8)$$
$$= 0.0054(2.6w - 35)^* \qquad (5\text{-}9)$$
$$= 0.009(\text{LL} - 10)^\dagger \qquad (5\text{-}10)$$

where w = moisture content
LL = liquid limit

From Eq. (5-8), we can derive Fig. 5-8, from which it may be seen that the value of $C_c/(1 + e_0)$ falls within a relatively narrow range from about 0.2 to 0.4.

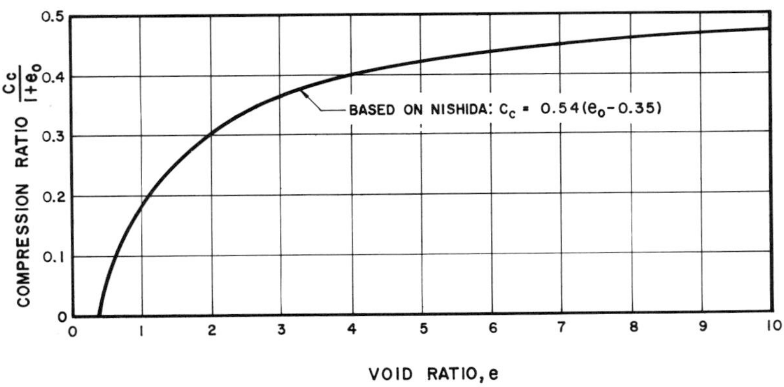

FIG. 5-8 *(After Nishida, Ref. 45.)*

CAUTION: The above discussion of settlement of a building founded on or above clay or silt strata has been presented in an ultrasimplified form. Neglected, for example, are effects such as rebound during excavation, history of previous loading (except as it relates to determination of the preconsolidation load), variation of C_c with loading increments, secondary compression, and sample disturbance. The effects of rigidity of the superstructure also are neglected. The computation, as described, provides an order-of-magnitude estimate of the probable settlement only. In many cases, this is enough. If the answer is an inch or two, as discussed in the following section, the proposed design is likely to be adequate. If the answer is several inches, the design merits careful review and is likely to need modification or special adaptation of the superstructure.

If the answer indicates an intermediate borderline condition or if the building is a sensitive type of construction, more sophisticated results

* Nishida [45].
† Terzaghi and Peck [5].

are needed. For this the services of a specialist are recommended (see also, Refs. 20 and 21).

In any event, whatever the type of building or the probable answers, the recovery of the undisturbed samples necessary for the consolidation tests and the execution of such tests should be performed only by or under the supervision of trained and experienced persons.

A sample computation of consolidation of a clay stratum, based on the simplifying assumptions described, is given in Sample Computation 5-1.

SAMPLE COMPUTATION 5-1
Consolidation of a Clay Stratum

STATEMENT OF PROBLEM

The condition to be analysed is indicated in the figure. Assume that the desired answer is the total settlement under the center of the building due to consolidation of the two clay strata. The compression of the sand strata is to be neglected. The building is assumed to be infinitely flexible.

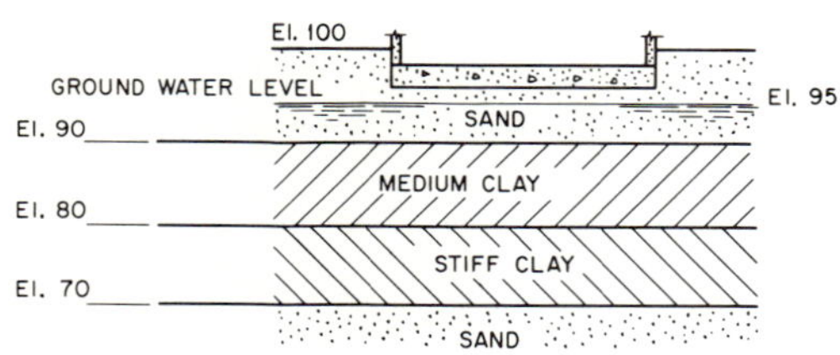

GIVEN

The following data are provided by laboratory test:

1. For sand strata:
   $\gamma_d = 120$ pcf
   $\gamma_s = 70$ pcf
2. For stratum of medium clay:
   $P_c = 1{,}200$ psf (average)
   $\gamma_s = 50$ pcf
   $C'_c = 0.22$
3. For stratum of stiff clay:
   $P_c = 2{,}500$ psf (average)
   $\gamma_s = 55$ pcf
   $C'_c = 0.18$
   $C'_r = 0.02$ (assuming slope = 10% of slope of virgin compression curve)

where $C'_c = \dfrac{C_c}{1 + e_0}$ (see Fig. 5-8)

$$C'_r = \frac{C_r}{1 + e_0}$$

PROCEDURE

Calculate p_0 from the weight of the overburden, as follows:

1. At middle of stratum of medium clay:

$$\begin{aligned} 5\text{ ft} \times 120\text{ pcf} &= 600\text{ pcf} \\ +5\text{ ft} \times 70\text{ pcf} &= 350\text{ pcf} \\ +5\text{ ft} \times 50\text{ pcf} &= \underline{250}\text{ pcf} \\ p_0 &= 1{,}200\text{ pcf} \end{aligned}$$

(stratum is normally consolidated because overburden pressure is equal to preconsolidation pressure)

2. At middle of stratum of stiff clay:

$$\begin{aligned} 5\text{ ft} \times 120\text{ pcf} &= 600\text{ pcf} \\ +5\text{ ft} \times 70\text{ pcf} &= 350\text{ pcf} \\ +10\text{ ft} \times 50\text{ pcf} &= 500\text{ pcf} \\ +5\text{ ft} \times 55\text{ pcf} &= \underline{275}\text{ pcf} \\ p_0 &= 1{,}725\text{ pcf} \end{aligned}$$

(stratum is overconsolidated because overburden pressure is less than preconsolidation pressure)

Values of Δp due to load from the building also are calculated at the middle of the compressible strata, using one of the several techniques described in Sec. 5C. Assume the results are:

1. At middle of stratum of medium clay:
   $\Delta p = 1{,}500$ psf
2. At middle of stratum of stiff clay:
   $\Delta p = 1{,}000$ psf

The settlement due to consolidation of the clay layers is computed by Eqs. (5-7*a*) and (5-7*b*), as follows:

1. In medium clay:

$$\Delta H = 120\text{ in.} \times 0.22 \log\left(\frac{1{,}200 + 1{,}500}{1{,}200}\right) \quad \text{(Eq. 5-7}b\text{)}$$
$$= 9.3\text{ in.}$$

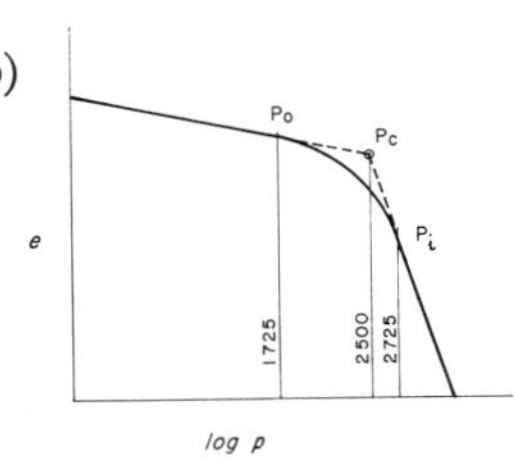

2. In stiff clay:

$$\Delta H = \Delta H_{p_0 \text{ to } p_c} + \Delta H_{p_c \text{ to } p_1} \quad \text{(see inset and Fig. 5-7)}$$

$$= 120\text{ in.} \times 0.02 \log\left(\frac{\overbrace{1725}^{P_0} + \overbrace{725}^{P_0 \text{ to } P_c}}{1725}\right) \quad \left(\begin{matrix}\text{Recompression} \\ \text{Eq. (5-7}a\text{)}\end{matrix}\right)$$

$$+\ 120\text{ in.} \times 0.18 \log\left(\frac{\overbrace{2500}^{P_c} + \overbrace{225}^{P_c \text{ to } P_i}}{2500}\right) \quad \left(\begin{matrix}\text{Virgin compression} \\ \text{Eq. (5-7}b\text{)}\end{matrix}\right)$$

$$= 1.2\text{ in.}$$

Total settlement $= 9.3 + 1.2 = 10.5$ in.

CASE 2 *Structure Founded on or above Partially Saturated Fine-grained (Clay or Silt) Strata:* The method described in Case 1 does not strictly apply to the analysis of consolidation of partially saturated soil strata. For such a case, there is no standard or usual approach. If the groundwater level can be relied upon to be more or less constant and if the rainfall is of moderately uniform distribution throughout the year so that the degree of saturation remains roughly constant, load-test procedures, as described in Sec. 5B.1*b*, Case 2, Method II, apply. Where such is not the case, again, the services of a specialist are recommended.

CASE 3 *Structure Founded on Coarse-grained (Sand or Gravel) Strata:* For structures founded on sand or gravel strata and for cemented soils such as hardpan or rock, it is conventionally assumed that the long-term consolidation aspects of the total settlement may be neglected.

b. Settlement Due to Shear Strain (Contact Settlement)

Theoretical solutions to determination of contact settlement can be made using the equations of Boussinesq. The expression for a uniformly loaded area is

$$\Delta H_c = q \frac{b(1 - \mu^2)}{E} I_p \tag{5-11}$$

where ΔH_c = contact settlement, in.
q = unit bearing pressure, psi
b = width of loaded area, in.
E = modulus of elasticity of soil, psi
μ = Poisson's ratio
I_p = shape factor (average of 0.95 for square loaded area and 0.85 for circular loaded area)

The problem with application of Eq. (5-11), or similar expressions, is in determining appropriate values for E and μ. Such determination usually requires load tests performed at the site to obtain values for E. Values for μ are approximated, a value of 0.5 commonly being assumed.

Determination of an appropriate E value is further complicated by the fact that E increases with depth (increased confining pressure), varies as the soil compresses under load, and is affected by the confining pressure of the foundation itself. As a result, the theoretical solution is of limited value, and, in practice, contact settlement usually is estimated from empirical relations or from load-test data, as follows. As with computation of the settlement due to long-term consolidations, calculation of the settlement due to shear strain must differentiate three cases depending on the type of supporting soil.

CASE 1 *Structure Founded in or above Fine-grained (Clay or Silt) Strata:* Shear strain settlements normally are neglected in the computation of total settlements of soft to medium (or loose to medium) clays or silts. The reason is not that such settlements are zero, but rather that they are likely to be small compared to the movements due to long term consolidation.

For stiff (or hard) or dense clay or silt strata, however, the shear strain settlement may constitute a major part of the overall displacement. Evaluation may be made by the use of load tests or using the method of Hough, both of which are described under Case 2.

CASE 2 *Structure Founded on Coarse-grained (Sand or Gravel) Strata:* In coarse-grained soils, shear strain constitutes the major part of the total settlement. The evaluation of settlement for such soils is based on (1) empirical relations such as that proposed by Terzaghi and Peck [5] or (2) on the extrapolation of data from *in situ* load tests, using relations proposed by Terzaghi and Peck [5], Housel [19], or Taylor [28].

Use of the method proposed by Hough [22] may be considered, subject to the cautions hereinafter described.

The three methods may be summarized, as follows:

I. *The Method of Terzaghi and Peck.* In their excellent text, "Soil Mechanics in Engineering Practice," [5] Terzaghi and Peck suggested that the settlement of shallow footings in sand can be related to the *N* value obtained from the standard penetration test. The suggested relation is shown in Fig. 5-9 and shows the correspondence of the *N* value to the bearing pressure which will produce a maximum total settlement of 1 in. The value of 1 in. was selected on the basis that, with a total settlement of 1 in., the amount of differential settlement was unlikely to exceed ¾ in. and that ¾ in. of differential settlement was a safe value for most buildings. This approach to the selection of allowable bearing pressures on sand soils has gained wide acceptance in engineering practice and will be utilized later in this text, subject to the following reservations.

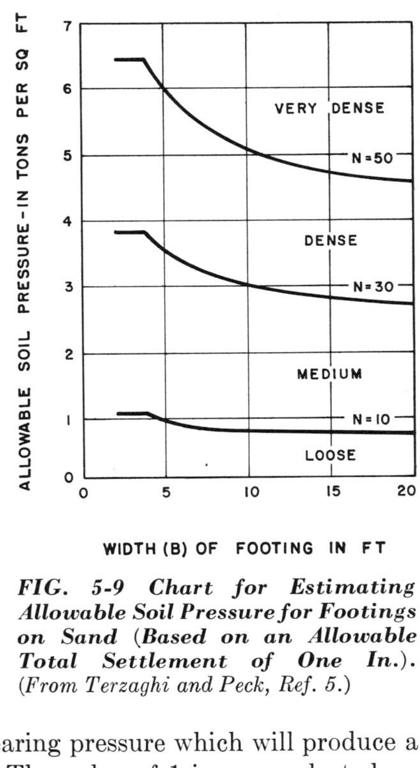

***FIG. 5-9** Chart for Estimating Allowable Soil Pressure for Footings on Sand (Based on an Allowable Total Settlement of One In.). (From Terzaghi and Peck, Ref. 5.)*

A. The approach is valid only for basically sandy soils (certain of the soils in groups SW, SP, and SM of the Unified Classification System).

It is not considered to be valid for gravel soils (groups GW, GP, GM, and GC), for cemented deposits such as hardpan (till) or rock, or for boulder deposits, because of the difficulty of obtaining a reliable N value in such soils.

B. The application to strata of fine sand should be limited, for reasons which will be discussed.

The method does not purport to permit of extrapolation to bearing pressures which will produce values of settlement different from 1 in.; i.e., it is not, per se, a method for calculating the amount of settlement.

II. *Use of Load Tests.* The settlement of a loaded area on a granular soil under a given intensity of bearing pressure can be related to the size of the bearing area by the expression proposed by Terzaghi and Peck [5]:

$$S_0 = S_1\left(\frac{2B}{B+1}\right)^2 \tag{5-12}$$

where S_1 = settlement of a 1- by 1-ft loaded area under a given loading intensity q

S_0 = settlement of a larger loaded area having a side dimension B

The procedure is to conduct a static soil load test, using a loaded plate, as described in Chap. 9, and using the results so obtained to extrapolate, by means of Eq. (5-12), to the performance of the larger loaded area.

The approach, as conventionally used, is somewhat questionable. First, it is applicable only to soils having no cohesive capacity, whereas most natural soils have some cohesion. Second, the usual plate load test is, say, 2 by 2 ft. Actual footings are several feet (or even tens of feet) in plan dimension. Therefore, the necessary extrapolation, which is proportional to the square of the size, gets rather wild. Lastly, again because of the size of the loaded plate, only a shallow depth of soil is actually stressed by the test load, further rendering the extrapolation unreliable.

Of course, if the size of the loaded area is made reasonably to approach the size of the actual footings, the worst of the above objections would be eliminated.

The matter of cohesive resistance also can be overcome, at least partly, by the use of load tests conducted on two or more plates of different sizes and extrapolating to the performance of the full-sized bearing area, using relations suggested by Housel [19] or Taylor [28].

The relation suggested by Housel is

$$P = Ap_c + P_r s \tag{5-13}$$

where P = total load on bearing area at a given settlement

A = contact area

p_c = compressive stress (*not* P/A) in pressure bulb
P_r = perimeter of contact area
s = unit shearing stress at perimeter (in effect, punching shear)

There are two unknowns in this expression, p_c and s. The procedure is to conduct two tests on different-size test plates. The load P required to produce the same total settlement of the two plates is determined, and the two resulting equations are solved for p_c and s. With this information, the size of bearing area required to support a given load without exceeding the same settlement can be established.

Again, however, in order to assure meaningful results, the size of the test plates should approximate the range of sizes of bearing areas to be used in the actual building, and the test must be carried to the point of producing the same settlement which is to be permitted in the proposed design. A sample computation is shown in Chap. 9.

Taylor suggests a similar relation, expressed as

$$\frac{q}{\rho} = C_1\left(1 + \frac{2d}{b}\right) + \frac{C_2}{b} \tag{5-14}$$

where $\frac{q}{\rho}$ = average strain in the soil (modulus of subgrade reaction)
C_1, C_2 = constants
d = depth of bearing area below adjacent grade
b = width of bearing area

Again, there are two unknowns evaluated by simultaneous solution of results from two tests on plates of different size.

The use of load tests to estimate settlements of foundations on granular soils is not a valid approach where the soil is a gravel or boulder formation, a cemented deposit, or a rock. The reasons are discussed in Chap. 9.

III. *The Method of Hough.* Hough [22] has noted that the load-settlement curves for plate load tests conducted on many types of soil, cohesive, cohesionless, and mixed, are of similar shape (Fig. 5-10) and that for low values of contact pressure (points 0 to a) settlement is roughly proportional to loading intensity. He concludes that, for usual working loads, settlement of structures founded in granular and mixed soils, as well as fine-grained deposits, is due primarily to one-dimensional compression.

If such is the case, Eq. (5-7) would apply, and Hough has suggested the use of Eq. (5-7) for the computation of settlements of cohesionless and mixed soils as well as saturated clay and silt deposits. He has suggested a relation between the value $(1 + e_0)/C_c$ and the N value as derived from the standard penetration test. This relation is shown in Fig. 5-11. Using this relation and Eq. (5-7) the amount of settlement for given conditions can be computed.

CAUTION: The applicability of Hough's method to the evaluation of settlements of structures founded in coarse-grained soils has been questioned on the basis that the assumption of settlement being due primarily to one-dimensional compression, is not valid for sands and other cohesionless strata. The validity of the assumption as applied to fine-grained soils is not in question.

The authors do not propose to comment on this debate. Mr. Hough's method is presented because it is potentially of considerable value for approximate analysis. It is not intended for detailed study. Plate load

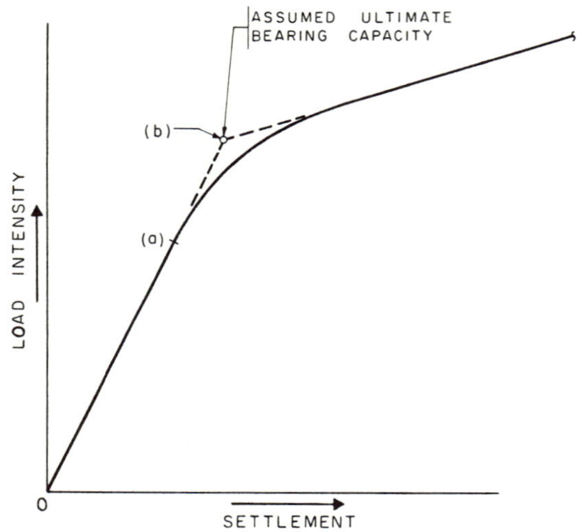

FIG. 5-10 Typical Relation between Load and Settlement (Plate Load Tests).

tests are used for this purpose. The method, if used at all, is to be used, like the other methods described—judiciously and in concurrence with, and checked by, the results indicated by other methods. Where doubt exists as to the validity of any one approach, the use of data from multiple load tests generally will give the most reliable results.

CASE 3 *Structure Founded on Mixed Soil Strata:* In the general case of a soil having both frictional and cohesive resistance, i.e., mixed soils, settlements may be estimated from plate load tests (using plates of two different sizes) or using the method of Hough where applicable. These methods are described under Case 2. The criteria proposed by Terzaghi and Peck also are applicable for types of mixed soils having basic characteristics similar to those of cohesionless deposits (see Sec. 5D).

c. Some Miscellaneous Considerations

1. *Initial Stress Condition* (p_0). Equation (5-7) contains the term p_0, defined as the average initial effective pressure under which consolidation

is taking place, and the term p_c, defined as the average preconsolidation pressure in the same stratum.

The value of p_0 usually is equal to the weight (submerged weight, if applicable) of the column of soil overlying the middepth of the stratum

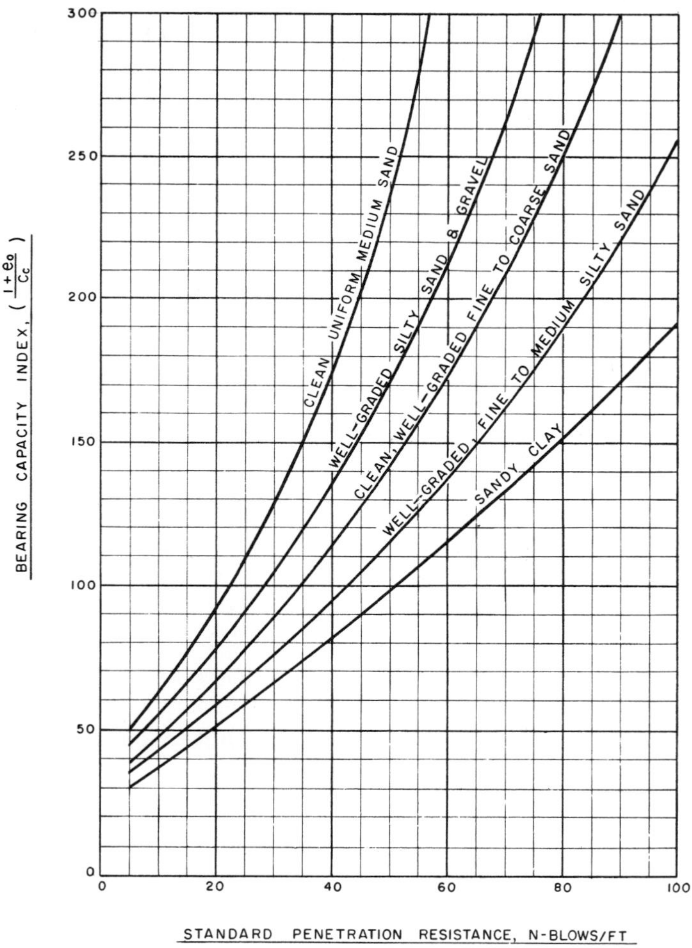

FIG. 5-11 Relation between N Value and Bearing-capacity Index $\left(\frac{1 + e_0}{C_c}\right)$. *(After Hough, Ref. 22.)*

being considered. However, this is *not* always the case. Some soils (such as those to which a weight of fill has been added recently or the groundwater level may have been lowered recently) may be underconsolidated, i.e., a condition of equilibrium has not yet been attained under the weight of the overburden. If such is the case, the preconsolidation pressure p_c will be less than the weight of the overburden. This will show up on the curve e versus log p (Fig. 5-7), and the value p_c should be used in Eq. (5-7*a*) rather than the value p_0.

The value of p_c (which may be substantially greater than the value of p_0) is determined from the curve e versus log p (Fig. 5-7) using the procedure given in chap. 2, of Ref. 5.

II. *Secondary Compression.* For fine-grained soils, the primary consolidation computed from Eq. (5-7*a* or *b*) will be followed by a secondary compression which is additive to the settlements resulting from the primary effects. The magnitude may be computed from

$$\Delta H_s = C_s H \log \left(\frac{t_{sl}}{t_{\text{prim}}} \right) \tag{5-15}$$

where ΔH_s = settlement due to secondary compression
C_s = coefficient of secondary compression
H = initial thickness of stratum
t_{prim} = time to completion of primary consolidation, yr; the value corresponding to U = 90 percent may be used
t_{sl} = required service life of proposed building, yr

Normally, the amount of the secondary compression occurring within the life of a building is sufficiently small, in comparison with the primary effects, that it may be neglected. It is seldom of importance except where the supporting soil is highly organic, micaceous, or for some highly compressible clays. If there is a doubt, however, the secondary effects can be estimated roughly, using a value of $C_s = 0.02$ and a value of t_{prim} from the consolidation test (see also Refs. 23 and 24).

III. *Swell.* In some cases, the ground supporting a building, instead of settling, will heave. This may be due to the relief of overburden pressure resulting from excavation or from a reduction in capillary stress in the soil due to an increase in moisture content.

The heave due to relief of overburden pressure may be estimated from the curve e versus log p.

Swelling due to changes in moisture content is best treated by preventing its occurrence. The foundation should bear on stable ground, as described in Chap. 3.

IV. *Accuracy of the Settlement Computation.* Settlements computed on the basis of Eqs. (5-6) or (5-7) are approximations only, but are likely to represent upper limit values. The major sources of error are:

1. Equations (5-6) and (5-7) assume one-dimensional consolidation of the soil strata. For a thin compressible stratum sandwiched between two rigid strata, this probably is nearly correct, but for a deep stratum, the effects of lateral deformation are significant. Skempton and Bjerrum [47] have proposed a series of correction factors for this condition. For normally consolidated or overconsolidated soil strata the correction

factors are less than unity, indicating that the actual settlement will be less than that calculated from Eqs. (5-6) and (5-7). It is only for highly sensitive soils that the correction factors exceed unity (the upper limit is about 1.1).

2. A certain amount of sample disturbance is likely to occur so that the sample, as tested, does not precisely conform to its condition in nature. In addition, the initial loading condition on the sample in its natural state is not likely to be precisely simulated in the laboratory test. The effect of these two conditions is to make the computed settlements larger than the observed values.

3. Other sources of error are in determining the correct stress-void ratio relation, the preconsolidation load, and the effects of stratification of the soil.

Sowers suggests that for homogeneous, normally consolidated soils, the total settlement may sometimes be computed with an accuracy of 10 percent but that for irregular strata or loading, the error may be 50 percent or more and up to 100 percent or more when the soils being considered are highly sensitive or heavily overconsolidated.

5B.2 Correction for Rigidity of Superstructure

The previous section presented several theoretical and empirical relations used to estimate probable settlements of buildings founded on various types of soil material. Except for the method of Terzaghi and Peck, which is based on limiting total settlement of the building to 1 in. and which presumes that, in such event, the differential displacement will not exceed ¾ in., each of the methods described presumes that the structure is completely flexible and will accommodate itself freely to the contours of displacement. In practice, the building has finite rigidity and tends to span any potential settlement craters, changing the stress pattern imposed on the soil and tending to even out the settlements, i.e., to reduce the amount of differential displacement to a value less than that indicated by theory. This, in fact, often is the basis for the use of mat or rigid-frame foundations when poor soil conditions are encountered. The effect is illustrated in Fig. 5-12.

A rigorous solution of the effect of rigidity on settlement can be made by treating the structure as a beam on an elastic support. Applicable equations are given in Ref. 29. The computational requirements are great, however, necessitating solution by computer. Further, the deformation of the soil takes place slowly, permitting the structure time to distort plastically (creep) so that conventional methods of analysis based on elastic action of the structure are not applicable. Accordingly, analysis of the effects of rigidity of the structure on the amount and distribution of settlements generally is not justified for routine problems, and recourse usually is made to one of the following devices.

1. *Empirical Rules Relating Actual to Theoretical Settlements.* As an example, Ref. 7 indicates that, for an ordinary multistory framed build-

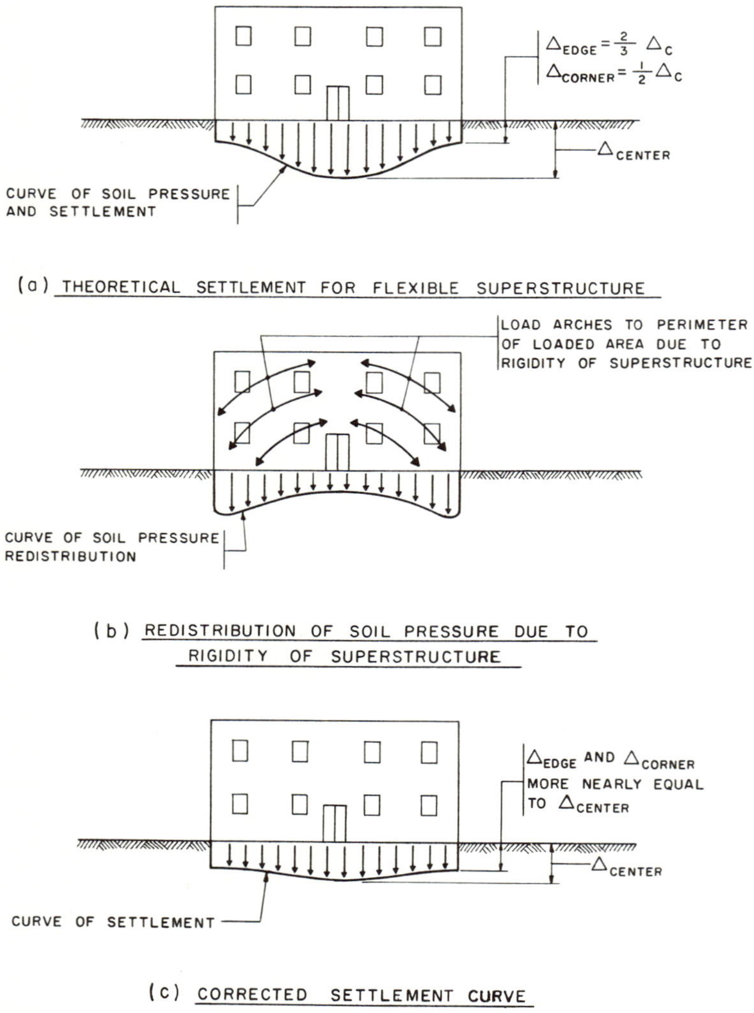

FIG. 5-12 Effect of Rigidity of Superstructure on Differential Settlement.

ing, the differential settlement is reduced to 80 to 90 percent of the theoretical value computed on the basis of complete flexibility.

Where rules of this type can be derived, and can be shown to be reliable, they are most valuable. The problem is that their development requires broad experience and actual measurements of postconstruction settlements, neither of which is easy to come by. Accordingly, it is common to neglect redistribution of reaction (and displacement) due to rigidity and to follow one of the following criteria.

2. *Limitations on Theoretical Angular Displacement.*
3. *Limitations on Theoretical Total Displacement.*

These two criteria are discussed in the following section.

5B.3 Allowable Settlement

The allowable settlement for a given building is limited by two considerations: (1) allowable total settlement and (2) allowable differential settlement, both for the building as a whole and between adjacent columns.

The total settlement seldom is important insofar as the building itself is concerned. Prentis and White [25], for example, cite the case of the School of Mines Building in Mexico City which successfully withstood over 4 ft of displacement. The famous Leaning Tower of Pisa still stands, although nearly 10 percent out of plumb. Krynine [18] cites the case of the National Theatre, also in Mexico City, which remained in use despite settlements in excess of 5 ft.

These facts notwithstanding, limitations on total settlement are used as criteria for design either (1) to prevent damage to utility lines entering the building and to preserve the grading around the building or (2) to limit indirectly the magnitude of differential settlements which may occur.

It is the differential settlement which is of primary concern and which causes damage to the building. This damage may be to the structural frame (in which case it is a very serious matter) or merely to the walls and partitions (in which case the resulting defects are more likely to be unsightly than dangerous). If the building houses equipment requiring careful alignment also, excessive differential movement must be avoided.

These problems are well known. The difficulty is to determine the amount of differential settlement which can be tolerated in a given case.

In theory, we should be able to calculate the amount of differential displacement which can be tolerated, at least in the structure. Structural theory has been developed to the point where this is possible.

In practice, however, the amount of computation necessary to make the determination, the problem of interaction between the building and the soil (effect of rigidity of the building), and the incidence and effects of inelastic redistribution of stress (plastic flow or the development of plastic hinges) preclude the execution of such analysis, except for very simple cases or for very large projects where the complex calculations may be warranted.

There is a similar problem in evaluating the amount of differential settlement which can be accommodated in the walls and partitions. Not only is the analysis time-consuming, but the constructions themselves do not conform to the uniform, isotropic, elastic continua assumed in deriving structural theory, nor does the construction approximate the shallow beams (strain proportional to distance from neutral axis) which the designer conventionally assumes.

As a result, allowable displacements (settlements) are approximated from rules of thumb, which take three general forms, as follows:

a. Limitations on Flexural Distortion (Maximum Deflection between Supports of Structural Frame)

Empirical limitations on allowable deflections between supports of the structural frame are in common use. For example, for members supporting walls or partitions, it is conventional to limit deflections to $L/360$ or $L/240$ (L is span length). The more restrictive value ($L/360$) is used when the walls or partitions are of masonry, glass, or other frangible material, and the value $L/240$ is used for walls sheathed in metal or similar nonfrangible finishes.

For steel or concrete frames, a limit of $L/150$ [26] to $L/180$ [30] is used. Such value would apply only to frames, however, and not to rigid elements such as trussed wind bents or to shear walls, where more restrictive limitations would apply.

For timber frames and members, the deflection which can be accommodated without "structural" damage will be greater than those indicated above for steel by a factor of, perhaps, 2.

For articulated structures (i.e. those with hinges), the structure itself need not necessarily be considered as having any limitation on allowable displacement.

b. Limitations on Slope

Limitations on slope are another approach to approximating allowable differential displacements. Feld [26], for example, reports the following data from the U.S.S.R. code of 1955.

Type of construction	*Maximum slope of deflected configuration from line (usually horizontal) of as-built configuration*
Brick dwellings:	0.003 for $L/H^* \leq 2$
	0.001 for $L/H = 8$
Crane runways:	0.003
Brick-clad column frame:	0.001 between columns

* L = length
H = height

Sowers [48] provides the following additional summary of slope limitations related to differential movement.

High, continuous brick walls	0.0005 to 0.001
One-story brick mill building (to prevent cracking of walls)	0.001 to 0.002
To prevent cracking of plaster (gypsum)	0.001
Reinforced-concrete building frame	0.0025 to 0.004
Reinforced-concrete curtain walls	0.003
Continuous steel frame	0.002
Simply supported steel frame	0.005

The Israel Standard Code of Practice [53] provides, as follows:

Differential settlement for—

Open steel and reinforced concrete frame structures	.002L to .005L
For rows of columns with concrete block walls	.0005L to .001L

c. Limitations on Total Settlement

A far more common approach than either of the above is to put an arbitrary limit on the total anticipated settlement of the structure. As previously noted, this device is used as an indirect means of controlling the amount of differential settlement which may occur.

Tschebotarioff [17], for example, suggests that no damage is to be expected in most buildings where the settlement does not exceed 2 to 3 in. and that with the total settlement so limited, the differential settlement will not exceed about 1 in.

Terzaghi and Peck [5] suggest a somewhat more conservative approach. Their bearing-capacity charts (Fig. 5-9) are based on limiting total settlement to 1 in. with the resulting probable differential not exceeding ¾ in.

Skempton and MacDonald [31] suggest a damage limit for conventional buildings having load-bearing walls, of 2 in. for buildings founded on sand and 3 in. for buildings founded on clay soils. The corresponding limitations on differential settlements suggested are 1¼ in. and 1¾ in.

Sowers [48] indicates that desirable limitation for total settlement of foundations on sand soils is 1 in. and on clay soils is 1½ in. The following limitations on total settlements derived from several sources also are indicated.

Masonry walled structures	1 to 2 in.
Framed structures	2 to 4 in.
Smokestacks, silos, mats	3 to 12 in.

Polshin and Tokar [49] suggest the following limitations on average settlement from the 1955 U.S.S.R. building code:

Buildings with plain brick walls	3 to 4 in.
Buildings with brick walls reinforced with reinforced-concrete or reinforced-brick belts	6 in.
Frame buildings	4 in.

d. Discussions and Conclusions

I. *Suggested Criteria.* Of the three criteria cited, limitation on flexural distortion, limitation on slope, or limitation on total settlement, only limitations relating to total settlement are independent of the rigidity of the superstructure and the complex condition of interaction which this rigidity introduces. Therefore, they are the most useful and most widely adopted.

The bearing-capacity charts of Terzaghi and Peck, for example, have gained wide acceptance for application to the design of foundations bearing on sand. Using these charts when applicable, foundations may be designed without any direct consideration for settlement, the settlement consideration being implicit in the chart relations. As previously noted, the charts appear to be somewhat conservatively derived, being predicated on a limitation of 1 in. on total settlement, whereas both Tschebotarioff and Skempton and MacDonald suggest limits of 2 to 3 in., with the Soviet Building Code providing for even more liberal values. The values of 2 to 3 in. seem quite reasonable except, perhaps, for sensitive buildings or for highly erratic soil conditions where the differential effects are likely to be pronounced.

Still, in keeping with the spirit of conservatism discussed in Chap. 1 and considering the wide acceptance of the criteria suggested by Terzaghi and Peck, the authors suggest a design limitation of 1 in. (corresponding to the bearing-capacity chart of Terzaghi and Peck) on settlement of buildings bearing on sand soils or other situations where the settlement is primarily due to relatively short term effects (principally shear strain). A more liberal limitation of 3 in. is suggested for foundations where the major displacement is produced by slow long-term effects (consolidation). The more liberal limitation represents, in part, the greater adaptability of a building to greater settlement where such settlement occurs slowly. It is also partly prompted by the difficulty of designing foundations bearing on clay or silt soils which are of reasonable size and cost under more restrictive criteria.

For intermediate cases, intermediate values may be interpolated, as judgment dictates.

II. *Exceptions*

A. *If the rigidity of the structure is neglected* and the soil is reasonably uniform, contours of differential settlement of a building can be derived without great effort by using the Newmark Chart (Fig. 5-16) to determine changes in bearing pressure and Eq. (5-6) or (5-7) to compute the amount of settlement. Under such circumstances, the slope or deflection limitations discussed in Secs. 5B.3a and 5B.3b can be applied as the controlling criteria. The approach is a conservative one in that the building always has some rigidity. Depending on the circumstances, it may give results which are more, or less, economical than the blanket limitations on gross settlement described above. It is often worth a try, but only if the soil is reasonably uniform so that there is some assurance against abrupt differentials.

B. *The amount of settlement* to be included in the limiting criteria is not the total amount of settlement which the building theoretically undergoes from the inception of construction. For example, in sand soils the settlement due to shear strain caused by the weight of the foundations is not included because such strain occurs while the foundation concrete is still plastic and therefore is able to adjust to the strain without developing stress. Similarly, short-term settlements (i.e., shear strains plus the amount of consolidation, if any, occurring during construction), which develop under the weight of elements added before rigidity is built into the structure, may be neglected in the computation of total settlement. If the building is without foundation walls, grade beams, or other stiffening elements, the settlement due to the weight of the first-floor framing can be neglected. Similarly, for a building such as a one-story factory (again, if the foundation itself is flexible), the settlement due to the weight of the roof framing can be disregarded. Significant savings are possible following this vein.*

* As an example, assume the building under design were a one-story factory. The roof load is 20 psf snow load and 50 psf dead load. The supporting soil is a sand showing an N value of 20 in the standard penetration test. Embedment of the footings is 3 ft. Column spacing is 40 by 50 ft. Column load (interior column) is 40 kips live load, plus 100 kips dead load = 140 kips.

The allowable bearing pressure for a sand having an N value of 20, from Fig. 5-9, is about 2 tsf. This value is dictated by considerations of settlement. The required footing size is 35 sq ft, or 6 by 6 ft.

The allowable bearing capacity as determined by the strength of the soil can be determined from Eq. (5-2). It is about 15,000 psf ($\phi = 32°$). Using a factor of safety of 2.5, the allowable bearing pressure would be 6,000 psf or 3 tsf if the settlement problem could, somehow, be bypassed.

Suppose we were to construct the footings, erect the frame using simple connections and temporary guys for wind bracing, place the roof, and release the guys as the permanent wind bracing was put in. Only then would the grade beams be poured and the walls and partitions be laid up. Under such a circumstance, all displacement of the footings under the dead load of the roof and roof structure would be accomplished before the roof structure was made rigid and could be disregarded. The allowable bearing pressure under such circumstance would be 3 tsf, not 2 tsf and the footing size would be 5 by 5 ft instead of 6 by 6 ft, a decrease of one-third in the amount of concrete, about one-quarter in the area of forms, and about one-half in the amount of

The settlement due to the weight of any fill added prior to final bolting up of the structure or hardening of the concrete framing also may be disregarded. This too can be of importance. For example, consider a case where the grade under the building is to be raised in order to drain away from the building or to form a loading dock. If the allowable bearing pressure, as limited by settlement, were low (say 1 tsf), the weight of the fill (about 600 psf) would be 30 percent of the total allowable bearing capacity.

At the opposite extreme, where the anticipated settlement results from long-term effects (consolidation of compressible strata), the design live load which is applicable in proportioning the superstructure may not be applicable in calculating settlement because it is not in place long enough to cause the requisite long-term effects. Design wind and earthquake loads definitely fall in this category and may be neglected in computing anticipated consolidation of compressible strata. As for the live load in the building, the usual live-load reductions applicable in the design of the superstructure probably compensate adequately for lack of their concurrence, and no further reductions applicable to the design of the foundations are recommended.

C. *Rigid Superstructure.* If the superstructure is deliberately designed to be rigid and to have sufficient strength to span any potential differential settlements, then, of course, no limitation need be placed on the amount of settlement which can be tolerated, other than to consider possible disruption of utility lines entering the building. This latter problem can be handled by running the lines in utility tunnels or using special flexible connections.

D. *Flexible Superstructure.* At the opposite extreme, if the superstructure is very flexible (the bottom of a steel tank is an example), again there need be no limitation on settlement (provided that the supports do not spread excessively) because the structure can accommodate itself to the settlement contours. Sometimes this flexibility can be attained without actually having flexibility in the superstructure. If the building is supported on three points, a tripod arrangement for example, large differential settlements can occur without inducing stresses in the superstructure, provided, of course, that the points of support do not spread apart as they settle. Another device is to construct the building as a series of small rigid abutting units with flexible jointing or space between units. These latter two devices (rigid superstructure or flexible

---

reinforcement. The postconstruction settlement of the footing would correspond to that caused by a load of 40 kips on a 5 by 5 ft footing bearing on 20-blow material. The bearing pressure is 1,600 psf. According to Terzaghi and Peck, 20-blow material can be loaded to about 4,000 psf before settlement of the footing will amount to 1 in. Therefore, the criterion that postconstruction settlement shall not exceed 1 in. will be satisfied.

or articulated superstructure) are a valuable alternate to the use of piling, particularly when light buildings are to be constructed over deep deposits of compressible materials.

Section 5C COMPUTATION OF ACTUAL BEARING PRESSURE

The bearing pressure at a point in a soil mass is the sum of the effects of:

1. The loads or surcharges due to fills or adjacent constructions
2. The weight of the overlying soil (including any reductions in soil weight resulting from excavation at or adjacent to the site)
3. Hydrostatic forces: negative (buoyancy), positive (desiccation or other capillary effects), and those due to seepage
4. Weight of the foundation and reactions from the superstructure, as described in Chap. 4

Normally, only items 1 and 4 are considered when checking allowable bearing capacity against the actual bearing pressure because, as will be discussed in Sec. 5D, items 2 and 3 usually produce a change in capacity equal to the change in pressure which they impose, giving a net zero effect. This is not always the case, however, and it is well to remember that items 2 and 3 are real loads, producing real pressure effects, and properly should be considered in the design. In particular, the reduction in pressure incident to excavation is important and is discussed in Sec. 5D.1 under Net Bearing Pressure.

Calculation of the soil pressure due to the weight of the overlying soil or due to hydrostatic forces (other than those due to seepage) is a matter of simple statics.

Calculation of seepage forces is not required in routine work and will not be treated here. Where such must be considered, the reader is directed to the excellent discussion in Ref. 17.

It is the calculation of the pressures imposed by the surface (surcharge) loads and by the reactions from the superstructure (items 1 and 4) which is the basic problem of the foundation designer and which will now be considered. This calculation involves consideration of two cases: (1) bearing pressure on soil at shallow depth under the bearing area and (2) bearing pressure on soil deep beneath the bearing area. For both cases, the theoretical evaluation is relatively complex, and for practical purposes, resort is made to approximations, as follows.

5C.1 Bearing Pressure on Soil at Plane of Contact with the Foundation

The theoretical solutions to this problem differentiate between foundations bearing on cohesionless (sand) soils and frictionless (clay) soils.

Differentiation also is made between flexible and rigid foundations. The results of these solutions are illustrated in Fig. 5-13.

In practice, the theoretical distributions seldom are used for design. Instead, it is assumed that the contact pressure between the soil and the

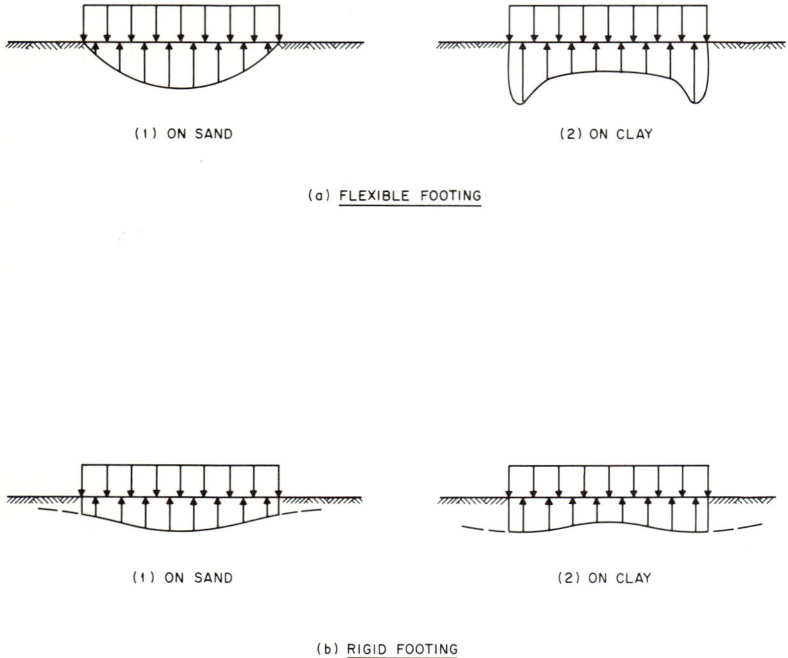

FIG. 5-13 Theoretical Distribution of Soil Reactions at Plane of Contact with the Foundation.

footing is uniform (P/A) or varies linearly $(P/A \pm Mc/I)$. These assumptions are illustrated in Fig. 5-6.*

The reasons for this disparity between theory and practice are as follows:

1. The distributions of pressure indicated in Fig. 5-13 would be modified by the effects of any surcharge confining the bearing area. The effect would be to make the distribution of contact pressure more uniform. Some surcharge effect almost always is present since footings are seldom placed directly on the surface of the ground.

2. Most foundations are neither absolutely rigid nor entirely flexible. As a result the actual pressure distribution at the contact face between soil

* The use of other criteria for dealing with eccentric loadings on foundations is discussed in Sec. 5E, Chap. Note 2.

and foundation would be intermediate between these shown in Fig. 5-13*a* and *b*.

3. Since both ultimate capacity and settlement involve strain in the soil to a considerable depth, the distribution of contact pressure does not materially influence the bearing capacity of the soil material. Rather, its importance is in determining the moments and shears in the transition structure of the foundation, i.e., the footings. For footings founded on granular soils, the contact pressure concentrates toward the center of the bearing area so that the assumption of uniform distribution results in moments and shears which are larger than the theoretical values and so is conservative. For footings founded on clay soils, in theory, the soil reaction is concentrated at the perimeter of the footings, and the assumption of uniform distribution gives moments and shears which are lower than theoretical values. Usually, this effect is neglected. However, some designers do increase the calculated moments and shears to approximate the effect.

4. Lastly, for footings, the design moments and shears are determined on the basis of procedures described in the Building Code of the American Concrete Institute (ACI-318). The ACI code stipulates an assumption of uniformly distributed soil reaction for concentric loading. This stipulation and the design procedure (particularly the locations of critical sections) appear to have an essentially empirical basis which is predicated on the assumption of uniformly distributed soil reaction.

EXCEPTION: An exception to the assumption of uniformly distributed soil reaction is made when the transition structure under design consists of combined footings or a mat. For such construction, the analysis of moment and shear sometimes is executed on the basis of varying subgrade contact pressure, using the equations developed for a beam or a plate on an elastic foundation [29, 52].

This latter problem has been considered by ACI committee 436. Their report [51] describes procedures for the design of combined footings and mats. The committee's report suggests that rigid mats or footings (or mats or footings supporting rigid structures which compel uniform settlements of the foundation and so approximate rigid footings) be designed for a uniform distribution of soil reaction and that flexible constructions be designed on the basis of variable subgrade reaction. Differentiation between rigid and flexible construction is made as follows:

> If the average of two adjacent spans $\bar{l}$ in a continuous strip having adjacent loads and column spacings that vary by not more than 20 percent of the greater value is less than $1.75/\lambda$ the footing can be considered rigid. . . .
>
> If the average of two adjacent spans $\bar{l}$, as limited above, is greater than $1.75/\lambda$ the footing shall be . . . [considered as flexible].

Evaluation of the factor λ can be made on the basis of the following formulas:

$$\lambda = \sqrt[4]{\frac{k_s b}{4E_c I}}$$

$$k_s = S k_{s'}$$

where k_s = coefficient of vertical subgrade reaction, kips per cu. ft.
$k_{s'}$ = basic value of coefficient of vertical subgrade reaction for a square area with width $b = 1$ ft., kips per cu. ft.
b = width of footing, ft.
S = size or shape factor for a footing on a particular type of soil
E_c = modulus of elasticity of concrete, kips per sq. ft.
I = moment of inertia of footing, ft.^4

For sandy soils the size factor S can be determined from the following formula:

$$S = \left(\frac{b + 1}{2b}\right)^2$$

with a limiting value of 0.25 for large footings.

For clay soils the shape factor S can be determined from the following formula:

$$S = \frac{n + 0.5}{1.5n}$$

where n is the ratio of the longer side to the shorter side of the footing. For extremely long footings, where n approaches infinity, S can be assumed as 0.67.

Values for $k_{s'}$ can be determined from the results of field tests performed on the subgrade of the proposed structure or they can be estimated on the basis of empirical values in 'Evaluation of Coefficients of Subgrade Reaction' by Terzaghi.

The procedures recommended by ACI committee 436 are simple (although the computations are somewhat lengthy) and are suggested as a standard of design for combined footings and for mats.*

* The solutions developed for beams on elastic foundations assume that the soil acts like a series of isolated springs; i.e., there is neither shear strain nor resistance between adjacent soil elements which undergo differential displacement. As a result, these solutions do not properly account for the stresses in the soil which exist at discontinuities in the loaded area, i.e., at the edges of the footing or mat. Therefore, some designers prefer to proportion the perimeter bays of a mat foundation on the basis of an assumed uniform distribution of soil reaction.

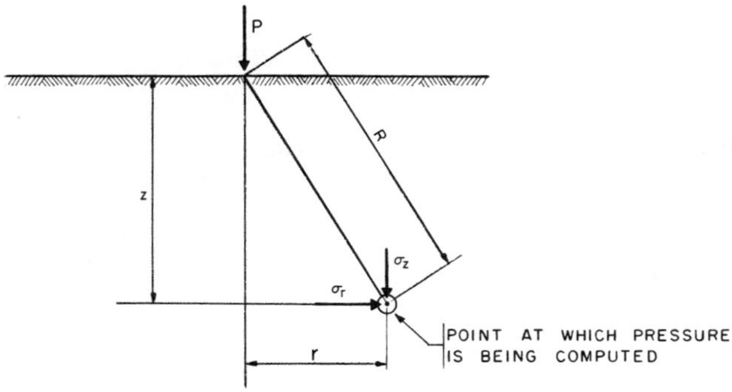

FIG. 5-14 Symbolism for Use in Boussinesq Equations.

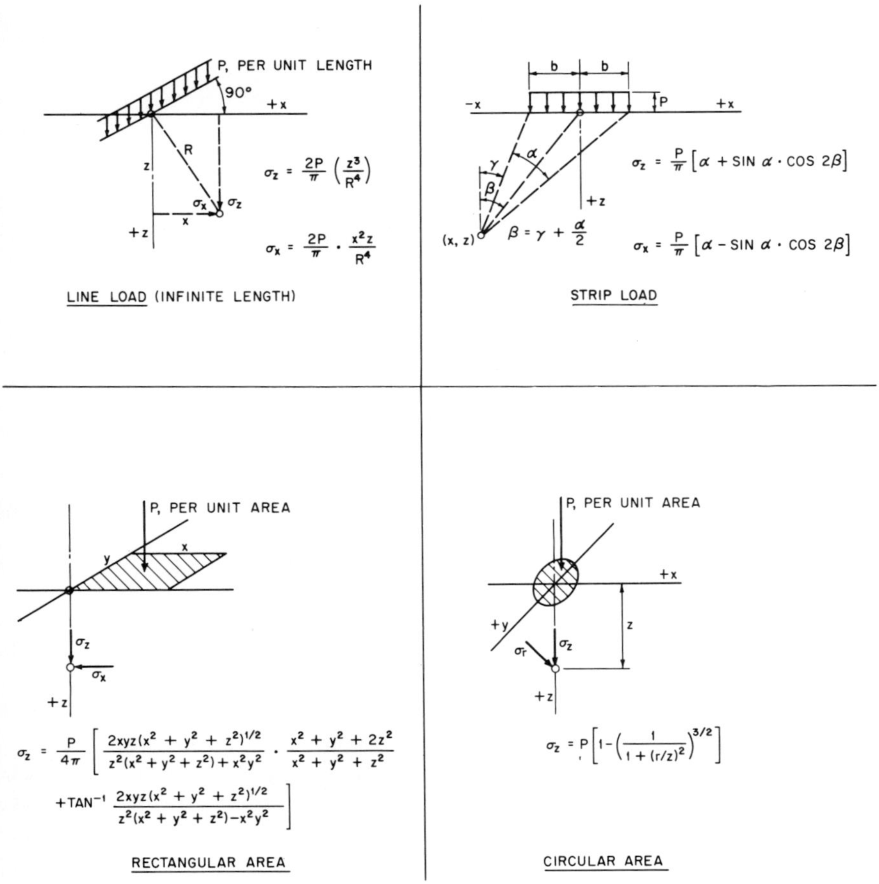

FIG. 5-15 Soil Pressures under Loaded Areas (by integration of Boussinesq Equations).

5C.2 Bearing Pressure on Soil below the Plane of Contact with the Foundation

a. Semi-infinite, Isotropic Soil Mass

Where subsurface conditions approximate those of a semi-infinite, isotropic continuum,* the equations of Boussinesq apply. These equations, for a point load, are

$$\text{Vertical stress: } \sigma_z = \frac{3P}{2\pi}\frac{z^3}{R^5} \tag{5-16}$$

$$\text{Horizontal stress: } \sigma_r = \frac{P}{2\pi}\left[\frac{3r^2z}{R^5} - (1 - 2\mu)\left(\frac{R - z}{Rr^2}\right)\right] \tag{5-17}$$

The symbolism is illustrated in Fig. 5-14.

* As a practical matter, such conditions are approximated only when a deep deposit of uniform soil, of uniform density (not increasing with depth), and without detectable stratification is encountered. Such a condition is not common.

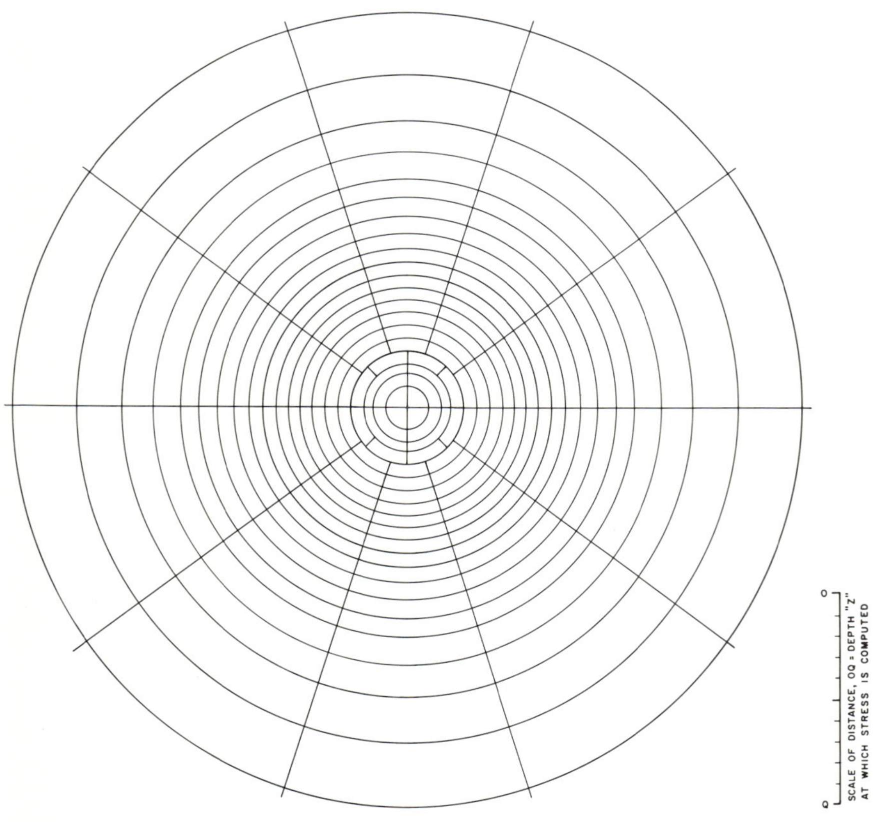

FIG. 5-16 Influence Chart for p_z. Vertical Stress on Horizontal Planes (Influence Value per Block Is 0.005). *(From Newmark, Ref. 42.)*

Equations (5-16) and (5-17) can be integrated to obtain expressions applicable to various line or area loads. Solutions for several simple cases are shown in Fig. 5-15.

For irregularly loaded areas, or for irregular loads, the influence chart developed by Newmark [42] as shown in Fig. 5-16 is used. Use of this chart is illustrated in Sample Computation 5-2.

SAMPLE COMPUTATION 5-2

Use of Newmark Chart

STATEMENT OF PROBLEM

Determine bearing pressure imposed on soil at point A, at level 50 ft below level of underside of foundation. Plan of building, as shown. Equivalent area loads:

Area A = 1,500 psf
Area B = 1,000 psf
Area C = 500 psf

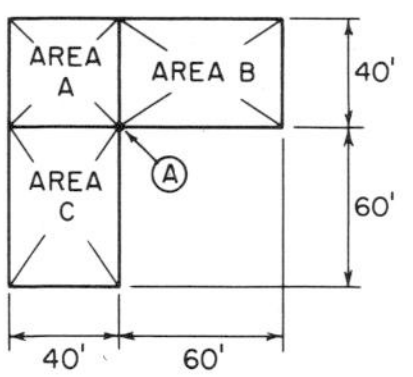

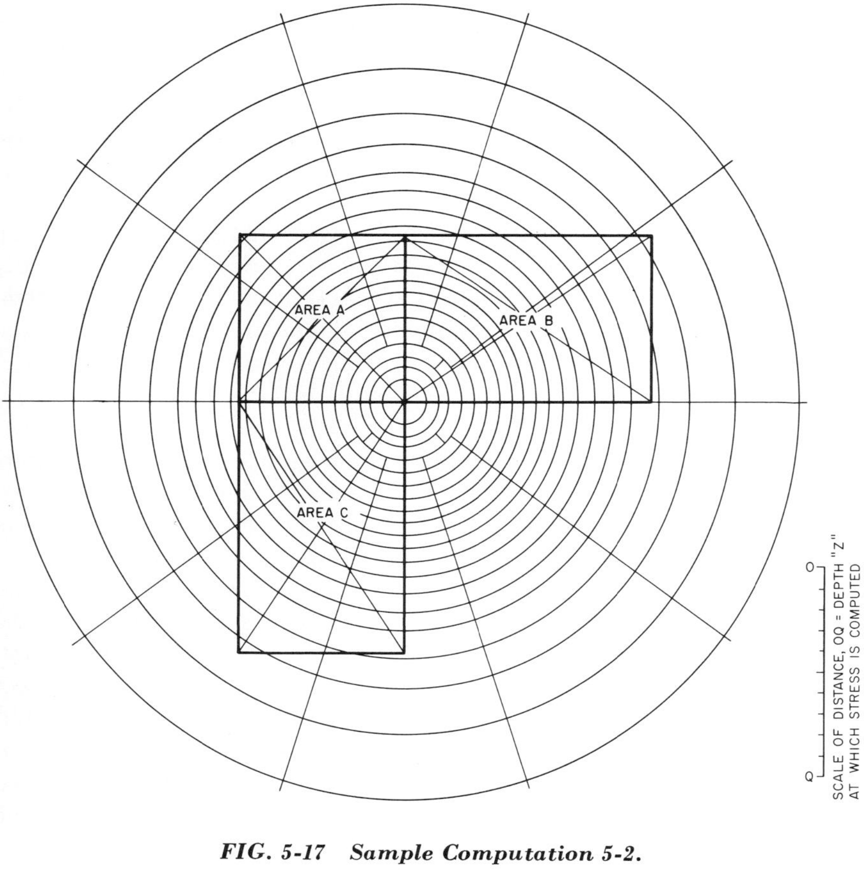

FIG. 5-17 Sample Computation 5-2.

PROCEDURE

STEP 1: Superimpose plan of building on Newmark Chart (Fig. 5-16) with point *A* over center of chart. Draw plan to scale such that distance *OQ* on chart is equal to 50 ft.

STEP 2: Count number of influence blocks encompassed by areas *A*, *B*, and *C* (see Fig. 5-17). For the case in point this number is:

Area *A*—29
Area *B*—34
Area *C*—34

STEP 3: Obtain pressure under point *A*, at depth of 50 ft by multiplying number of influence blocks by the influence value of 0.005 and multiplying the result by the imposed pressure. For the case in point, the value is:

Area *A*—29 blocks × 0.005 × 1,500 = 218
Area *B*—34 blocks × 0.005 × 1,000 = 170
Area *C*—34 blocks × 0.005 × 500 = 85
Total pressure = 473 psf

b. Layered or Anisotropic Soil Mass

In nature, soils seldom approximate the semi-infinite, isotropic continuum assumed by Boussinesq. The soil usually is layered, with elastic properties differing from layer to layer. Within a given layer, the elastic properties frequently will differ between horizontal and vertical planes, with rigidity in the horizontal direction usually being greater than that in the vertical direction.

A limited number of solutions have been derived for cases involving layered or anisotropic soils (see Refs. 33 and 34). Except for the limiting cases where Burmister's influence charts [34] apply (two-layer system, base layer rigid, rectangular loaded area), the available solutions are too complex and time-consuming to be of practical value.

c. Sixty-degree Approximation

It has been noted that soil strata seldom approximate the isotropic, semi-infinite continuum assumed by Boussinesq and that the soil conditions in the specific instance under consideration usually do not conform to the specific layering systems for which there are theoretical solutions available for computing the bearing pressures, at depth. In addition to these difficulties, the theoretical solutions (Boussinesq, et al.) assume the foundation to be flexible, whereas in practice, it is likely to have considerable rigidity. Thus, for practical problems, the theoretical solutions have limited use, and recourse often is made to the so-called 60° approximation.

In this approximation, computation of the vertical pressure in the soil at any depth below the foundation is made on the assumption that the load is spread uniformly at an angle at 60° with the horizontal (Fig. 5-18*a*). Naturally, the area considered as supporting a given foundation does not extend beyond the intersection of 60° planes projected downward from adjacent foundations (Fig. 5-18*b*). Provided that the depth (D)

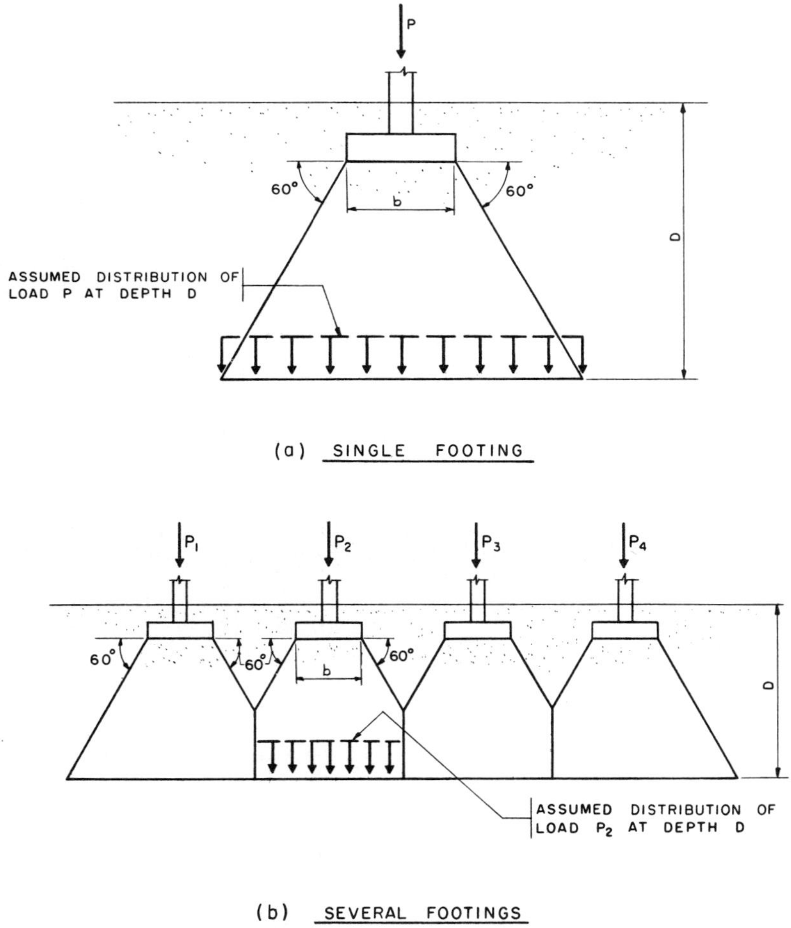

FIG. 5-18 Sixty-degree Approximation for Pressure Distribution below Bearing Area.

at which the bearing pressure is being computed is not too large compared to the width of the loaded area (b) (say 2:1 or 3:1), this approach gives results which closely approximate average pressures as computed by the Boussinesq equations.

The approximation has gained wide acceptance. It appears in several texts [9, 11] and in several building codes including those of Boston and Los Angeles.*

Using the 60° approximation, the pressure distribution at any depth is presumed to be uniform. Variations in bearing stress between the center and edges of the bearing area are neglected.

Since allowable bearing pressure usually is controlled by considerations of allowable settlement rather than of ultimate capacity, if the 60° approximation is used, it is important that the allowable settlement be established as a limitation on total displacement (see Sec. 5B.3). In such a case, it is only the maximum bearing pressure (or a reasonable approximation thereof) which is of interest, and not the variations therein. This is one of the reasons that limitations on total displacement are more useful than are limitations on angular displacement or flexural distortion.

Section 5D APPLICATION TO PRACTICAL PROBLEMS

The procedure for estimating allowable bearing pressures presented in this Section consists of calculating a base value applicable to the type of soil encountered and then of correcting for certain effects, such as embedment and width of the bearing area, depth of the bearing stratum, infrequent loading conditions, etc. Evaluation of base pressure values is discussed in Sec. 5D.2. The correction factors are described in Secs. 5D.3 through 5D.6. *The corrections are cumulative.*

Before proceeding to the details, however, two introductory considerations must be developed: (1) *net*-versus-*gross* bearing pressure and (2) the classification system to be used to describe the soil strata which are encountered.

5D.1 Introductory Considerations

a. Net Bearing Pressure

It is of the utmost importance to recognize that allowable bearing pressures as discussed in this text and as prescribed in most building codes are allowable *net* bearing pressures, i.e., the amount of pressure that the soil can support over and above the stabilized weight of overburden which it is already supporting.†

* Some engineers [9] assume a distribution at 63.5° with the horizontal, instead of 60°. At 63.5°, the planes defining the distribution of the applied load slope at 2:1 with the horizontal, and the computation is somewhat simplified.

† The proof of this statement relates to the previous discussion about allowable bearing pressure being controlled by two considerations, namely, ultimate bearing capacity and settlement. Consider settlement first. If the overburden has been in place long enough to have come to a stable condition with the bearing stratum, any settlement of the bearing stratum which would be caused by its weight will have been

As an illustration, assume a foundation consisting a belled piers bearing on a hardpan, which has a presumptive bearing capacity of 6 tsf as shown in Fig. 5-19. Assume the weight of the overburden to be 120 pcf and the depth of overburden to be 100 ft. The surface of the hardpan is subject to a pressure of 6 tsf due to the weight of the overburden. If the overburden pressure of 6 tsf were to be included in the

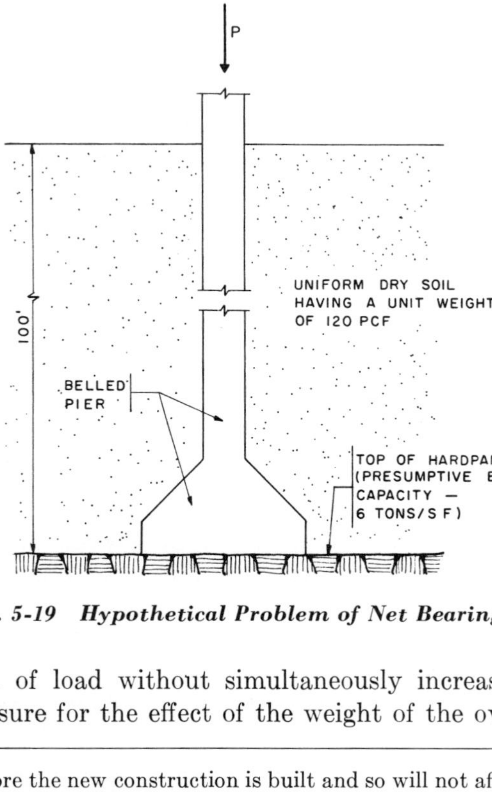

FIG. 5-19 Hypothetical Problem of Net Bearing Pressure.

computation of load without simultaneously increasing the allowable bearing pressure for the effect of the weight of the overburden, the soil

---

completed before the new construction is built and so will not affect the new construction. Of course, if a stable condition has not been reached, additional settlement of the supporting soil will occur and would affect the proposed construction. Therefore the allowable net bearing pressure is increased by the amount of *stabilized* overburden pressure and not, necessarily, by the total overburden pressure.

With regard to ultimate capacity, refer to Eq. (5-1). The second term $\Sigma \gamma_d D_f N_q$ represents the increase in ultimate strength of the soil due to the effects of confining pressure. Confining pressure includes surcharge or overburden pressure. Referring to Fig. 5-1 it may be noted that the value of N_q always is equal to or greater than 1.0. Therefore, if a pound of overburden pressure is added to a soil stratum, the ultimate strength of that stratum is correspondingly increased by at least 1 lb or, since the value of N_q, except for purely frictionless soils, is more than 1.0, by several pounds. Accordingly, insofar as ultimate bearing capacity is concerned, the weight of the overburden contributes an increase in strength equal to or greater than the pressure increment which it engenders.

could not support any foundation at all, as the hardpan surface would be overloaded by the weight of the concrete shaft and bell alone. Similarly, if the hardpan surface were at a depth of 110 ft below grade, it would be overloaded just by the weight of the overburden. Neither of these conclusions is correct, of course. A soil having a presumptive bearing capacity of 6 tsf is a soil which can support 6 tsf more than the weight of the overburden which it already supports, provided, however, that said weight has been in place long enough to have become stabilized (stabilized overburden pressure).

This reference to *stabilized* overburden pressure is an important qualification, because the stabilized overburden pressure is not always equal to the total overburden pressure, as indicated by the following considerations.

I. *Stabilized Overburden Pressure.* The basic criterion in establishing the amount of stabilized overburden pressure existing at a given point in the soil mass is that the weight of the overburden has reached a condition of stability with respect to the supporting bearing stratum. In effect, this means that the weight must have been in place long enough so that it is supported by intergranular forces within the bearing stratum and not by excess pore water pressure.

If the bearing stratum consists of medium to coarse granular soils, the stabilized overburden pressure may be assumed to be equal to the full weight of the overburden less any amount which is stripped and not replaced (see Sec. 5E, Chap. Note 6). If the bearing stratum is rock, the stabilized overburden pressure is assumed to be zero (see Sec. 5E, Chap. Note 7). If the bearing stratum consists of fine-grained materials (fine sands, clays, silts), then an intermediate condition prevails, and the portion of the total weight of the overburden which has attained a condition of stability should be determined from the laboratory consolidation test (curve e versus log p), again as described in Sec. 5E, Chap. Note 6.

CAUTION: In Chap. 4 the design approximation of omitting the weight of the soil overlying the foundation (the weight of the overburden) from the computation of load and of treating the allowable bearing pressure as the net pressure without increase for the effects of stabilized overburden pressure was described. It was stated that, while this device in most cases gives results which are numerically correct, in some cases it does not. It was suggested that a preferable practice is to include the weight of the overburden or fill, if any, which overlies the foundation as part of the load to be supported and to take the allowable bearing pressure as the gross allowable value, which is the *net* allowable pressure (derived as hereinafter described) plus the stabilized overburden pressure, if any.

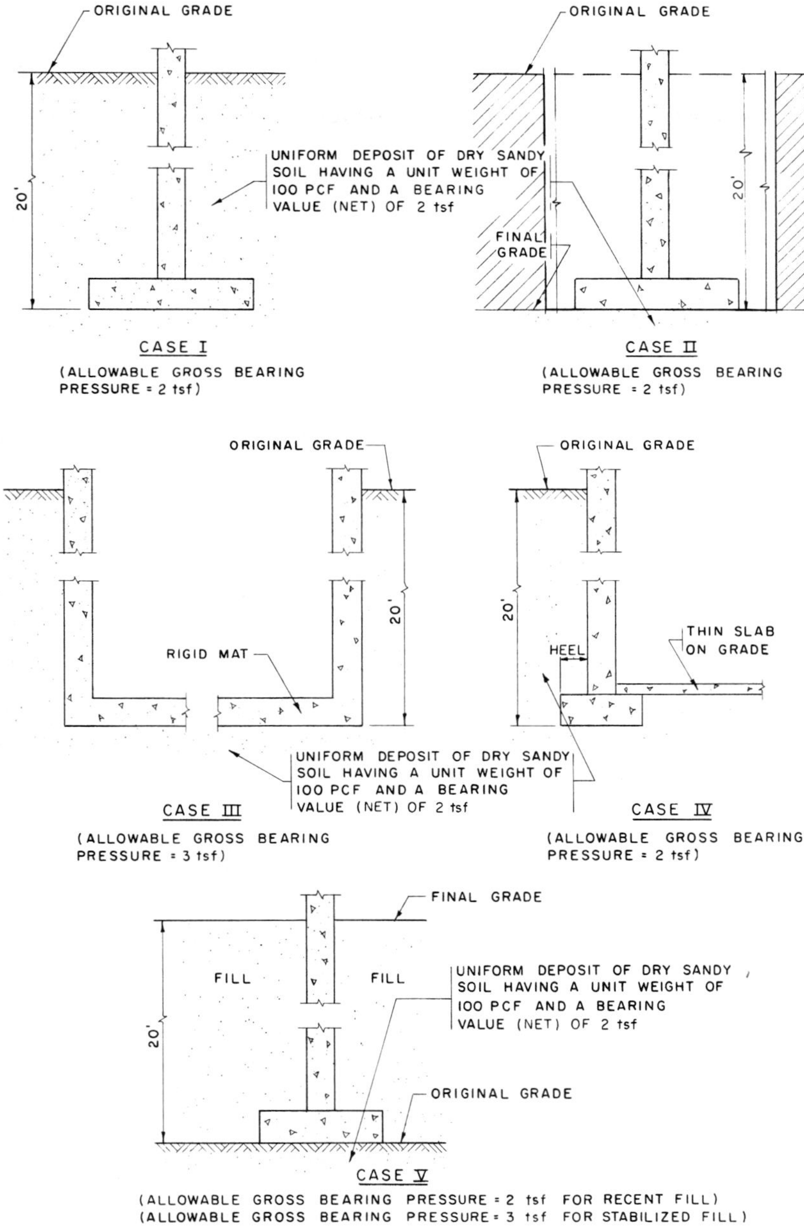
ORIGINAL GRADE
UNIFORM DEPOSIT OF DRY SANDY SOIL HAVING A UNIT WEIGHT OF 100 PCF AND A BEARING VALUE (NET) OF 2 tsf
20'
CASE I
(ALLOWABLE GROSS BEARING PRESSURE = 2 tsf)
ORIGINAL GRADE
FINAL GRADE
20'
EXCAVATED
CASE II
(ALLOWABLE GROSS BEARING PRESSURE = 2 tsf)
ORIGINAL GRADE
RIGID MAT
20'
CASE III
(ALLOWABLE GROSS BEARING PRESSURE = 3 tsf)
ORIGINAL GRADE
HEEL
THIN SLAB ON GRADE
20'
UNIFORM DEPOSIT OF DRY SANDY SOIL HAVING A UNIT WEIGHT OF 100 PCF AND A BEARING VALUE (NET) OF 2 tsf
CASE IV
(ALLOWABLE GROSS BEARING PRESSURE = 2 tsf)
FINAL GRADE
FILL
FILL
UNIFORM DEPOSIT OF DRY SANDY SOIL HAVING A UNIT WEIGHT OF 100 PCF AND A BEARING VALUE (NET) OF 2 tsf
20'
ORIGINAL GRADE
CASE V
(ALLOWABLE GROSS BEARING PRESSURE = 2 tsf FOR RECENT FILL)
(ALLOWABLE GROSS BEARING PRESSURE = 3 tsf FOR STABILIZED FILL)

FIG. 5-20

The validity of this point and its importance can be illustrated by reference to several typical cases as shown in Fig. 5-20.

Case I This figure represents an isolated footing and pier embedded below the original grade level. The load to be supported by the footing includes the weight of the foundation plus the weight of the soil overlying the footing. The gross allowable bearing pressure includes the weight of the overburden. This may be expressed as:

1. $\overbrace{\text{Foundation reaction + weight of overburden}}^{\text{load}} = \overbrace{\text{net allowable bearing pressure + stabilized overburden pressure}}^{\text{resistance}}$
2. Stabilized overburden pressure = weight of overburden

The weight of the overburden therefore appears on both sides of the equation, in the load and in the resistance, and, in effect, cancels out. For this case, then, the approximation of omitting the weight of the overburden from the calculation of load and resistance gives results which are numerically correct.

Case II This figure represents the same isolated footing shown in Case I, but with the overburden excavated and the footing at the excavated surface (or at shallow depth). For this case the overburden pressure remaining after the completion of the construction is zero and the aforementioned approximation of omitting the weight of the overburden from the calculation of load and resistance again gives results which are numerically correct:

1. $\overbrace{\text{Foundation reaction + weight of overburden}}^{\text{load}} = \overbrace{\text{net allowable bearing pressure + stabilized overburden pressure}}^{\text{resistance}}$
2. Weight of overburden = zero
3. Stabilized overburden pressure = weight of overburden = zero

Case III This figure represents a floating rigid-box or mat foundation. For this case, the mat is rigid and fully confines the underlying soil. The effective confining pressure on the soil underlying the base of the mat is equal to the weight of the 20 ft of overburden overlying this level. Except for the effects of rebound, this has the same effect as if the mat were overlain by 20 ft of soil. For granular soils (sands and gravels), rebound is considered to be negligible. Therefore, the allowable bearing pressure would be 3 tsf (2 tsf net allowable value plus 1 tsf overburden pressure). At the same time, the weight of the overburden on the foundation is zero, i.e., the applied pressure due to the weight of the overburden is less than the confining pressure from the same source. Accordingly, the above described approximation would not give correct results for this case nor for any case where the adequacy of the foundation is predicated on relief of overburden pressure. The applicable equations are:

load		resistance
1. Foundation reaction + weight of overburden	=	net allowable bearing pressure + stabilized overburden pressure

2. Weight of overburden = zero
3. Stabilized overburden pressure = effective confining pressure ≠ zero

If the bearing stratum were a silt or clay, the same reasoning would apply except that the effects of rebound would not be negligible, and correction must be made, as described in Sec. 5E, Chap. Note 6.

An extreme example of the situation shown in Case III would be if the soil had zero net allowable bearing pressure; i.e., it was a pure fluid. The construction shown could be used by limiting the bearing pressure to 1 tsf, which is the effective confining pressure. The structure simply would float.

Case IV This is another case in which the subject approximation does not give correct results. The slab on grade is assumed to be of insufficient rigidity to confine the underlying soil. Accordingly, the confining pressure due to the overburden effectively is zero, whereas the weight of the soil on the heel of the basement wall is a positive pressure.

Case V This case is intended to illustrate the difference in effect between the case (same as Case I) where the overburden pressure has reached a condition of equilibrium with the supporting soil (stabilized overburden pressure) and the case where it has not reached that state. The weight of the fill exists as a load in either case, but the allowable gross bearing pressure is different. The appropriate equations are:

load		resistance
1. Foundation reaction + weight of overburden	=	net allowable bearing pressure + stabilized overburden pressure

2. Weight of overburden ≠ zero
3. Stabilized overburden pressure = zero ≠ weight of overburden

II. *Economic Significance.* The economic significance of this matter of net versus gross bearing capacity is important. It is most pronounced where the presumptive bearing capacity of the soil is relatively low. For example, assume a footing embedded 4 ft (frost depth) in a loose sand having a presumptive bearing capacity of 2 tsf and a unit weight of 100 pcf. The allowable bearing pressure would be 2.2 tsf (2 tsf + 400 psf overburden) if treated as a net pressure, but only 2.0 tsf if the net pressure effect were neglected. The difference in area of the footing would be 10 percent, with an even greater difference in volume and in reinforcement due to the reduced moment lever arm engendered by the reduced size.

b. Soil Classification System

Allowable bearing capacity depends primarily on the type of soil being loaded. Gravel, for example, behaves differently from sand. Both behave differently from silt or clay soils. Therefore, an essential step in determining allowable bearing capacity is to classify the soil which is to constitute the bearing stratum.

Several excellent classification systems have been developed. The one which will be used for purposes of this text is the Unified Classification System as adopted by the U.S. Army, Corps of Engineers, and by the Bureau of Reclamation. The system and the classification notation (GM, GW, SM, CL, etc.) are shown in Fig. 2-1.

Equivalent classifications under other systems may be substituted. The important thing is to recognize that soils, in nature, generally occur as mixtures of particle fractions and that it is the behavior of the whole soil and not of an individual fractions which is of interest. For example, a sand is a sand because it behaves like a sand, and this does not necessarily imply that sand is the volumetrically dominant fraction in the aggregate soil mass (although this usually is the case).

The classifications under the unified system will be supplemented for purposes of this text by the following definitions:

1. *Fine Sand.* Soils of group SM containing more than 50 percent (by weight) of particles passing the no. 60 mesh sieve.

2. *Hardpan.* A glacial till (classification groups GM, GC, or SW) sufficiently cemented to be difficult to remove by picking.

5D.2 Basic Allowable Bearing Pressures

a. Rock

I. *Base Values.* At the present (1968) state of the art, no generally accepted rational method for determining the allowable bearing capacity of rock has been developed. Computation based on Eq. (5-1), assuming zero friction angle and value of cohesion c equal to one-half the unconfined compressive strength q_u, sometimes is attempted (see Ref. 48, for example), but would be applicable only if the rock under and adjacent to the bearing area of the footing or pier were without defects such as cracks, seams, or weathered zones. Sometimes, this is the case. More often, it is not.

A rational analysis, based on the principles of rock mechanics or on load tests, sometimes also is attempted when a large project such as a dam or other major structure is contemplated. For the general run of building foundations, however, the standard of practice is to use the provisions of the applicable (or most nearly applicable) building code.

Some typical values of allowable bearing pressures on rock as given in various building codes are summarized in Table 5-6, and the suggested procedure is for the designer to interpolate the case under consideration from the classifications and descriptions in this table. Unfortunately, the variation in values is extreme, and the designer is forced to do the best he can.

Equally frustrating is that the indicated bearing values are limited to those for materials classified as either "in sound condition" or in "shattered or broken condition."

In nature, rock often is neither sound nor shattered or broken. It is something in between. Moreover, the term "sound" and the additional terms "soft" and "disintegrated," which appear in Table 5-6, are not well defined. As a result, when the bearing area is uncovered the engineer is likely to be left in a considerable dilemma not just to identify the type of rock (which is relatively easy) but to define its condition. When this happens, there is apt to be some difference of opinion as to allowable bearing capacity, particularly between the design engineer and the building department inspector. This leads to difficulties.

To alleviate the problem, in conjunction with their task of advising and assisting in the development of a foundations code for the City of New York, the Code Committee of the Soils and Foundations Group, Metropolitan Section ASCE [38] developed the following descriptive definitions relating to the condition of rock:

> *Sound Rock:* Rock which rings when struck with a pick or bar; does not disintegrate after exposure to air or water; breaks with a sharp, fresh fracture; in which cracks are unweathered and less than ⅛″ wide, generally no closer than 3 ft. apart; core recovery with a double tube, diamond core barrel is generally 85%, or greater, for each 5 ft. run.
>
> *Medium Rock:* Characteristics as for sound rock except that cracks may be ¼″ wide and slightly weathered, generally spaced no closer than 2-ft. apart; core recovery, as above, generally is 50%, or greater.
>
> *Intermediate (between medium and soft) Rock:* Rock which gives dull sound when struck with a pick or bar; does not disintegrate after exposure to air or water; in which broken pieces may show weathered faces; may contain fractured or weathered zones up to 1″ wide spaced not closer than 12″ apart; and wherein core recovery, as above, generally is 35%, or greater.
>
> *Soft (or disintegrated) Rock:* Any rock which slakes on exposure to air or water, or a rock which gives a dull sound when struck with a pick or bar and wherein core recovery, generally, is less than 35%, but standard penetration resistance in sampling is more than 50 blows per foot.

TABLE 5-6 ***Allowable Bearing Pressure on Rock***

Allowable Bearing Pressure before Increase for Embedment, tsf

Description	Building Code of City of Baltimore (1955)	Building Code of City of Pittsburgh (1947)	Building Code of City of St. Louis (1961)	Building Code of City of Los Angeles (1959)	Uniform Building Code (1961)	BOCA (1960)	National Building Code (1955)	Building Code of City of Cleveland (1951)	Building Code of City of San Francisco (1962)	Building Code of City of Boston (1962)	Building Code of City of Kansas City (1961)	Building Code of City of Washington, D.C. (1961)
Massive bedrock in sound condition.........	100											
Massive igneous rock in sound condition......	....	....	....	....		....	....	....	....	100		
Massive crystalline bedrock, including granite, diorite, gneiss, traprock, hard limestone, and dolomite......	....	....	100	....		100	100					
Solid bedrock....	....	....	....	....	20% of crushing strength	....	....	....	....	....	....	30
Massive crystalline bedrock in sound condition..........	....	....	....	10								
Slate in sound condition......	....	....	....	....		....	....	....	....	50		
Foliated rock, including bedded limestone, schist, and slate, in sound condition......	....	....	40	4		40	40					

TABLE 5-6 *Allowable Bearing Pressure on Rock (Continued)*

Description	Building Code of City of Baltimore (1955)	Building Code of City of Pittsburgh (1947)	Building Code of City of St. Louis (1961)	Building Code of City of Los Angeles (1959)	Uniform Building Code (1961)	BOCA (1960)	National Building Code (1955)	Building Code of City of Cleveland (1951)	Building Code of City of San Francisco (1962)	Building Code of City of Boston (1962)	Building Code of City of Kansas City (1961)	Building Code of City of Washington, D.C. (1961)
Laminated rocks, such as schists, in sound condition	35											
Solid limestone											20	
Solid rock		25										
Sedimentary rock, including hard shales, sandstones, and conglomerates			25	3	20% of crushing strength	25	15	25				
Medium rock		15										
Shales in sound condition	10											
Shale		8								10		
Shale and sandstone, unweathered											10	
Rock in natural beds									5			
Residual deposits of shattered or broken bedrock	10									10		
Soft or broken bedrock			10			10						
Soft rock		8										10
Disintegrated rock												5

Using these definitions one may attempt to interpolate between the values given for sound and for soft rock.

The reason for the wide variation in bearing values between building codes for what, by description, would be the same material is not known.

CAUTIONS:

1. The indicated values of allowable bearing pressure on rock would apply only where the strata are level, or nearly so, where the rock is sufficiently cemented or massive that separating of the strata is not a problem, where the bearing area has ample lateral support, i.e., it is not next to a cut or on a steep slope, and where the bearing area is not underlain by cavities or by heavy fissures or seams.

2. When dealing with soft rock, if core recoveries are less than 20 percent, the material should be treated, not as a rock, but as a soil (classification may be as described in Fig. 2-1).

II. *Correction Factors.* The base values for bearing capacity of rock may be increased for embedment (Sec. 5D.3, but subject to the stipulations noted on page 171).

b. Hardpan

I. *Base Values.* As in the case of rock, to date (1968) no generally accepted criteria for determining allowable bearing pressure on hardpan have been developed. Again, selection of an appropriate value is made by interpolation in a table of typical values (Table 5-7). Fortunately,

TABLE 5-7 ***Allowable Bearing Pressure on Hardpan***

Code	*Allowable Bearing Pressure before Increase for Embedment, (tsf)*
Building Code of Baltimore City (1955)	10
Building Code of St. Louis (1961)	10
Building Code of City of Pittsburgh (1947)	8
Public Buildings Administration Structural Engineering Design Manual	10
Building Code of City of Boston (1962)	10
National Building Code (1955)	10
U.S. Navy BuDocks Design Manual DM7	10
BOCA (1960)	10

however, in contrast to the wide range of applicable values which obtain for rock, the range of values utilized in representative building and design codes is very limited. There seems to be a good consensus on 10 tsf.

II. *Correction Factors.* The base value may be increased for:

1. Stabilized overburden pressure (Sec. 5D.1)
2. Embedment (Sec. 5D.3)

c. Gravel Soils (soil groups GW, GP, GM, and GC)

I. *Base Values.* The base value of allowable bearing pressure on gravel (or almost any granular soil) is a function of the compaction and grading of the soil.

A reproducible numerical measure of the tightness of the grading of a gravel soil has not yet been developed. Grading usually is estimated in the field by visual examination. For laboratory purposes some evaluation can be made from the coefficients of uniformity and curvature determined from the sieve test,* but the reliability of these coefficients as a measure of grading suitable for evaluating bearing capacity is questionable.

Evaluation of relative compaction of a gravel soil also is an unsolved problem. Because of lack of uniformity of the soil and the large size of the particles with respect to the sampling equipment, any attempt at an accurate reproducible numerical evaluation is nearly hopeless. Therefore, the suggested approach to estimating bearing capacity of gravel soil is the device of bracketing the extremes and interpolating between them.

First, we must establish the extremes. Values of allowable bearing capacity of gravel soils appearing in several prominent building codes and other sources are given in Table 5-8. It may be noted that except for Los Angeles and San Francisco, where earthquake is a special consideration, the range is from about 4 to 10 tsf. Therefore, extremes are suggested as follows: for loose, poorly graded gravel soils—4 tsf; for compact, well-graded gravel soils—10 tsf.

For what constitutes a compact, well-graded soil and what constitutes a loose, poorly graded soil, the following guides may be used:

1. Compaction is best measured by in-place density tests. The density so determined is then converted to a relative density D_r, from the following equation:

$$D_r = \frac{e_0 - e}{e_0 - e_{\min}}$$

where e_0 = void ratio of soil in its loosest state
$e_{\min}$ = void ratio of soil in its densest state
e = void ratio of soil in the field

The values e_0 and $e_{\min}$ are determined in the laboratory.

The relative density is a measure of the compaction of the soil.

If the execution of in-place density tests is not feasible, the relative compaction may be judged (very approximately because of the difficulty of sampling gravel) from the resistance to penetration of the standard

* For determination of coefficient of uniformity and coefficient of curvature, see Sec. 5E, Chap. Note 3.

TABLE 5-8 ***Allowable Bearing Pressures on Gravel Soils***

Allowable Bearing Pressure, before Increase for Embedment, on Bearing Areas 3 ft or Greater in Width, tsf

Description	Building Code of City of San Francisco (1962)	Building Code of City and County of Denver	Building Code of City of Los Angeles (1959)	Building Code of City of Baltimore (1955)	Building Code of City of Cleveland (1951)	Building Code of City of St. Louis (1961)	BOCA (1960)	Building Code of City of Pittsburgh (1947)	Public Bldgs. Adminstn. Struc. Engineering Div. Manual	Building Code of City of Boston (1962)	York New City (1949)	U.S. Navy BuDocks Design Manual
Exceptionally compact gravels. Compact, partly cemented gravel					10	10	10					
Very compact												8
Compact gravel	1.5			5	6				5			6
Compact, well-graded gravel			3									
Gravel						6	6	4		5		
Loose gravel	1.0	3		4	4	4	4		4			4
Loose, poorly graded gravel			2									
Compact, silty or clayey gravel			3									
Loose silty or clayey gravel			1.5									

sampling spoon. To avoid being misled by the presence of large pieces of gravel, the N value used should be the average of the lower one-third of the recorded values, with not less than 15 values being recorded. With this stipulation, the authors suggest that a compact material may be considered as one exhibiting a penetration resistance of 80 or more blows per foot and that a loose material is one exhibiting a penetration resistance of 20 blows or less per foot.

2. Grading may be measured (in relative terms) from the coefficients of uniformity and curvature taken from the size distribution curve from the sieve test. A coefficient of uniformity greater than 4 coupled with a coefficient of curvature between 1 and 3 should be considered as indicative of a well-graded material [37]. Otherwise the material should be considered to be poorly graded.

The above criteria will serve to establish the desired extremes, i.e. loose, poorly graded material (4 tsf) and compact, well graded material (10 tsf). The reader is left to his own devices to interpolate between these extremes.

II. *Correction Factors.* The values of allowable bearing pressure on gravel soils, determined as above, may be increased for:

1. Stabilized overburden pressure (Sec. 5D.1)
2. Embedment (Sec. 5D.3)
3. Limited depth of the bearing stratum (Sec. 5D.5)

III. *Correction for Size of Bearing Area (See also Sec.* 5D.4). Where the bearing area is less than 3 ft in dimension or the embedment is less than 2½ ft, or both conditions prevail, the ultimate bearing capacity of a gravel soil should be checked by the methods indicated in Sec. 5A, since there is a danger of shear failure in the soil. Where the bearing area is both 3 ft or larger in dimension and embedded 2½ ft or more, the danger of shear failure can be neglected, provided that the bearing values are kept within the indicated limits of 4 and 10 tsf, plus increases for stabilized overburden pressure and embedment (see Sec. 5E, Chap. Note 4).

Other than this verification of safety against shear failure for narrow footings, no correction of allowable bearing capacity need be made because of the size of the bearing area. Theoretically, the larger the bearing area, the greater the settlement at a given intensity of bearing pressure. However, for granular soils, the effect is relatively small and too refined a correction considering the gross approximations inherent in the above described determination.

d. Sands (soil groups SW, SP and SM containing few, or no, particles retained on a #4 sieve)

I. *Base Values.* As with gravels, the allowable bearing pressure on sands also is a function of compaction and of grading. However, we are not faced with quite the same uncertainty as with gravel soils. For sands we have a widely recognized measure of compaction, the N value obtained from the standard penetration test.

As previously discussed, the N value is the parameter used by Terzaghi and Peck [5] to develop their chart (Fig. 5-9) of allowable bearing capacity for sands. This chart, which has an empirical basis, relates bearing capacity to penetration resistance (N) and is predicated on limiting settlement of footings bearing at the indicated pressure to a total of 1 in. Note that the chart also relates allowable bearing capacity to size of footing for each N value.

The chart, although somewhat conservative, has given satisfactory results in practice, has gained wide acceptance, and is recommended for use, subject to the following modifications and amplifications:

A. The *value of penetration resistance* (N) will vary over the site with location and with depth. The value to be used in entering the charts may be determined by averaging the measured resistance within the *significant depth* of the foundation, i.e., within that depth of soil below the bearing area which contributes significantly to the settlement of the foundation. For sands this may be taken as a depth below the proposed footing level equal to the width of the footing.* Such averages should be taken for each boring, and the lowest of the average values so obtained should be the value used in entering the chart. In some cases, the variation in average value is large. When this occurs, there are two problems:

1. Loss of economy if all of the footings are proportioned for an inordinately low bearing value representing one or a few borings
2. Possible differential settlement if all footings are proportioned for the same bearing pressure on a soil having an excessively variable profile

The solution is that the bearing pressure under the footings should be varied in consonance with capacity of the soil. For this purpose, the area under the building is zoned according to the average N value, using whatever increments seem reasonable, and the soil bearing pressure correspondingly zoned.

B. The allowable bearing capacity of *fine sand* (for definition see page 154), may be determined in the same manner described above except where the materials are in a loose state (N equal to 10, or less) and are subject to vibratory load. Under such circumstances fine sand is a treacherous material, sensitive to disturbance and subject to spontaneous liquification (sudden loss of strength under impact load). Determination of its bearing capacity requires great caution.

* Sowers [48] indicates that the penetration resistance should be averaged for a depth of 1.5 rather than 1.0 times the width of the footing.

C. For mat foundations, wherein the mat has sufficient rigidity to bridge local soft spots in the subgrade and so to reduce differential settlements between columns, Ref. 11, indicates that the allowable bearing pressure may be taken as $(N - 3)/5$, or about twice the values indicated in Fig. 5-9.

The hypothesis that the mat tends to iron out the differential settlements doubtless is correct. Its actual capability would depend on the provision of excess reinforcement and section as required to resist the additional stresses incident to preventing said displacements. The problem is that mat foundations are not in sufficiently common use for an adequate breadth of experience to have been developed to confirm or disaffirm the validity of the hypothesis in actual cases. As a result many foundation designers (rightly or wrongly) do not take advantage of the indicated increase, but proportion mat foundations bearing on sand for the bearing pressures indicated in Fig. 5-9.

D. N values notwithstanding, experience seems to indicate that the bearing capacity of sands *need not be* taken as *less than 2 tons per square foot,* unless some special consideration of impact or vibratory loading (earthquake, for example) exists.

E. *N values* for design purposes should not be taken as greater than, say, 60. The reason is that the standard penetration test is not a reliable measure of compaction at the extreme ranges (say, less than 20 or more than 60).

II. *Correction Factors.* The values of allowable bearing pressure on sands, determined as above, may be increased for

1. Stabilized overburden pressure (Sec. 5D.1)
2. Embedment (Sec. 5D.3)
3. Limited depth of the bearing stratum (Sec. 5D.5)

III. *Correction for Size of Bearing Area.* Where the bearing area is less than 3 ft in dimension or the embedment is less than 3 ft, or both conditions prevail, the ultimate bearing capacity of the soil should be checked by the methods indicated in Sec. 5A, since there is a danger of shear failure in the soil. Where the bearing area is both 3 ft or larger in dimension and embedded 3 ft or more, the danger of shear failure can be neglected provided that the bearing pressure does not exceed, say, 6 tsf, plus increases for stabilized overburden pressure and embedment (see Sec. 5E, Chap. Note 4).

In connection with the aforementioned procedure for determining the allowable bearing capacity for sand soils, since so much depends on the N value, every reasonable precaution should be taken to assure that the N values are reliable. This is one of the reasons why inspection of borings is so important. Good borings are not nearly as common as they should be. The reader is referred to Chap. 2 for a discussion of some of the more important points to watch for when performing the standard penetration test.

e. Gravelly Sands (soil groups SW, SP, and SM containing a substantial percentage of particles retained on a #4 sieve)*

I. *Base Values.* A soil in this classification is neither a sand nor a gravel and, theoretically, neither of the above indicated criteria (paragraph *c*, Gravel Soils or paragraph *d*, Sands) would apply. The problem, again, is that of obtaining a meaningful N value when the soil contains a substantial amount of very coarse particles. However, within the range of base values for sands indicated in paragraph *d* (2 to 6 tsf), use of the N value criterion presented therein will give reasonable results if care is observed to take enough samples (see paragraph *c*, Gravel Soils); to use lowest average values in determining N; and to carefully inspect the samples and the variation in blow count for each 6 in. increment of sampling to detect the presence and influence of any pieces of gravel, eliminating any tests which seem to show extraordinarily high values. Alternatively, the method used for estimating bearing capacity of gravel soils may be adopted, but lowering the range of allowable bearing values to, say, 3 tsf for loose, poorly graded soils and 8 tsf for compact, well graded deposits.

II. *Correction Factors.* The values of allowable bearing pressure on gravelly sands may be increased for

1. Stabilized overburden pressure (Sec. 5D.1)
2. Embedment (Sec. 5D.3)
3. Limited depth of the bearing stratum (Sec. 5D.5)

f. Clay and Silt Soils (soil groups SC, CL, CH, ML, and MH in the medium to stiff or dense state)

I. *Base Values.* Consideration in this section will be limited to inorganic silts and clays. The bearing capacity of organic deposits (OH, OL, and Pt soils and of soft clays and loose, saturated silts) is considered in Sec. 5D.2*g*.

The physical characteristics and the behavior of inorganic clay and silt soils tend to approximate the physical properties which are conventionally assumed for purposes of mathematical analysis. As a result, the allowable bearing capacity can be determined on the basis of a

* "Substantial" is left undefined. It is a matter of judgment, and would depend on the actual size of the coarse particles. In general, however, if the individual particles are not especially large, a content of 10 to 20 percent of coarse particles would not be "substantial."

rational analysis in which the factor of safety is computed and compared to the desired value as described in Sec. 5A, and the amount of probable settlement is calculated and compared to the value presumed to be allowable, as described in Sec. 5B. Reference, therefore, is made to Secs. 5A and 5B for details of the procedures and the computational requirements.

Values of ϕ (friction angle), c (cohesion), and C_c (compression index) are required to make these computations, and for this purpose, undisturbed samples and laboratory test data are required.

This rational analysis is, by far, the preferable approach. There are, however, some shortcut methods which are in use. For example, the building code of the city of Boston gives the allowable bearing capacity of soft and medium clay as $1.50q_u$ for round and square footings and $1.25q_u$ for rectangular footings having length-width ratios greater than 4. Linear interpolation is permitted between these limits. As can be derived by reference to Eqs. (5-2) and (5-3), this device is equivalent to assuming a surface footing, purely cohesive soil, and a safety factor of 2 against shear failure. The values of allowable bearing pressure so obtained provide safety against shear failure only and do not guarantee a satisfactory limitation on settlement. The settlement problem must be analyzed.

The other common approximation is to use the nominal values specified in building codes. This is a dangerous practice. The values tabulated in building codes are more properly approximate upper limits than specific permissive values, and they definitely do not guarantee satisfactory limitations on settlement.

II. *Correction Factors.* The base values may be increased for

1. Stabilized overburden pressure (Sec. 5D.1)

The values may *not* be increased for

1. Embedment
2. Limited depth of the bearing stratum

III. *Caution re: Fissured Clays.* Laboratory test criteria, as described, are of doubtful application in estimating the bearing capacity of stiff, fissured clays. Determination of the probable behavior of such soil requires considerable judgment and experience.

g. Miscellaneous Soil Materials

Rock, hardpan, gravel, sand, inorganic silt, and inorganic clay soils constitute the gamut of soil types normally considered as bearing materials. The remaining types of soil which are of common occurrence are fills and organic deposits (including peat, organic silts, and organic clays).

Normally these latter soil types (and sometimes soft clays and loose silt deposits as well) are not considered as bearing materials; i.e., they are assumed to have zero bearing capacity. Similarly, soils containing lenses of such soil types or soils which are underlain by such strata are assumed to have zero bearing capacity. Zero bearing capacity means that when such soils are encountered, piling, piers, or other construction must be used to bypass the foundation reactions around the soft compressible material into bearing in a more satisfactory stratum.

The reason for this disrepute of fills, organic deposits, soft clays, and loose silts is that they tend to be highly compressible, and have an unsavory record as a source of difficulties when used for the support of structures. Also, there has been some overreliance perhaps on that adage, "When in doubt, drive piles."

In point of fact, fills, organic, and soft deposits can be used as bearing materials, if properly handled. Some of the means of handling are as follows:

1. *Fills.* To start, two types of fills must be differentiated; controlled and uncontrolled. They are entirely different. A controlled fill is one in which the material is placed under supervision, properly compacted, and with the material itself properly selected. In general, the placement of a controlled fill follows the procedures applicable to the construction of a satisfactory embankment. A typical specification for controlled fill is contained in Sec. 5E, Chap. Note 5.

A fill conforming to the provisions of Sec. 5E, Chap. Note 5 is a good bearing material. It corresponds, roughly, to a soil of groups GW, GP, SW, or SP, i.e., a gravel or a sandy material, and may be treated as such. The bearing value can be determined as described in Secs. 5D.2*c*, 5D.2*d*, and 5D.2*e*.

An uncontrolled fill is an entirely different matter. It could be a good bearing material or it could be worthless. The problem is that an uncontrolled fill may contain practically anything. Garbage, tin cans, automobile bodies, baby carriages, and paper are common inclusions. At the present state of the art of foundation design, no generally accepted means has been developed for determining the bearing capacity of automobile bodies, baby carriages, garbage, and paper. The usual approach is to assume that they have zero bearing capacity. Therefore, the key to successful utilization of an uncontrolled fill as a bearing material is to satisfy oneself that it does not contain such inclusions, at least not under the footings. This is the essential purpose of the soil exploration program for fill suggested in Chap. 2 (page 36). If as a result of the indicated program, it appears that the fill is free of unsatisfactory inclusions (again, at least under the footings), then a nominal bearing capacity, say, 1 or 2 tsf, may be assigned to the material. If such inclusions do show up in the explorations, then either the building should be supported on piles or piers bearing below the fill strata or a special design should be adopted which is rigid enough to bridge the soft spots or is sufficiently articulated to conform to the settlement contours without creating a condition of distress.

II. *Soft and Organic Deposits.* In theory, organic deposits are merely very weak and very compressible clays or silts, and they can be treated as such. The allowable bearing capacity for such soils (and for soft clays and loose silts) can be computed and the settlements calculated as for more satisfactory clay or silt soils. As a practical matter, however, the values of allowable bearing pressure so obtained are apt to be

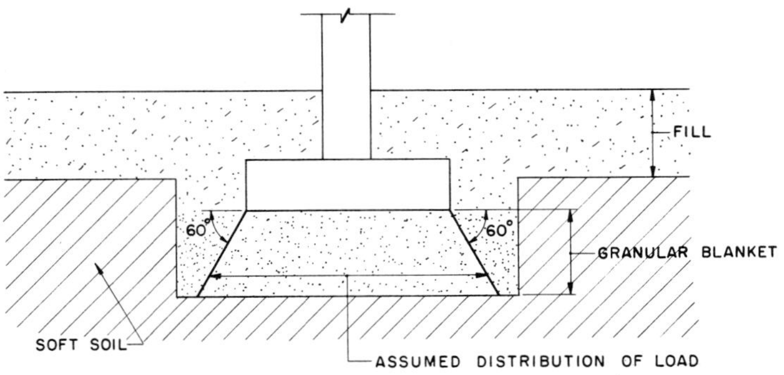

FIG. 5-21 Use of Granular Blanket to Distribute Footing Reaction.

very low. Values less than ¼ tsf are common. For most designs, a bearing capacity of ¼ tsf has the same effect as no presumptive bearing capacity at all, in that it is more economical to support the foundation on piles or piers than on shallow footings. Occasionally, however, a case is encountered in which the cost of piling or piers would be so high that recourse must be made to founding the building on, in, or above the organic materials.* Where this is the case, there are two choices: (1) to spread out the load sufficiently to reduce the applied pressure to a value equal to or less than the allowable or (2) to strengthen the soil.

The first device can be accomplished by making the footings large enough. Alternatively, and usually more economically, a blanket of granular material can be spread between the footings and the soft soil (Fig. 5-21).

* A case commonly encountered is that of a small one-story industrial building being built in marginal land. The columns support only the roof. The column spacing is, say, 20 by 40 ft. The total roof load is 40 psf. Each column carries only 16 tons (8 tons for the exterior lines). The use of piling would cost several times the cost of spread footings designed for a nominal bearing capacity (say, ½ to 1 tsf).

The second device, in general, involves some form of precompression technique. Use of a surcharge (with or without sand drains) is the most common of these. The surcharge usually consists of an earth or rock fill, although water loads (or the equivalent in drawdown of the groundwater level) have been used. Jacking also has been used and is virtually standard practice in underpinning. Aldrich [39] summarized the "state of the art" in precompression technique, noted its application to organic silt and peat soils and to miscellaneous fills, and cites instances of its application to buildings three to five stories in height, as well as its more common use for foundation preparation for one-story structures.

In many cases, the soil already has been precompressed by natural forces such as desiccation or glaciation. If this is the case, it will show up in the consolidation test (pressure-voids ratio curve). The effect is the same as if the soil had been artificially surcharged.

In this connection, a word of caution is in order. The basic techniques and computations involved in checking a proposed design intended to bear in or above strata of soft material, such as the organic deposits indicated, are relatively simple. In practice, however, there are likely to be complications.

To start with, there is the problem of getting good reliable strength data with which to check the safety against shear failure in the subsoil. In general, triaxial-test data are required for this purpose. Obtaining satisfactory test samples and making proper interpretation of triaxial-test data are not simple matters.

Secondly, a careful watch must be kept during filling or surcharging (if either is required) to assure against shear failure in the soft soil. To this end, settlement plates and piezometers should be installed and read. The results should be compared to the calculated effects, and corrective measures taken if things appear to be going awry.

Lastly, time almost always is a problem. There is continuing pressure to cut down on the surcharge period either by stripping early or by increasing the height. Neither is likely to be a good idea. Stripping too early reduces the effectiveness of the surcharge. Increasing the height increases the chances of shear failure in the soil which is supposed to be consolidated.

5D.3 Increase for Effect of Embedment of Bearing Area

The allowable bearing pressure for any soil is increased if the bearing area is embedded below the surface of the ground. The reasons are related to the two basic considerations in bearing capacity, ultimate strength and settlement, and may be derived as follows:

1. *Ultimate Strength.* Refer to Eq. (5-1). The second term represents the effect of confining pressure of the surrounding soil on the bearing

area. From Fig. 5-1 it may be noted that N_q is positive and is equal to or greater than 1.0. Therefore, as the bearing area is embedded deeper below the ground surface, the term $\gamma D N_q$ increases, and the ultimate-bearing capacity increases.

2. *Settlement.* As the depth of embedment of the bearing area increases, the settlement due to a given increase of applied pressure decreases. This will be apparent from consideration of Eq. (5-7):

$$\Delta H = H \frac{C_r \text{ or } C_c}{1 + e_0} \log \frac{p_0 + \Delta p}{p_0} \tag{5-7}$$

In this expression, as the depth of embedment increases, the value of p_0 increases. Therefore, for a given increase in applied pressure (Δp), the value $\log [(p_0 + \Delta p)/p_0]$ decreases, and the settlement H also decreases. Of course, Eq. (5-7), strictly speaking, is applicable only for cases of one-dimensional compression. Fox [35], however, has derived similar results for cohesionless soils (lateral displacement of soil during compression considered).

In practice, the increase in allowable bearing pressure due to embedment is determined as follows:

a. Clay and Silt Soils

As discussed in Sec. 5D.2, the bearing capacity of clay and silt soils may be estimated in two ways: (1) by rational analysis and (2) from prescription values as stipulated in various building codes.

As indicated, the method of rational analysis is preferable, and where used, the bearing capacity is determined from Eqs. (5-1) and (5-7), i.e., directly from theoretical considerations. In Eqs. (5-1) and (5-7), the influence of embedment of the bearing area is included in the basic computation, and *no further increase in bearing pressure due to the effects of embedment* should be taken.

Similarly, where bearing capacity is based on prescription values as stipulated in building codes, if such values are considered as net bearing pressures, the increase in allowable bearing pressure which may be taken due to the weight of the overburden (Sec. 5D.1) corresponds to an assumption of $N_q = 1.0$ (see Sec. 5E, Chap. Note 6). Unless N_q exceeds 1.0, no further increase for embedment of the bearing area should be taken. Since clays and silt soils conventionally are assumed to have zero frictional resistance ($N_q = 1.0$), the increase in allowable bearing pressure for the effects of embedment usually is taken to be zero.

However, if the assumed friction angle ϕ is greater than zero, an increase in allowable bearing pressure for embedment would be justified, and is used. The Uniform Building Code, for example, provides for an increase of allowable bearing pressure of 5 percent per foot of embedment for soft clays, 10 percent for medium clays, and 20 percent for hard clays. It is emphasized, though, that such increases, where applicable, relate only to prescription values and do not apply where allowable bearing pressures are determined on the basis of analysis.

b. Granular Soils

Allowable bearing capacity of granular soil deposits (sand, gravels, hardpan, and the like—groups G and S of the Unified Classification

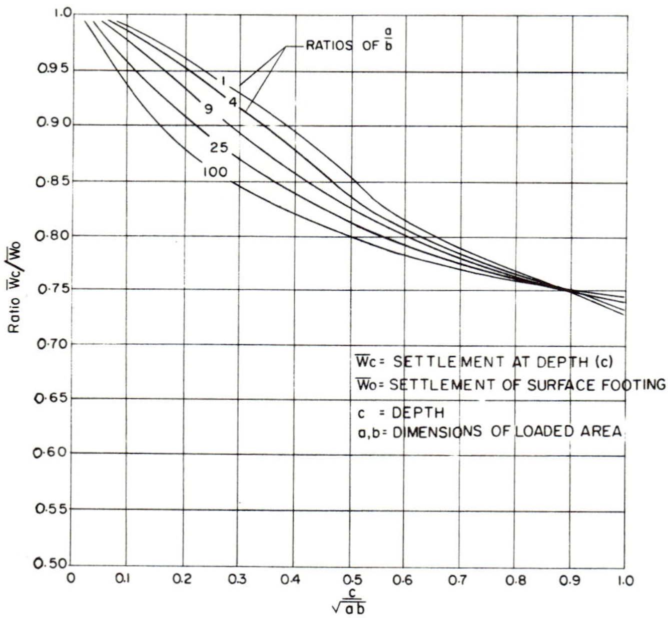

FIG. 5-22 Decrease in Settlement of Loaded Area with Increased Embedment (Cohesionless Soil). *(From Fox, Ref. 35.)*

System) generally is limited by considerations of settlement and not of ultimate strength and is derived on an empirical rather than on a theoretical basis. As described above, as the embedment of the bearing area increases, the settlement under a given loading intensity decreases. Therefore, for granular soils, the allowable bearing pressure *may be increased for embedment.*

The amount of such increase is a matter of debate. Again citing the aforementioned sources, the Uniform Building Code permits increases of 20 to 25 percent per foot of embedment for foundations bearing on sand soils, whereas the Boston building code permits an increase of only 5 percent per foot of embedment.

On a theoretical basis, Fig. 5-22, which is taken from Fox [35], shows the approximate rate of decrease of settlement with depth for granular soils. It varies from 3 to 4 percent per foot of embedment for square footings to about 8 to 10 percent per foot of embedment for narrow footings. Therefore, it appears that the 5 percent figure adopted by the Boston Code is a conservative value and is recommended for use.

EXCEPTION: The increase in allowable bearing capacity for embedment does not apply for narrow, shallow footings where allowable bearing capacity is controlled by considerations of ultimate shear strength (see Sec. 5D.4).

c. Rock

For foundations bearing on rock, increase of allowable bearing pressure with increased depth of embedment also *is justified*. This is because the prescription values hereinbefore indicated necessarily are for surface bearing areas. Again, the amount of such increase is a matter of debate. Settlement, of course, is not a problem. The increase must be arbitrarily selected. Again quoting the Boston building code, an increase of 10 percent per foot of embedment is stipulated. Design Manual DM7, U.S. Navy [7] gives a corresponding value. The value is certainly conservative and is recommended for use.

EXCEPTION: The indicated allowable increase for embedment would not apply if the rock was shattered by blasting, expanded due to slaking, or altered during, or incident to, construction. In such case, the classification of the rock is changed and lower base values would apply.

If the rock is not so affected, as, for example, where the footing excavation is line drilled or at the bottom of a drilled hole for a drill-in caisson, the increase would apply as described above.

d. Measuring Depth of Embedment

Theoretically, the depth of embedment used to determine the allowable increase in bearing pressure due to embedment would be measured from the upper surface of the bearing stratum and would be increased for the strength and initial pressure contributed by the weight of the overburden above the top of the bearing stratum.

In practice, this is seldom done, the depth of embedment being measured either as the depth of the bearing area below the level of frost penetration, or, if the top of the bearing stratum is below the frost level, as the depth below the top of the bearing stratum, whichever gives the lesser value (Fig. 5-23*a*). The strength and initial pressure contributed by the overburden is neglected.

For foundations bearing on rock, the measurement conventionally is made as the depth of embedment in excess of some nominal value, say

1 ft, with the first foot of embedment being neglected. The frost depth requirement, of course, does not apply (see Fig. 5-23b).

There are several reasons for this disparity between theory and practice. The disallowance for the depth of frost penetration relates to the

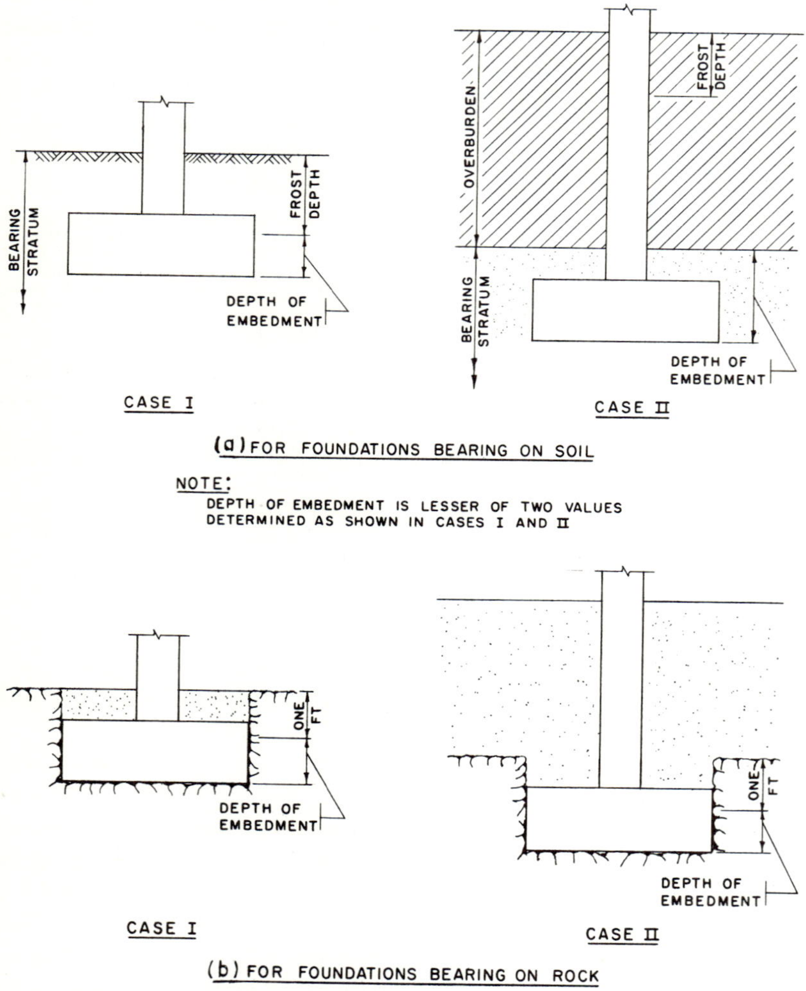

FIG. 5-23 Measurement of Depth of Embedment.

potential loss of strength due to freezing and thawing as discussed in Chap. 3, which, of course, does not apply to rock. The disallowance for the first foot of depth in rock relates to the improbability of finding the rock surface around the footing to be level and sound, particularly the latter. The matter of measuring from the top of the bearing stratum, which, in effect, neglects the contribution of the weight of the overburden to strength and rigidity of the bearing stratum, appears to be a concession to the difficulties inherent in evaluating such contribution. It is a conservative approach and, considering that part of the effect of that contribution already has been discounted by establishing the allowable bearing pressures as net pressures, is not particularly penalizing.

e. Upper Limits

In theory, the allowable bearing pressure could be increased indefinitely as the depth of embedment of the bearing area increases. However, since the reduction of settlement with depth follows a logarithmic curve, this increase cannot be made indefinitely at the usually stipulated straight line rates (5 percent per foot, etc.). Therefore, it is customary to establish an upper limit to the amount of permissible increase. Allowable increases of anywhere up to double or triple the basic allowable values are permitted by the provisions of different building codes.

5D.4 Decrease for Effect of Size of Loaded Area

As the size of the loaded area increases, for a given bearing pressure, the settlement increases. The primary reason is that the bulb of pressure affects a greater depth of soil, and so the total strain in the underlying soil mass is greater. On the other hand, as the size of the loaded area increases, for a given bearing pressure, the foundation becomes increasingly safe against shear failure of the soil [see Eq. (5-1)]. Thus, the size of the loaded area affects both determinants of allowable bearing capacity —settlement and safety against shear failure—and the allowable bearing capacity for foundations, as determined in Sec. 5D.2, should be corrected for the effects of the size of the loaded area, as follows:

a. Foundation Bearing on Rock

No area correction is required. The indicated evaluation of allowable bearing capacity for rock is so gross an approximation that refinements relating to size of bearing area are not warranted.

b. Foundations Bearing on Hardpan or Gravel Formations

Area correction is required only for shallow knife-edge bearing areas (areas less than 3 ft wide and embedded less than 2½ ft). For shallow narrow footings, check for safety against shear failure using Eqs. (5-1) to (5-3) (see also Sec. 5E, Chap. Note 4).

With regard to settlement, if the bearing capacity is determined in accordance with Sec. 5D.2, the gross settlement will be sufficiently small that the refinement of correcting for size of the bearing area is not warranted.

c. Foundations Bearing on Sand

Again area correction is required only for shallow, knife-edge footings (areas less than 3 ft wide and embedded less than 3 ft). As in the case of foundations bearing on hardpan or gravel soil, shallow narrow footings bearing on sand soils should be checked for safety against shear failure using Eqs. (5-1) to (5-3). For wider or deeper bearing areas, the possibility of shear failure in the soil is not the controlling consideration.

With regard to settlement, the bearing capacity charts of Fig. 5-9 already reflect the necessary relation between settlement and size.

d. Foundations Bearing on Silt or Clay Soils

If bearing capacity is determined by rational analysis as described in Sec. 5D.2, the effects of size of the bearing area will have been considered, and no further correction is required.

If the bearing capacity is taken from prescription (code specified) values, the effects may be important, however. For example, the Boston building code contains a provision to the effect that the prescribed allowable bearing values for clay soils apply only to pressures directly under individual footings, walls, or piers and that the area load (average load over an entire bay or major portion of the building) should not exceed one-half the prescribed values. The reader is reminded that prescription values for allowable bearing pressures on clay or silt soils tend to represent limitations imposed by considerations of ultimate strength and do not necessarily consider limitations based on satisfactory performance with regard to settlement.

e. Balancing Settlements

Most foundation design is predicated on using a fixed allowable bearing pressure for a given soil stratum without regard to size of the bearing area; i.e., all footings are designed for 2 or 3 tsf or some other fixed value of bearing capacity. On the other hand, most foundation designers are aware of the inverse relation between size of footing and settlement under a given intensity of bearing pressure as described in the introduction to this section.

As a result, a practice has arisen wherein the designer proportions the footings for a fixed allowable bearing pressure and then either slightly oversizes the larger footings or undersizes the smaller ones. The idea is to balance the settlements and minimize the amount of differential displacement.

The principle is correct. For footings bearing on rock or hardpan, the displacements at usual design pressures are so small that as stated above, the refinement probably is not warranted, but for gravel soils the device is appropriate. For sand soils, if the designer prefers to use fixed values of allowable bearing pressure rather than the variable relation shown in Fig. 5-9, the device is an approximation of that variable relation.

For foundations bearing on clay or silt soils, the authors much prefer to see allowable bearing pressures derived on the basis of analysis, but the principle of oversizing the larger footings still is valid.

5D.5 Increase for Effect of Limited Depth of Bearing Stratum

An additional increase in allowable bearing pressure may be taken for the effects of limited depth of the bearing stratum, if the foundation bears on or in sand or gravel soils.

It has been noted that allowable bearing pressure is limited by considerations of ultimate bearing capacity (shear failure) and of settlement. Shear failure requires development of a failure surface (Fig. 5-24*a*). Values of allowable bearing pressure as determined in Sec. 5D.2 are predicated on potential failure along that surface which is most critical. If the bearing stratum is underlain by denser, stronger material at a shallow depth below the footing, the critical failure surface cannot be developed (Fig. 5-24*b*). For such a case the factor of safety against shear failure would be increased, and the allowable bearing capacity as limited by ultimate strength of the soil would be increased.

In a similar vein, if the depth to harder material is less than the depth required to develop the full bulb of pressure (Fig. 5-24*c*), then the settlement of the footing due to strain in the soil mass will be decreased, as compared with the amount of settlement which would occur if the bulb of pressure were to be fully developed. This also corresponds, in effect, to an increase in allowable bearing pressure.

To illustrate the point, consider a hypothetical case in which a footing bears on 3-ton sand, and 40-ton rock is only a few inches below the subgrade of the footing. Both theoretically and intuitively the allowable bearing capacity under such a footing is closer to 40 tsf than 3 tsf.

The concept is simple, although not always recognized. The problem is to determine how much increase in allowable bearing pressure is warranted by the condition. The determination is made as follows. First, the factor of safety against shear failure should be determined by drawing several possible failure planes and balancing the static forces of load, soil weight, friction, and cohesion in a manner similar to that employed for slope stability computations (Ref. [17], page 178 et seq., for example). The factor of safety should not be less than that prescribed in Sec. 5A.2.

Next, the effect of reduced settlement should be estimated. If we assume that the depth of soil below the bearing area which contributes significantly to settlement of the footing is equal to the width of the foot-

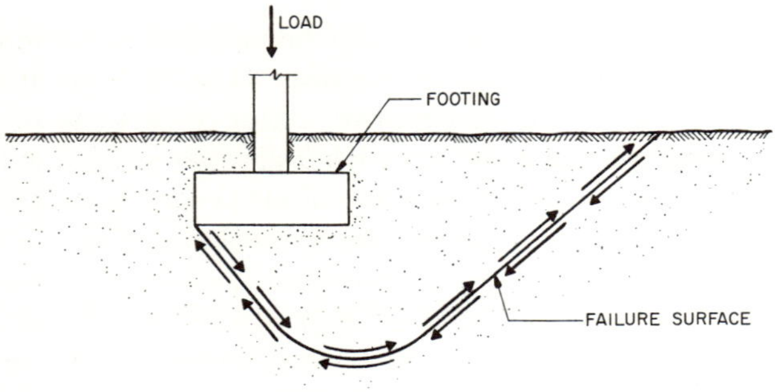

(a) CONVENTIONAL REPRESENTATION OF FAILURE SURFACE UNDER FOOTING

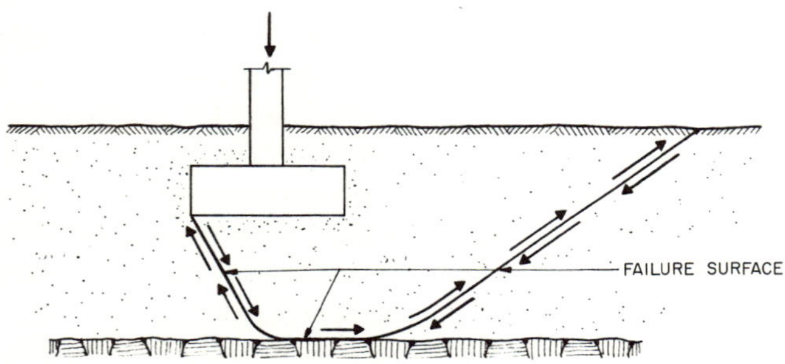

(b) FAILURE SURFACE FOR LIMITED DEPTH OF BEARING STRATUM

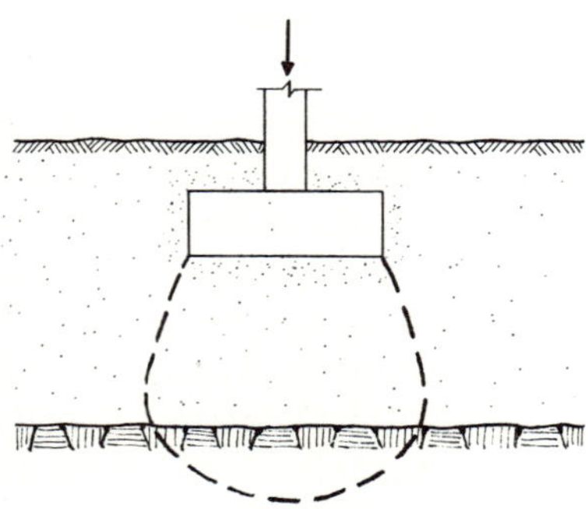

(c) RESTRICTED DEVELOPMENT OF BULB OF PRESSURE

FIG. 5-24 Correction for Limited Depth of Bearing Stratum (Part 1).

ing, we may conservatively approximate the allowable bearing capacity as controlled by considerations of settlement using the equation shown in Fig. 5-25.

Certain restrictions must be placed on application of this increase in allowable bearing pressure, however. First, it applies only if the bearing stratum is underlain by a substantially denser and stronger material and, second, if the denser and stronger material occurs within the significant depth below the footing, a depth which, as noted above, may be

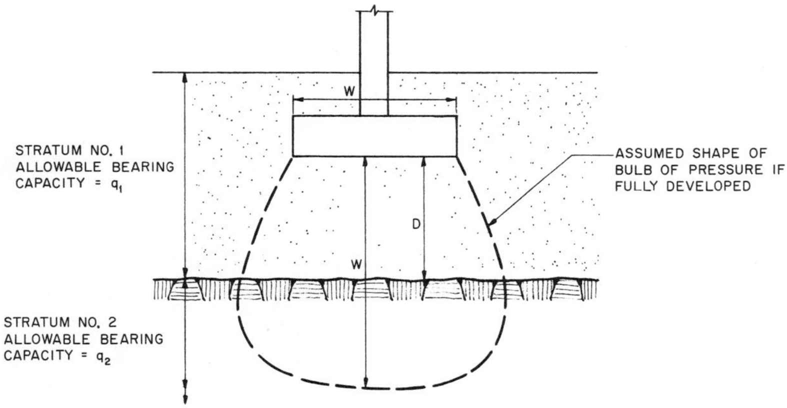

$$q_{ALLOWABLE} = \sqrt{\frac{W}{D}}\, q_1 \leq q_2$$

FIG. 5-25 Correction for Limited Depth of Bearing Stratum (Part 2).

taken as equal to the width of the footing. It is emphasized, however, that this increase in allowable bearing pressure should be applied only for foundations bearing on, or in, sand or gravel soils and not for foundations bearing on, or in, silt, clay or rock deposits. The reasons are that for clay or silt, the effect of variable depth already is reflected in the indicated analysis and that for rock, strength characteristics simply are not well-enough known to warrant the refinement.

5D.6 Increase for Infrequent Loading Conditions

The allowable bearing capacity, as determined in Secs. 5D.2 through 5D.5, must be less than the ultimate bearing capacity by the safety factor indicated in Sec. 5A.2.

In this connection, it is noted that it is common practice in design to vary the required safety factor to suit the design loading condition. The more improbable the loading condition, the smaller the required safety factor. In structural design, for example, an overstress factor of one-third commonly is permitted for members subject to wind or to earthquake load. In highway-bridge design, overstress factors up to one-half (50 percent) are permitted for combinations of loading involving dead plus earth pressure, plus buoyancy, plus wind, plus ice pressure.

This same principle applies in the design of foundations, and allowable bearing pressures may be increased for combined loading conditions. However, if the procedures set forth in this text are followed, this already has been done in computing the loads (see Chap. 4), and no further increase in allowable bearing pressure by virtue of infrequent occurrence of loading conditions should be taken.

5D.7 Miscellaneous Considerations

a. Effect of Inundation of Bearing Stratum

According to theory (see Ref. 5, article 54), the allowable bearing pressure on a given sand stratum, if submerged, would be less than the value for the same sand stratum, if dry.

This fact is widely known. Unfortunately, there has been considerable misunderstanding of its implications with regard to practical design of foundations, and some clarification is warranted.

First, if the design bearing pressure is taken from the N value and if the N value is taken from a penetration test performed on inundated material, the N value (at least in some circumstances) represents the density (and presumably the foundation modulus) of the inundated soil.

Secondly, while the theoretical decrease in ultimate bearing capacity appears to be correct and must be considered, as previously indicated, allowable bearing pressure on sand soils is limited by considerations of settlement, not of ultimate strength and, at the usual permissible intensities of bearing pressure on sands, there does not appear to be any substantial difference in settlement for a given intensity of bearing pressure between the dry and inundated condition. Parsons [40] for example, on the basis of many load tests, reports that inundation during the period of test does not substantially affect the results.

By way of explanation of this paradox it is noted that the basic relationship used to determine settlement of a bearing area on sand is derived from the fact that the settlement involves lateral as well as vertical displacement of the soil material. Therefore, confining pressure (to the extent that such lateral displacement is prevented) would affect settlement. A decrease in confining pressure, such as that due to submergence of the soil, should cause an increase of settlement for a given bearing area under a given load.

On the other hand, if the settlement of a bearing area on sand was due only to one-dimensional displacement of the sand particles, confining pressure would affect settlement only as it might affect the p_0 value.

Hough [9, 22] has long contended that, at usual design values of bearing pressure, the settlement of a bearing area on sand is due, primarily, to one-dimensional displacement and that it is only at higher loading intensities that lateral displacement of the sand particles becomes a significant consideration. This is offered as a tentative explanation for the observed phenomenon.

b. Where the Foundation Bears on Soil Material of Varying Bearing Capacity

This situation is encountered on virtually every job. For example, the density of the bearing stratum may vary so extensively that some footings are founded on, say, 2-ton soil and others on what would be 3- or 4-ton material. In more extreme cases, the soil material underlying the building may vary to the extent that it is impractical to found the entire building on or in the same bearing stratum.

In either event, founding the building on or in soil materials of varying bearing capacity accentuates the problem of differential settlement, and the design bearing pressure should be varied from zone to zone of the building to conform to the varying soil capacity.

There is, of course, a practical limit to this. Designing each footing for a different bearing capacity is unrealistic, as would be varying the allowable bearing capacity by ultrarefined increments (say, 2.0 tsf, 2.1 tsf, etc.). A decision must be made concerning the increments of bearing capacity into which the building should be zoned. The authors have the following suggestions:

I. *For Foundations Bearing on Rock or Hardpan.* Zone in accordance with changes in soil classification only, and for rock, only in accordance with changes in geological classification. The reason is that for rock or hardpan the vertical displacement of the foundation under bearing pressures of the intensities prescribed in Secs. 5D.2*a* and 5D.2*b* are so small that variations of material within a given soil or geologic classification may be neglected. Only gross variations involving change of classification need be considered.

II. *For Other Soil Materials.* Zone for increments of allowable bearing pressure equal to 30 percent of the least value of allowable bearing pressure represented at the site.

Section 5E CHAPTER NOTES

1 Capacity of Drilled Piers

Drilled pier construction is illustrated in Fig. 5-26. Little has been published on the matter of the capacity of such construction. There is general agreement that the bearing capacity of the soil under the pier base should be computed using the techniques described in Sec. 5D, perhaps with some variation in the constants N_ϕ, N_q, or N_γ where Eq. (5-2) or (5-3) are used. There is considerable difference of practice, however, with regard to evaluating the additional capacity contributed by the soil surrounding the shaft. Those standards of practice which have been described in the literature are as follows:

1. *The Building Code of the City and County of Denver* (1963) permits utilization of friction capacity of cast-in-place friction piles cast in drilled or bored holes (which would be a drilled pier without an enlarged

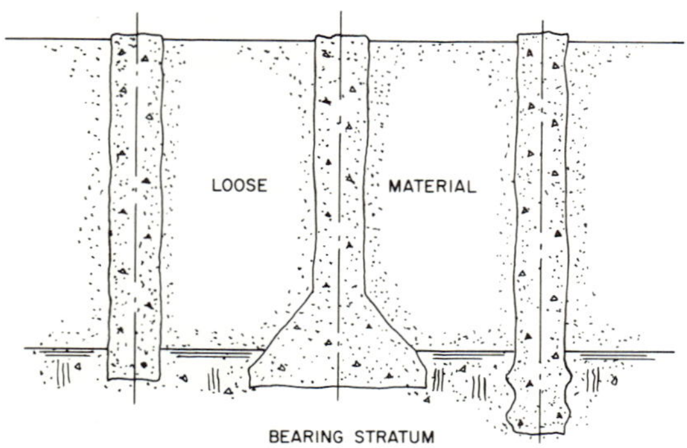

FIG. 5-26 Configurations of Drilled Piers.

base) at one-sixth the bearing capacity of the material confining the shaft, with a maximum value of 500 psf unless justified by special investigation. It is further specifically provided for such cases that frictional resistance and bearing resistance shall not be assumed to act simultaneously.

2. *The Building Code of the City of Los Angeles* (1958) contains a like provision, but without a specific limitation relating to the simultaneous occurrence of frictional and bearing resistance.

3. *Skempton* [43] in summarizing the results of numerous load tests on piling embedded in the clay of London reports an average value of adhesion (C_a) between clay and shaft, at ultimate load, of 0.45 times the shear strength and indicates, contrary to common American practice, that the adhesive resistance along the shaft, computed as C_a times the area of the shaft, should be added to the end bearing resistance. A factor of safety of 2.5 is used for design.

4. *Teng* [44] suggests computation of the frictional resistance along a shaft embedded in granular soils from the expression

$$S_f = 0.7\gamma d D^2 \tan\phi \qquad (5\text{-}18)$$

where S_f = total ultimate skin friction, lb
γ = average unit weight of soil—use submerged weight for soil below water level, pcf
d = diameter of caisson shaft, ft
D = depth of caisson from ground surface to top of bell, ft
ϕ = angle of internal friction of the soil

This expression is based on the assumption that the lateral confining pressure around the shaft is equal to the at-rest pressure K_0. The assumed value of K_0 is 0.45.

For cohesive soils, the adhesive resistance is indicated to be equal to the cohesion of the soil, taken as one-half the unconfined compressive strength. A maximum adhesive resistance of 1,000 to 1,500 psf is indicated.

2 Bearing Capacity of Foundations under Eccentric Loading

a. Conventional Approach

Section 5A (page 114) describes the conventional procedure for estimating the bearing capacity of an eccentrically loaded shallow foundation. This procedure is to calculate (or estimate) the allowable bearing pressure as though the foundation were concentrically loaded, and then to proportion the construction so that the maximum bearing pressure, calculated as $P/A + Mc/I$, does not exceed the allowable value for concentric loading (see Fig. 5-6). The resulting variation of allowable total load with increased eccentricity for a square footing is shown as the solid line in Fig. 5-27. For an eccentricity equal to one-sixth the width of the bearing area (middle third) the load capacity is reduced 50 percent. At an eccentricity of one-half the width of the bearing area, the foundation, theoretically, becomes unstable, and the capacity is zero.

b. Approach of Meyerhof

The conventional approach assumes a linear distribution of reaction of the soil supporting the bearing area. At ultimate loads, this assumption is not even approximately true, and Meyerhof [32] has proposed that the condition of eccentric loading be treated by assuming that the contact pressure distribution is the same as for a centrally loaded foundation, i.e., a rectangular block distribution, but with an effective width of the foundation reduced from the value B to a value B' such that

$$B' = B - 2e \tag{5-19}$$

where B = width of footing
B' = effective width of footing
e = eccentricity

The bearing capacity of the footing may then be computed as

$$Q = qB' \tag{5-20}$$

and

$$q = cN_c + \frac{\gamma B' N_\gamma}{2} \tag{5-21}$$

N_c and N_γ are the bearing-capacity factors from Fig. 5-1.

Equation (5-21) corresponds to Eq. (5-1) except for the introduction of the term B' in lieu of B, representing the width of the footing.

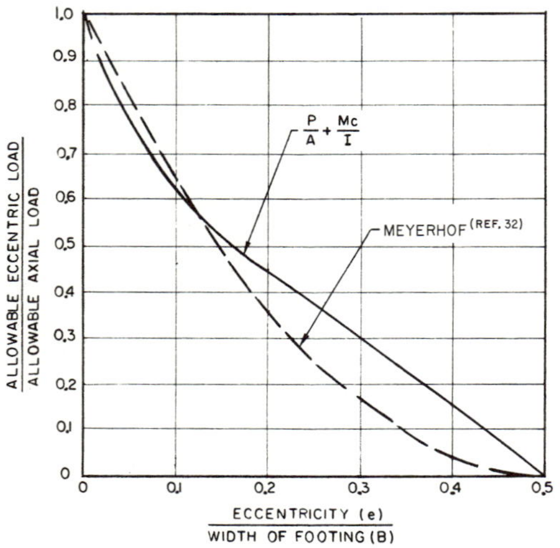

FIG. 5-27 Relation between Allowable Axial Load and Allowable Eccentric Load on Square Footings as Computed by Different Procedures.

The effect of Eqs. (5-20) and (5-21), for a strip footing founded on a cohesionless soil, is to reduce the bearing capacity to an amount equal to B'^2/B^2 times the basic allowable capacity Q. For a load at the limit of the kern section (middle third) the reduced value is $\frac{4}{9}$ or 45 percent, which corresponds, roughly, to the results derived from the conventional expression $P/A + Mc/I$. For an eccentricity equal to $0.5B$, again the footing becomes unstable and the effective width B' and the bearing capacity are equal to zero.

Equations (5-19) to (5-21) relate to strip footings having eccentricity about one major axis. For footings of other shape or for eccentricity about two axes, Meyerhof suggests application of the above procedure by finding the minimum contact area A', with straight boundary across the base (see Fig. 5-28), such that its centroid coincides with that of the load.

Using this contact area, the ultimate load capacity is given by the expression

$$Q = qA' \tag{5-22}$$

where q is determined from Eq. (5-2) or (5-3).

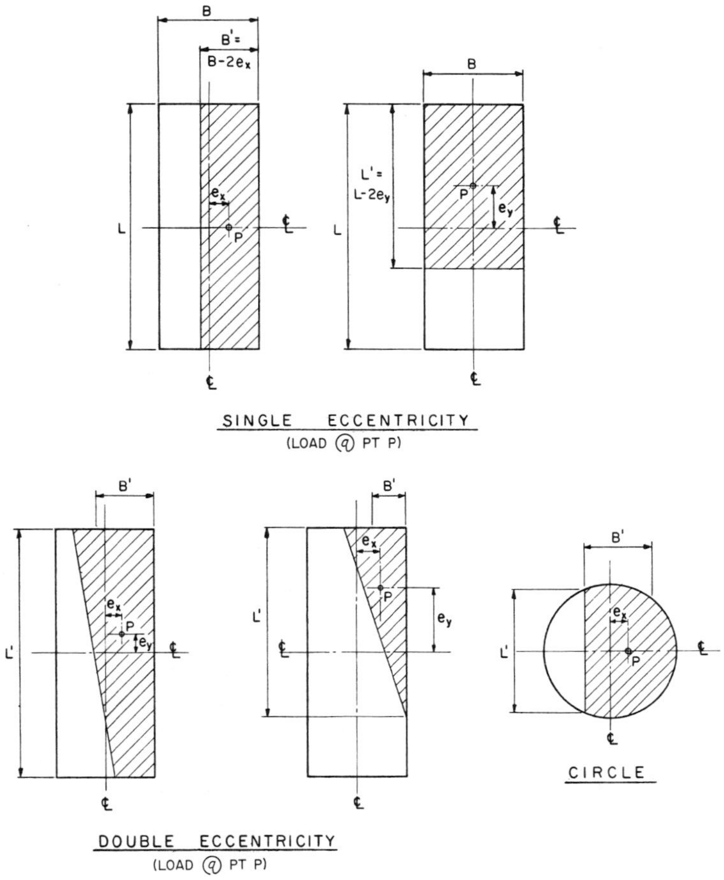

FIG. 5-28 Effective Contact Area of Foundations with Eccentric Load. *(After Meyerhof, Ref. 32.)*

The variation of allowable total load with increased eccentricity for a square surface footing bearing on a cohesionless soil is shown as the dashed line in Fig. 5-27. Comparison of the solid and dashed lines in Fig. 5-27 shows that the conventional approach is more conservative than the approach of Meyerhof for low values of eccentricity (within the kern) and somewhat less conservative for larger eccentricities.

c. *Ultimate Strength Design Concept*

The expression $P/A \pm Mc/I$ represents an assumption of linear variation of reaction between the foundation and the supporting soil. It corresponds to the working stress concept used in the design of concrete columns.

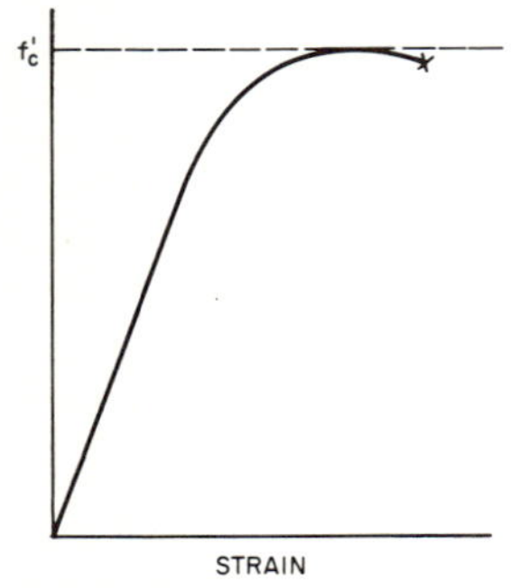

(a) STRESS-STRAIN CURVE FOR CONCRETE COMPRESSION TEST

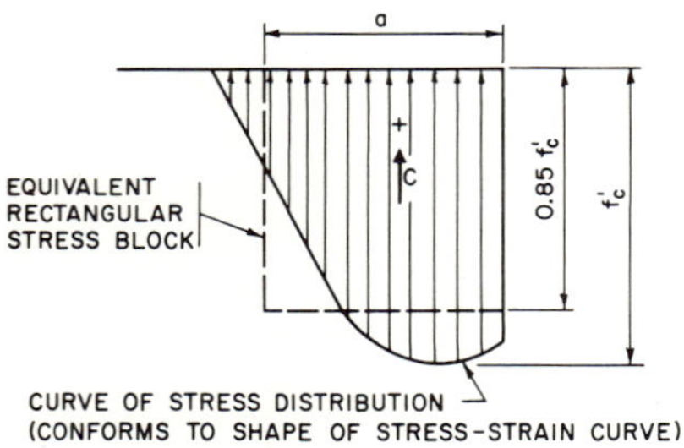

(b) STRESS DISTRIBUTION AT ULTIMATE LOAD

FIG. 5-29 ***Shape of Stress-strain Curve for Concrete Compression Test and Compression Stress Block for Concrete in Flexure at Ultimate Load.***

The approach of Meyerhof with its assumption of a rectangular stress block (stress of uniform intensity) corresponds to the Whitney [41] concept of ultimate strength design. This is the modern approach to the design of concrete members now rapidly supplanting the working stress concept because of its advantages with regard to consistency of safety factors, economy, and its more direct relation to ultimate or true strength of the materials.

The concept of ultimate strength design as used for concrete design is predicated on the fact that the stress-strain (load-deformation) relation for concrete (Fig. 5-29*a*) creates a condition of stress distribution at ultimate load as shown in Fig. 5-29*b*. For ultimate strength analysis, it is necessary to know the actual magnitude of the maximum compressive stress, the magnitude of the resultant (C), and the location of the resultant. These are derived from three parameters obtained from tests, and once they are known, the stress-strain shape may for convenience be shown as an equivalent stress rectangle having the same properties.

The load-deformation curve for a loaded footing (or for a pile) has a similar shape to the load-deformation curve for concrete. The stress distribution at ultimate load, as assumed by Meyerhof, also is similar to the equivalent rectangular stress block. It may be concluded that the ultimate strength design techniques used for proportioning eccentrically loaded concrete compression members are applicable, at least in principle, to the design of footings and pile groups. In fact, the approach of Meyerhof corresponds to such techniques with the exception that, unlike concrete design, the allowable stress intensity (bearing pressure) varies with the eccentricity [see Eq. (5-20)].

For piling, however, the allowable stress intensity does not vary with eccentricity, and the ultimate strength design concept, if applied, provides considerable economy in proportioning eccentrically loaded pile groups. This matter is discussed in Chap. 6.

3 Coefficients of Uniformity (C_u) and Curvature (C_c)

The coefficient of uniformity is a measure of the size range of the particles in a given soil and is the ratio of the 60 percent-finer-than size (D_{60}) to the 10 percent-finer-than size (D_{10}). *Coefficient of curvature* (C_c) is the ratio of the square of the 30 percent-finer-than size (D_{30}) to the product of D_{60} by D_{10}.

4 Potential for Shear Failure in Granular Soils

Several building codes contain a requirement to the effect that the nominal allowable bearing pressures on granular soils relate to footings 3 ft or more in least dimension, but should be reduced in proportion to the ratio of the actual width of the footing to the 3-ft dimension, if the width is less than 3 ft. The purpose of this provision is to avoid the possibility of shear failure under knife-edge bearing areas.

For bearing areas 3 ft or larger in least dimension (and embedded 2½ to 3 ft) and bearing on granular soil, shear failure is not a problem provided that the bearing values do not exceed the nominal allowable values derived as indicated in the text. For example:

a. Gravel Soils

For a loose poorly graded gravel soil, assume $\phi = 35°$, $c = 0$, $\gamma_a = 110$ pcf. For $B = 3$ ft, $D = 2.5$ ft, substituting in Eq. (5-2), $q_{ult} = 8.5$ tsf (FS $= 2.13$ based on nominal allowable value of 4.0 tsf). For a compact well-graded gravel soil, assume $\phi = 40°$, $c = 0$, $\gamma_d = 120$ pcf. For $B = 3$ ft, $D = 2.5$ ft, $q_{ult} = 19.5$ tsf (FS = 1.95 based on nominal allowable value of 10 tsf).

b. Sand Soils

For a loose soil ($N = 10$), assume $\phi = 28°$, $c = 0$, $\gamma_d = 95$ pcf. For $B = 3$ ft, $D = 3$ ft, $q_{ult} = 3.6$ tsf (FS = 1.80 based on nominal allowable value of 2.0 tsf). For a compact soil ($N = 60$), assume $\phi = 36°$, $c = 0$, $\gamma_d = 110$ pcf. For $B = 3$ ft, $D = 3$ ft, $q_{ult} = 10.4$ tsf (FS = 1.73 based on nominal allowable value of 6.0 tsf).

When the bearing area of a footing on a sand or gravel soil is narrower than 3 ft, or the depth of embedment is less than the above values (2½ to 3 ft), the possibility of shear failure should be checked by computation. The aforementioned device of reducing the bearing pressure in proportion to the width of the footing is convenient in this regard, but is quite a conservative approach where the footing is embedded more than a foot or two.

5 Specification for Controlled Fill

a. Materials

Materials for controlled fills shall consist of natural or washed soils free of organic matter and conforming to the following limitations on gradation:

1. Maximum size of particles—4 in.
2. Retained on ¾-in. sieve—30 percent maximum
3. Passing no. 100 sieve—45 percent maximum
4. Passing no. 200 sieve—12 percent maximum

The grading shall be determined in accordance with ASTM standard specification C117 and C136.

b. Preparation of Subgrade

Before placing controlled fill, the surface to receive the fill shall be stripped of all organic growth, timber, rubbish, and debris, and all surface boulders larger than 6-in. size shall be removed. The surface to receive the fill then shall be rolled with a roller with a minimum weight of 20 tons and exerting a pressure of not less than 300 lb per inch width, until all weave has been eliminated.

Where pavement is present within a depth of 2 ft below the surface which is to receive the fill, such pavement shall be broken up to provide for drainage of the fill.

Where the surface to receive fill slopes in excess of 1 on 3, the surface shall be benched before placing the fill.

c. Placing and Compacting Fill

Controlled fill shall be placed and compacted at a moisture content approximately 3 percent drier than the optimum moisture content. It shall be placed in uniform layers not more than 12 in. thick (after compaction). Each layer shall be compacted to not less than 95 percent of the maximum density determined in accordance with ASTM standard specification D1557.

Fill shall not be placed when frozen, or placed on frozen or wet subgrade. At the close of each day's work or at the inception of periods of rain, the surface shall be sealed by rolling with a smooth wheel roller.

6 Stabilized Overburden Pressure (Soils)

The calculation of stabilized overburden pressure is based on the curve e versus log p from the laboratory consolidation test. If the preconsolidation load is equal to, or exceeds, the weight of the overburden, the stabilized pressure is equal to the weight of the overburden, less the effects of rebound, which may be all or part of the weight of excavation removed during the construction of the building. If the preconsolidation load is less than the weight of the overburden, the stabilized pressure is equal to the preconsolidation load, again less the effects of rebound.

The effects of rebound can be evaluated from the slopes of the virgin and recompression branches of the compression curve, as illustrated in Fig. 5-30. If the bearing stratum consists of fine-grained materials, this difference in slope is large, and the effects of rebound are significant. If the bearing stratum consists of granular free-draining soils such as medium and coarse sands and gravels, however, there is no detectable difference in slope of the virgin and recompression branches of the compression curve, and the effects of rebound are approximated by taking the stabilized overburden pressure as the full overburden pressure existing *at the completion of construction;* i.e., the weight of any overburden which is stripped and not replaced is not included unless the effect of such weight is simulated and maintained in the final construction by confining the soil under the bearing area, as illustrated in Case III of Fig. 5-20.

Where the stabilized overburden is not the same over the entire bearing area, the least value (i.e., the overburden pressure measured from the level of the bearing area to the lowest level of the adjacent grade) is used in the bearing-capacity computation. Also, where the groundwater level is above the level of the bearing area, the computation of stabilized overburden pressure should be made on the basis of the submerged (buoyant) weight of the soil below the phreatic surface.

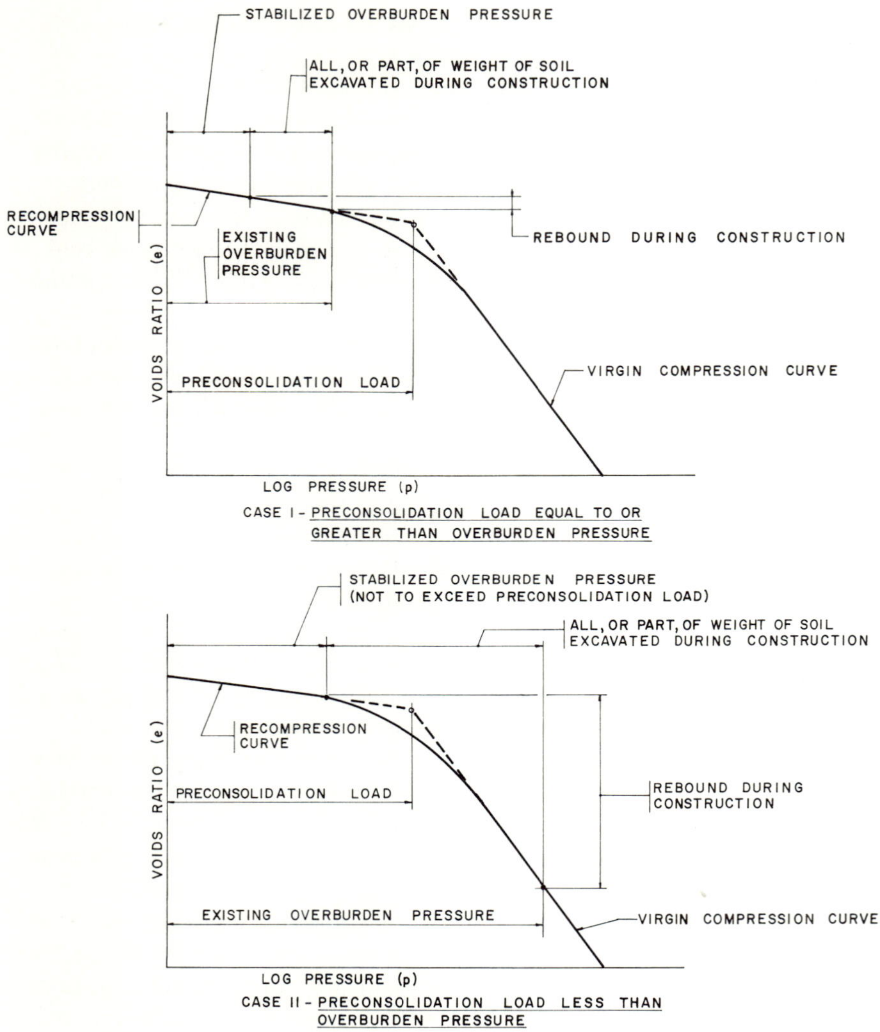

FIG. 5-30 Stabilized Overburden Pressure (Soils).

7 Stabilized Overburden Pressure (Rock)

For foundations bearing on rock (other than decomposed rock) stabilized overburden pressure is normally assumed to be zero. The probable reason is that the weight of the overburden is so small compared with the bearing capacity of the rock that it can be neglected. Besides, conventional values for bearing capacity of rock are not based on rational considerations, so that the refinement of adding the stabilized overburden pressure is not justified.

REFERENCES

1. Terzaghi, Karl: "Theoretical Soil Mechanics," John Wiley & Sons, Inc., New York, 1943.
2. Meyerhof, G. G.: The Ultimate Bearing Capacity of Foundations, *Geo technique,* 1951.
3. Hansen, Brinch: A General Formula for Bearing Capacity, *Geoteknisk Institute Bulletin* 11, Copenhagen, 1961.
4. Balla, Arpad: Bearing Capacity of Foundations, *Journal of the Soil Mechanics and Foundations Division, Proceedings of the American Society of Civil Engineers,* vol. 88, no. SM5, October, 1962.
5. Terzaghi, Karl, and R. B. Peck: "Soil Mechanics in Engineering Practice," 2d ed., John Wiley & Sons, Inc., New York, 1967.
6. Meyerhof, G. G.: Influence of Roughness of Base and Ground Water Condition on the Ultimate Bearing Capacity of Foundations, *Geotechnique,* 1955.
7. Design Manual DM7, "Soil Mechanics," U.S. Navy, Bureau of Yards and Docks, Washington, D.C.
8. Henry, F. D. C.: "The Design and Construction of Engineering Foundations," McGraw-Hill Book Company, New York, 1956.
9. Hough, B. K.: "Basic Soils Engineering," The Ronald Press Company, New York, 1957.
10. Paaswell, G.: "Retaining Walls," McGraw-Hill Book Company, New York, 1920.
11. Peck, Ralph B., Walter E. Hanson, Thomas H. Thornburn: "Foundation Engineering," John Wiley & Sons, Inc., New York, 1953.
12. "Structural Engineering Manual," Public Buildings Administration, Federal Works Agency, Washington, D.C.
13. Dunham, C. W.: "Foundations of Structures," 2d ed., McGraw-Hill Book Company, New York, 1962.
14. Terzaghi, Karl: Stability and Stiffness of Cellular Cofferdams, *Transactions of the American Society of Civil Engineers,* vol. 110, p. 1083, 1945.
15. White, Ardis, James A. Cheney, and C. Martin Duke: Field Study of a Cellular Bulkhead, *Transactions of the American Society of Civil Engineers,* vol. 128, pt. I, p. 463, 1963.
16. White, Lazarus, and E. A. Prentis: "Cofferdams," Columbia University Press, New York, 1950.
17. Tschebotarioff, G. P.: "Soil Mechanics, Foundations, and Earth Structures," McGraw-Hill Book Company, New York, 1951.
18. Krynine, D. P.: "Soil Mechanics," McGraw-Hill Book Company, New York, 1941.

19. Housel, W. S.: A Practical Method for the Selection of Foundation Based on Fundamental Research in Soil Mechanics, *University of Michigan Engineering Research Bulletin* 13, October, 1929.
20. Lambe, T. William: Methods of Estimating Settlement, *Journal of the Soil Mechanics and Foundations Division, Proceedings of the American Society of Civil Engineers*, vol. 90, no. SM5, September, 1964.
21. Seed, H. Bolton: Settlement Analyses: A Review of Theory and Testing Procedures, *Journal of the Soil Mechanics and Foundations Division, Proceedings of the American Society of Civil Engineers*, vol. 91, no. SM2, March, 1965.
22. Hough, B. K.: Compressibility as the Basis for Soil Bearing Value, *Journal of the Soil Mechanics and Foundations Division, Proceedings of the American Society of Civil Engineers*, vol. 85, no. SM4, August, 1959.
23. Lo, K. Y.: Secondary Compression of Clays, *Journal of the Soil Mechanics and Foundations Division, Proceedings of the American Society of Civil Engineers*, vol. 87, no. SM4, August, 1961.
24. Wahls, H. E.: Analysis of Primary and Secondary Consolidation, *Journal of the Soil Mechanics and Foundations Division, Proceedings of the American Society of Civil Engineers*, vol. 88, no. SM6, December, 1962.
25. Prentis, E. A., and L. White: "Underpinning," Columbia University Press, New York, 1931.
26. Feld, Jacob: Tolerance of Structures to Settlement, *Journal of the Soil Mechanics and Foundations Division, Proceedings of the American Society of Civil Engineers*, vol. 91, no. SM3, May, 1965.
27. Anderson, Paul: "Substructure Analysis and Design," The Ronald Press Company, New York, 1948.
28. Taylor, D. W.: "Fundamentals of Soil Mechanics," John Wiley & Sons, Inc., New York, 1948.
29. Hetenyi, M.: "Beams on Elastic Foundations," The University of Michigan Press, Ann Arbor, Mich., 1947.
30. "Building Code Requirements for Reinforced Concrete," ACI 318-63, American Concrete Institute.
31. Skempton, A. W., and D. H. MacDonald: Allowable Settlement of Buildings, *Proceedings of the Institution of Civil Engineers*, pt. III, vol. 5, 1956.
32. Meyerhof, G. G.: The Bearing Capacity of Foundations under Eccentric and Inclined Loads, *Proceedings, Third International Conference on Soil Mechanics and Foundation Engineering*, vol. 1, 1953.
33. Mehta, M. R., and A. S. Veletsos: Stresses and Displacements in Layered Systems, *University of Illinois, Structural Research Series*, no. 178, 1959.
34. Burmister, D. M.: Stress and Displacement Characteristics of a Two-layer Rigid Base Soil System: Influence Diagrams and Practical Applications, *Proceedings, Highway Research Board*, 1956.
35. Fox, E. N.: The Mean Elastic Settlement of a Uniformly Loaded Area at a Depth below the Ground Surface, *Proceedings, Second International Conference on Soil Mechanics and Foundation Engineering*, 1948.
36. Building Code of the City of Boston, 1962.
37. "Earth Manual," U.S. Department of Interior, Bureau of Reclamation, p. 26. Denver, 1951.
38. Building Code Committee, Soil Mechanics and Foundations Division, Metropolitan Section, ASCE. Members: Messrs. J. D. Parsons, Cove Sullivan, Martin Kapp, Gordon Fletcher, Salvatore DeSimone, and Reuben Samuels.
39. Aldrich, H. P., Jr.: Precompression for Support of Shallow Foundations, *Journal*

of the Soil Mechanics and Foundations Division, Proceedings of the American Society of Civil Engineers, vol. 91, no. SM2, March, 1965.
40. Parsons, J. D.: chairman, Building Code Committee, Soil Mechanics and Foundations Division, Metropolitan Section, ASCE, personal communication.
41. Whitney, Charles S.: Plastic Theory of Reinforced Concrete Design, *Transactions of the American Society of Civil Engineers,* vol. 107, p. 251, 1942.
42. Newmark, Nathan M.: Influence Charts for Computation of Stresses in Elastic Foundations, *University of Illinois Bulletin,* vol. 61, no. 92, June, 1964.
43. Skempton, A. W.: Cast-in-situ Bored Piles in London Clay, *Geotechnique,* vol. 9, no. 4, 1959.
44. Teng, W. C.: "Foundation Design," Prentice-Hall, Inc., Englewood Cliffs, N.J., 1962.
45. Nishida, Y.: Brief Note on Compression Index of Soil, *Proceedings of the American Society of Civil Engineers,* vol. 82, no. SM3, July, 1956.
46. Button, S. J.: The Bearing Capacity of Footings on a Two Layer Cohesive Subsoil, *Proceedings, Third International Conference on Soil Mechanics and Foundation Engineering,* vol. 1, 1953.
47. Skempton, A. W., and L. Bjerrum: A Contribution to the Settlement Analysis of Foundations on Saturated Clay, *Geotechnique,* vol. 7, no. 4, December, 1957.
48. Sowers, G. F.: Shallow Foundations, in G. A. Leonards (ed.), "Foundation Engineering," McGraw-Hill Book Company, New York, 1962.
49. Polshin, D. E., and R. A. Tokar: Maximum Allowable Non-uniform Settlement of Structures, *Proceedings, Fourth International Conference on Soil Mechanics and Foundation Engineering,* 1957.
50. *Proceedings, Conference on Soil Shear Strength,* American Society of Civil Engineers, 1961.
51. "Suggested Design Procedures for Combined Footings and Mats," Report of Committee 436, *Proceedings of the American Concrete Institute,* vol. 63, no. 10, October, 1966.
52. Malter, Henry: Numerical Solutions for Beams on Elastic Foundations, *Journal of the Structural Division, Proceedings of the American Society of Civil Engineers,* vol. 84, no. ST2, March, 1958.
53. Israel Standard Code of Practice, The Standards Institution of Israel, Ramat Aviv, Tel Aviv.

CHAPTER 6
Allowable Pile Loads

The allowable load on a foundation pile is limited by two considerations:

1. The capacity of the pile to transmit load to the soil
2. The capacity of the soil to support the load transmitted from the pile

The design load is the smaller of the two values, divided by a suitable factor of safety.

The first consideration (capacity of the pile) is a matter of structural strength and is discussed in Sec. 6A. The second (capacity of the soil) relates to the allowable soil bearing pressure, is treated in Sec. 6B, and involves considerations of ultimate strength capacity of the soil and of settlement, in a similar vein to the discussion of Chap. 5 (Allowable Soil Bearing Pressure). Accordingly, the reader should be familiar with the text of Chap. 5, before proceeding to Sec. 6B.

Sections 6A and 6B deal with the usual consideration of downward vertical load. The capacity of piling to resist lateral loads and uplift forces also is important and is discussed in Secs. 6C and 6D respectively.

Section 6A STRUCTURAL STRENGTH OF THE PILE

Table 6-1 summarizes allowable stress values in piling of various materials and types as prescribed by a number of authorities. As may be noted, two basic approaches to determining the structural strength of a pile are in use: (1) fixed values of allowable unit stress and (2) calculation of the capacity of the pile as a "short column." The essence and basis

TABLE 6-1 ***Typical Values of Allowable Unit Stress for Design of Piling***

A. Cast-in-place concrete piling

Source	Allowable unit stress, psi: Shell	Allowable unit stress, psi: Reinforcing steel	Allowable unit stress, psi: Concrete	Remarks
National standards				
1. National Research Council Survey (1962) [3]	Varies: zero–24,000 psi. Generally limitation as to minimum thickness of metal included where structural capability is considered	Varies: 3,500–24,000 psi. AASHO [6], AISC [5], and ACI [4] specifications also referenced. Value of $0.35F_y$ also indicated	Varies: 600–3,000 psi. Also expressed as function of f_c' with values varying from 0.20 to $0.45f_c'$	
2. "Pile Foundations" reference bulletin issued by American Iron and Steel Institute (1963)	$0.35F_y$	$0.40F_y$	$0.225f_c'$	Shells to have minimum thickness of 0.10 in.
3. National Building Code (1967)			$0.225f_c'$	
Regional codes				
4. BOCA (1963)	$0.50F_y$	$0.34F_y$	$0.225f_c'$	Designed as short column. Indicated values are for tied and composite columns (ACI-318) [4].
5. Southern Standard Building Code (1961)			330 psi for piles with temporary shells; 850 psi for piles with permanent shells	Ostensibly piles are designed as short columns. Maximum load values are stipulated, however, which limit stresses to the values indicated.

TABLE 6-1 ***Typical Values of Allowable Unit Stress for Design of Piling*** **(*Continued*)**

A. Cast-in-place concrete piling

Source	Allowable unit stress, psi: Shell	Allowable unit stress, psi: Reinforcing steel	Allowable unit stress, psi: Concrete	Remarks
		Regional codes		
6. Uniform Building Code (1961)		$0.34F_y$	$0.225f'_c$	
7. Building Code of City of Boston (1962)	8,500		$0.225f'_c$ but not to exceed 900 psi	Minimum wall thickness of 0.20 in.
8. Building Code of City of St. Louis (1961)	$0.50F_y$	$0.34F_y$	$0.225f'_c$	Designed as short column. Indicated values are for tied and composite columns (ACI-318) [4].
9. Building Code of City of Baltimore (1955)	See remarks	See remarks	$0.20f'_c$ (uncased) $0.25f'_c$ (cased) $0.30f'_c$ (conical)	Based on transformed section.
10. Los Angeles Building Code (1958)	9,000	$0.34F_y$	$0.225f'_c$	
11. Building Code of City and County of Denver (1962)	9,000	$0.40F_y$	$0.225f'_c$	
12. Building Code of City of Chicago (1963)			$0.40f'_c$	
13. Building Code of City of Buffalo (1965)	12,000	nf_c	$0.25f'_c$	Minimum wall thickness of ⅛ in.

B. Precast concrete piling

Source	Allowable unit stress, psi: Reinforcing steel	Allowable unit stress, psi: Concrete	Remarks
	National standards		
1. National Research Council Survey (1962) [3]	Varies: 3,500–24,000 psi. Also referenced as $0.35F_y$	Varies: 460–3,000 psi. Also expressed as function of f'_c with values varying from $0.225f'_c$–$0.45f'_c$	

TABLE 6-1 ***Typical Values of Allowable Unit Stress for Design of Piling (Continued)***

B. Precast concrete piling

Source	Allowable unit stress, psi: Reinforcing steel	Allowable unit stress, psi: Concrete	Remarks
National standards			
2. "Pile Foundations" (1963)	$0.40F_y$	$0.225f'_c$	
3. National Building Code (1967)		$0.225f'_c$	
Regional codes			
4. BOCA (1963)	$0.34F_y$	$0.225f'_c$	Designed as short column. Indicated values are for tied columns (ACI-318) [4].
5. Southern Standard Building Code (1961)			Ostensibly piles are designed as short columns. Maximum load value is stipulated, however as: load (in tons) = 2.2 × side dimension (in inches).
6. Uniform Building Code (1961)	$0.34F_y$	$0.225f'_c$	
Local codes			
7. Building Code of City of Boston (1962)			Maximum allowable load of 50 tons on 12 × 12 pile of f'_c = 4,000 psi. Increase proportionately to area to a maximum 90-ton load.
8. Building Code of City of St. Louis (1961)	$0.34F_y$	$0.225f'_c$	Designed as short column. Indicated values are for tied columns (ACI-318) [4].
9. Building Code of City of Baltimore (1955)	$0.40F_y$	$0.225f'_c$	Designed as reinforced concrete columns having $L/d = 10$.
10. Los Angeles Building Code (1958)	$0.34F_y$	$0.225f'_c$	
11. Building Code of City and County of Denver (1962)	$0.40F_y$	$0.225f'_c$	
12. Building Code of City of Chicago (1963)			Designed as column. Assumed unbraced length not specified.
13. Building Code of City of Buffalo (1965)	nf_c	$0.25f'_c$	

TABLE 6-1 ***Typical Values of Allowable Unit Stress for Design of Piling (Continued)***

C. Timber piling

Source	Allowable unit stress, psi	Remarks
National standards		
1. National Research Council Survey (1962) [3]	Varies: 400–1,550	1. Allowable stress is a function of species. 2. Stipulated limitations on maximum loads vary from 20–25 tons for various agencies reporting.
2. "Pile Foundations" (1963)	Southern pine, Douglas fir (coast region), oak, 1,200 Norway pine, Douglas fir (mountain region), western hemlock, 850	
3. National Building Code (1967)	Cedar, Douglas fir (mountain region), western hemlock, northern pine, 850 Cypress, Douglas fir (coast region), hickory, oak, southern pine, 1,200	
Regional codes		
4. BOCA (1963)	1.0c where c is allowable bearing stress in compression parallel to grain	Designed as short column.
5. Southern Standard Building Code (1961)	Southern pine, Douglas fir, oak, 800 Norway pine, cedar, spruce, 600	
6. Uniform Building Code (1961)	60% of basic stress for clear material as set forth in ASTM specification D2451	Not to exceed 1,000 psi.
Local codes		
7. Building Code of City of Boston (1962)	Expressed as allowable total load rather than as allowable unit stress. Allowable load: 12 tons for piles of spruce, Norway pine, or similar woods, 10-in. butt, 6-in. tip, increased 3 tons for each 1 in. increase in butt and tip dimension to maximum of 24 tons. 16 tons for piles of oak, southern yellow pine, or similar woods, 10-in. butt, 6-in. tip, increased 4 tons, as above, to maximum of 30 tons.	
8. Building Code of City of St. Louis (1961)	1.0c, where c is allowable bearing stress in compression parallel to grain	Designed as short column.

TABLE 6-1 ***Typical Values of Allowable Unit Stress for Design of Piling (Continued)***

C. Timber piling

Source	Allowable unit stress, psi	Remarks
Local codes		
9. Building Code of City of Baltimore (1955)	500 (computed at midlength)	
10. Los Angeles Building Code (1958)	60% of basic stress for clear materials as set forth in ASTM specification D2451	Not to exceed 1,000 psi.
11. Building Code of City and County of Denver (1962)	60% of basic stress for clear material as set forth in ASTM specification D2451	Not to exceed 1,000 psi.
12. Building Code of City of Chicago (1963)		Design as column. Assumed unbraced length not specified.
13. Building Code of City of Buffalo (1965)	Cedar, western hemlock, Norway pine, spruce, 600. Douglas fir, hickory, oak, southern pine, 800	

D. Steel pipe piling

Source	Allowable unit stress, psi			Remarks
	Pipe	Reinforcing steel	Concrete	
National standards				
1. National Research Council Survey (1962) [3]	Varies: Zero–20,000 psi. Also expressed as $0.35F_y$	Varies: 6,000–20,000 psi. Also expressed as $0.35F_y$	Varies: 600–4,000 psi. Also expressed as function of f'_c with values varying from $0.20f'_c$ to $0.45f'_c$	Also expressed in terms of following column formulas: 1. $18{,}000 - 70\dfrac{L}{r} \leq 9{,}000$ psi 2. $17{,}000 - 0.485\dfrac{(L)^2}{r}$
2. "Foundation Piling" (1963)	$0.35F_y$	$0.40F_y$	$0.25f'_c$	
3. National Building Code (1967)	$0.35F_y$		$0.225f'_c$	1. Minimum shell thickness = 0.10 in. 2. F_y not to exceed 36,000 psi.

TABLE 6-1 ***Typical Values of Allowable Unit Stress for Design of Piling (Continued)***

D. Steel pipe piling				
	Allowable unit stress, psi			
Source	Pipe	Reinforcing steel	Concrete	Remarks
		Regional codes		
4. BOCA (1963)	$0.50F_y$	$0.34F_y$	$0.225f'_c$	Designed as short column. Indicated values are for tied and composite columns (ACI-318) [4].
5. Southern Standard Building Code (1961)	About 6,000		About 500	Ostensibly piles are designed as short columns. Maximum load values are stipulated, however, which limit stresses to values indicated.
6. Uniform Building Code (1961)	9,000		$0.225f'_c$	
		Local codes		
7. Building Code of City of Boston (1962)	8,500		$0.225f'_c \leq 900$	15,000 psi on steel core.
8. Building Code of City of St. Louis (1961)	$0.50F_y$	$0.34F_y$	$0.225f'_c$	Designed as short column. Indicated values are for tied and composite columns (ACI-318) [4].
9. Building Code of City of Baltimore (1955)	7,500		$0.25f'_c$	Reduction for long piles.
10. Los Angeles Building Code (1958)	9,000		$0.225f'_c$	
11. Building Code of City and County of Denver (1962)	9,000		$0.225f'_c$	
12. Building Code of City of Chicago (1963)	12,000		$0.40f'_c$	
13. Building Code of City of Buffalo (1965)	12,000	nf_c	$0.25f'_c$	

TABLE 6-1 ***Typical Values of Allowable Unit Stress for Design of Piling (Continued)***

E. Steel H piling

Source	Allowable unit stress, psi	Remarks
National standards		
1. National Research Council Survey (1962) [3]	Varies: 6,000 to 20,000 psi. Also expressed as $0.35F_y$ or by reference to AISC Specification [5]	1. Also expressed by column formula: $\frac{18{,}000}{1 + L^2/18{,}000R^2} \leq$ 9,000 psi 2. Also differentiates between friction and end-bearing piles.
2. "Foundation Piling" (1963)	$0.35F_y$	
3. National Building Code (1967)	$0.35F_y$	1. Minimum thickness of metal = ⅜ in. 2. F_y not to exceed 36,000 psi.
Regional codes		
4. BOCA (1963)	20,000	
5. Southern Standard Building Code (1961)	12,000	
6. Uniform Building Code (1961)	12,000	
7. National Building Code of Canada (1965)	15,000	
Local codes		
City*	**Allowable unit stress, psi**	**Remarks**
Allentown, Pa.	17,000	
Camden, N.J.	17,000	
Detroit, Mich.	17,000	
Erie, Pa.	17,000	
Hartford, Conn.	17,000	
Providence, R.I.	17,000	
South Bend, Ind.	17,000	
St. Louis, Mo.	17,000	
Waterbury, Conn.	17,000	
Atlanta, Ga.	12,000	
Norfolk, Va.	12,000	
Seattle, Wash.	12,000	

TABLE 6-1 ***Typical Values of Allowable Unit Stress for Design of Piling (Continued)***

E. Steel H piling

City*	Allowable unit stress, psi	Remarks
Local codes		
Charlotte, N.C.	12,000	
Greensboro, N.C.	12,000	
Winston-Salem, N.C.	12,000	
Chicago, Ill.	12,000	
Kansas City, Mo.	12,000	
Los Angeles, Calif.	12,000	
Washington, D.C.	12,000	
Phoenix, Ariz.	$0.35F_y$	
Scranton, Pa.	20,000	
Oakland, Calif.	12,000	
Pasadena, Calif.	12,000	
San Diego, Calif.	12,000	
Santa Ana, Calif.	12,000	
Tucson, Ariz.	12,000	
Newark, N.J.	20,000	
Philadelphia, Pa.	12,000	
Pittsburgh, Pa.	12,000	
San Francisco, Calif.	12,000	
Wichita, Kans.	12,000	
Boston, Mass.	7,500	
Baltimore, Md. (1955)	8,000	
Denver, Colo. (1961)	9,000	
New York, N.Y. (1965)	9,000	
Buffalo, N.Y. (1965)	12,000	

F. Requirements expressed in general terms (for all types of pile)

National Building Code of Canada (1965)	1. Design as intermediate or long column in air, water, peat, soft clay, or very soft clays. 2. Design as short column in other soils. 3. Column capacity to be taken as 75% of capacity of superstructure column if proved by load test, or 50% of said capacity if not so proved.

* As of 1964, except when specified otherwise; courtesy American Iron and Steel Institute.

of these two approaches, and some of the problems of application, are as follows:

6A.1 Fixed Values of Allowable Pile Stress

As indicated in Table 6-1, opinions on allowable stresses in piles vary widely, even for exactly the same material in the same type of pile. This is a curious situation. The reason usually given is that the differences represent the effects of variations in local soil conditions. However, this is only an attempt at rationalization. Variations in soil conditions do occur, but they hardly account for the extremes indicated in Table 6-1. Why, for example, does the concrete in a cast-in-place pile in Boston have only 55 percent of the stress value of that same pile in Chicago? Why should the steel in an H pile designed under the Boston or New York codes have less than half the capacity of that same steel, in the same pile, designed under the BOCA code? Why do standards of a city differ so widely from those of regional or national standards used in adjacent municipalities? There are two reasons:

a. The Values Represent Opinion Derived from Experience

The first reason is that the stress values indicated in Table 6-1 do not have a rational basis, but rather represent an expression of experience. As stated in Chap. 1, the interpretation and application of experience to design is a subjective matter which depends on the interpreter, his background, his age, personality, and often his professional and economic position as well. Among different persons using the same pile, under the same conditions, some will have had a satisfactory experience, and others a poor one. Those who have good fortune and are bold tend to act more liberally. Those who are cautious or have had unfortunate experiences are conservative.

As anyone familiar with the development of design standards will recognize, this personal factor has an important influence in deriving code provisions, and the indicated standards reflect this influence.

b. Capacity of the Pile versus Capacity of the Soil

The second reason, bluntly, is that allowable unit stress values for materials in piling often are set artificially low in order to limit the *total* load on the pile and thereby, indirectly, to limit the load intensity imposed on the soil. This type of indirect regulation is common in foundation design practice and, generally, works well. However, in the case of piling it is grossly wasteful, since the resulting pile section often is many times stronger than it need be.

The fact of the matter is that the structural strength of the pile and the capacity of the soil, for purposes of design, are separate considerations and are best treated separately.

For example, consider a closed-end pipe pile driven to rock. Assume that the design load on such a pile was 100 tons and that this value represented the capacity of the rock to receive load. Suppose the allowable unit stress values for steel and concrete in piling were 9,000 psi for the steel and $0.25f'_c$ for the concrete. To support the 100-ton load would require the use of a 10¾-in.-OD pipe of ½-in. wall thickness (extra strong pipe), filled with 3,000 psi concrete (capacity = 101 tons). Now suppose that the designer were satisfied that the steel in the pipe could support, instead of 9,000 psi, a stress of 12,000 psi without structural failure of the filled-pipe section (a stronger grade of steel might be used, for example). The load per pile still would be limited to 100 tons (the capacity of the rock), but the wall thickness of the pipe could be reduced from ½ in. (extra strong) to ⅜ in. (standard weight). The second pile can probably be driven as readily as the first, without damage to the shell, and the saving would be 14 lb of metal or (at 1968 prices) over a dollar per foot of pile.

As will be shown in Sec. 6A.2 there is little doubt that a pipe material could be found which would take the higher stress. The question which bothers most designers is—Can the ground take it? The answer is—how does the ground know the difference? In both cases, the load is 100 tons, the OD of the pile is 10¾ in., and the concrete is the same. All the ground "knows" is that it has in each case been pierced by a 10¾-in.-diam solid shaft. It certainly cannot differentiate whether that shaft has a ⅜-in. shell and a 10-in. concrete core or a ½-in. shell and a 9¾-in. core.

Also, take the case of a short H pile driven through soft soil to end bearing on rock. The bearing capacity between the steel and the rock can be increased by adding reinforcing plates (or shoes) to the bottom of the pile. The entire length of pile need not be reinforced to accomplish this increase, however, because the strength of the pile exceeds the contact bearing capacity of the rock.

The point being made is one which has been demonstrated many times (see Refs. 1 to 3, for example). In a given soil formation, pile load capacities, insofar as the capacity of the ground is concerned (and with certain limited exceptions), depend on the size, shape, and length of the pile, the energy exerted in driving, and the mode of installation, and not on the pile material or on the stresses therein. It is the structural capacity of the pile which is a function of pile material and pile stress. Structural capacity and the capacity of the soil are separate considerations. The pile material should be stressed to whatever value the pile material (under the given conditions of lateral restraint) can support, and the soil, in turn, should be stressed to whatever its capacity may be.

The importance of this distinction is a matter of economical design. Also, piling material is continually being improved. Higher-strength steels (A36, A242, A441) and stronger concretes (5,000 to 10,000 psi) are now available. New design concepts and techniques are in vogue.* Improvements are being made on the problem of stress grading of timber piling. Plastic pipe or extrusions may find future use for pile material, particularly as a solution to problems of potential corrosion. Soil, however, remains as nature created it. Its properties are not getting any better. Therefore, if use is to be made of these new materials and advantage taken of their improved strength and properties of durability, each component (pile and soil) must be treated for what it, separately, is worth, and the design must be so detailed that each component is capable of supporting the total load most efficiently.

In order to do this, the concept of fixed values of allowable unit stress for piling gradually is being abandoned, and a more rational approach is being substituted. This new approach is to treat the pile as a column. The section embedded in the ground is treated as a short column.† The section projecting above grade, if any, is treated as an extension of the building columns. The practice has gained relatively wide acceptance and has been adopted by several prominent local codes and by all of the regional and national standards (see Table 6-1).

Basically, the new approach is a good one. There are problems in its application, however, which are not widely recognized and which preclude any simple use of column formulas for the design of piling. These problems are discussed in the following section (6A.2).

6A.2 Rational Design of the Pile Section (The Pile as a Column)

A pile is a column, at least to the extent that it carries compressive load (and excluding tension piles).

The design of a column is a function of six basic parameters:

1. The strength (usually yield strength, F_y) of the material

* For example, the basic allowable stress in short compression members of steel conforming to ASTM specification A7 was increased from 17,000 psi in the 1946 AISC design standard to 19,770 psi in the 1963 edition, an increase of 16 percent.

Similarly, in concrete, compression members in modern practice are designed by ultimate strength criteria rather than according to the older concept of working stress. The savings are appreciable. For example, for a 16- by 16-in. concrete section, fully laterally braced, with minimum steel (four no. 7 bars) the capacity under the working stress provisions of the 1956 ACI Code [4] is 274,000 lb. Under the ultimate strength provisions of the 1963 ACI Code, the capacity is 326,000 lb. For a 12- by 12-in. section with four no. 6 bars, the corresponding values are 157,000 and 185,000 lb. The increases amount to 19 percent. Both sections are in common use for precast concrete piling.

† A short column is not defined. Presumedly it would be a column which is sufficiently short and stiff that failure takes place by yield of the column material rather than by buckling.

2. The stiffness of the material (as measured by the modulus of elasticity E and the radius of gyration r)
3. The equivalent unbraced length KL
4. The amount of load P acting on the column
5. The eccentricity of that load
6. The cross sectional area A of the column

Residual stresses in the column section also are important.

The relationships between these parameters and allowable unit stress have been studied extensively, and column formulas have been derived expressing the design capacity of column sections, as follows:

a. Steel Sections [5]

I. Compression on gross section of axially loaded compression members

$$F_a = \frac{[1 - (KL/r)^2/2C_c^2]Fy}{\text{FS}} \tag{6-1}$$

or
$$F_a = \frac{149{,}000{,}000}{(KL/r)^2} \tag{6-2}$$

where F_a = axial stress permitted in the absence of bending moment
K = effective length factor
L = actual unbraced length
r = governing radius of gyration
C_c = Euler ratio (column slenderness ratio dividing elastic and inelastic buckling) equal to $\sqrt{2\pi^2 E/F_y}$
FS = factor of safety $= \dfrac{5}{3} + \dfrac{3(KL/r)}{8C_c} - \dfrac{(KL/r)^3}{8C_c^3}$
F_y = specified minimum yield point for the type of steel being used

Equation (6-1) applies where the ratio KL/r is less than the value C_c. Where KL/r is greater than C_c, Eq. (6-2) applies. Equations (6-1) and (6-2) are represented graphically in Fig. 6-1, for ASTM A7 structural steel (F_y = 33,000 psi).

II. For combined axial load and bending, an interaction-type formula is specified.

When $f_a/F_a \leq 0.15$,

$$\frac{f_a}{F_a} + \frac{f_b}{F_b} \leq 1.0 \tag{6-3}$$

When $f_a/F_a > 0.15$,

$$\frac{f_a}{F_a} + \frac{C_m f_b}{(1 - f_a/F'_e)F_b} \leq 1.0 \tag{6-4}$$

where f_a = computed axial stress
F_a = axial stress permitted in the absence of bending moment

f = computed compressive bending stress
F_b = compressive bending stress permitted if bending moment alone existed
C_m = coefficient dependent on column curvature caused by applied moments

$$F'_e = \frac{149{,}000{,}000}{(KL_b/r_b)^2}$$

K = effective length factor, in plane of bending
L_b = actual unbraced length in plane of bending
r_b = radius of gyration about axis of concurrent bending

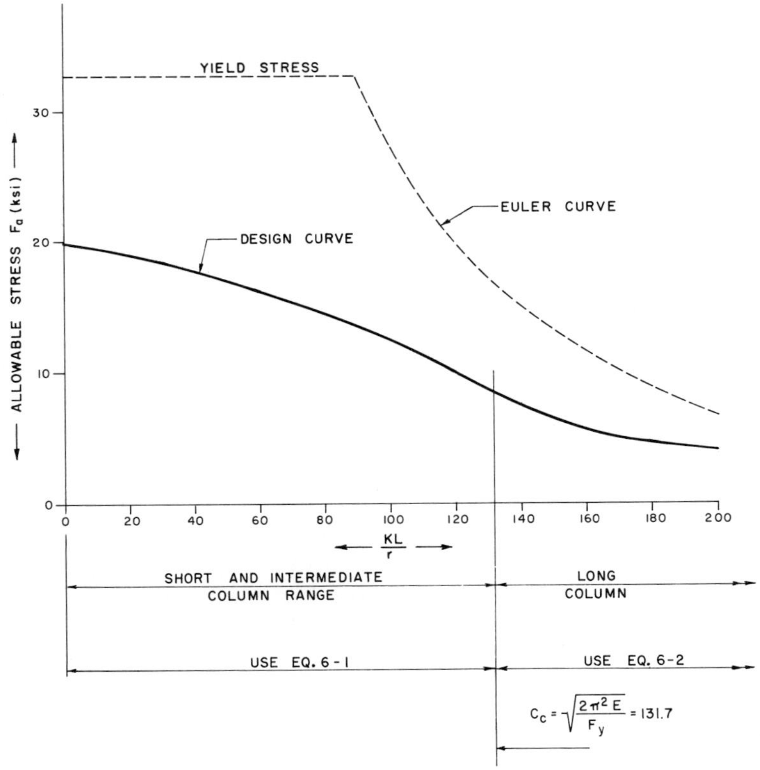

FIG. 6-1 Graphical Representation of Design Equations for Axially Laoded Columns of ASTM A7 Structural Steel (F_y Equals 33,000psi).

b. Reinforced Concrete Sections [4]

I. *Working stress design:*

1. Axially loaded, spirally reinforced columns:

$$P = RA_g(0.25f'_c + f_s p_g) \qquad (6\text{-}5)$$

2. Axially loaded tied columns:

$$P = 0.85RA_g(0.25f'_c + f_s p_g) \tag{6-6}$$

where P = allowable axial load
R = reduction factor for length
A_g = gross area of column
f'_c = compressive strength of concrete
f_s = allowable stress in column vertical reinforcement
p_g = ratio of area of vertical reinforcement to gross area of column A_g

The expressions for R are:

1. Columns bent in double curvature:

$$R = 1.32 - \frac{0.006h}{r} \leq 1.0 \tag{6-7}$$

2. Columns bent in single curvature:

$$R = 1.07 - \frac{0.008h}{r} \leq 1.0 \tag{6-8}$$

where h = actual unbraced length
r = governing radius of gyration of gross concrete area

II. *Ultimate strength design.* No simple formulations are applicable for ultimate strength design of reinforced concrete sections. Instead, it is assumed that all sections are subject to some eccentricity of loading (0.05 × diameter for round columns and 0.10 × side dimension for rectangular sections), and the design is predicated on interaction diagrams (see Sec. III below). In general, for the same load capacity, slightly smaller sections are required by ultimate strength design procedures than would be needed to support the same load were the capacity calculated in accordance with working stress formulations.

III. *Combined axial load and bending.* Design is based on interaction diagrams such as that indicated in Fig. 6-2.

c. *Timber Sections* [7]

I. *Axially loaded sections:*

$$f_a = \frac{3.619E}{(L/r)^2} \leq c \tag{6-9}$$

where f_a = allowable axial stress
E = modulus of elasticity of wood of species used
L = actual unbraced length
r = governing radius of gyration
c = allowable compressive stress, parallel to grain

II. *Combined axial load and bending:*

$$\frac{f_a}{F_a} + \frac{f_b}{F_b} \leq 1.0 \qquad (6\text{-}10)$$

where the symbols have the same significance as in Eq. (6-3).

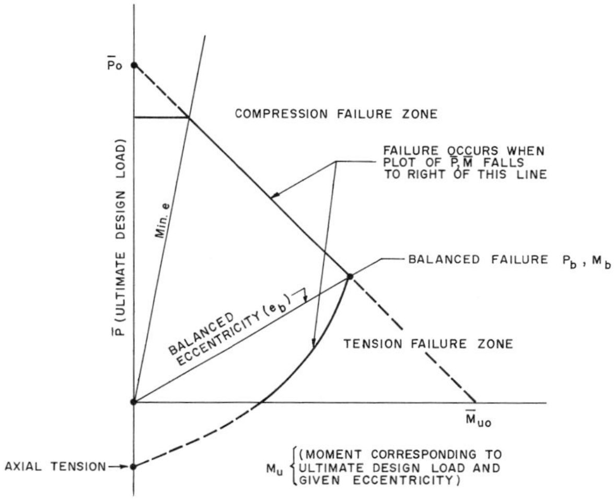

FIG. 6-2 Idealized Interaction Diagram—Eccentrically Loaded Reinforced Concrete Columns.

Application of these formulas [Eqs. (6-1) to (6-10)] to the design of the pile column is as follows.

6A.3 Application of Column Formulas to Design of the Pile Column

As indicated in Sec. 6A.2, the application of conventional column formulas requires evaluation of the parameters F_y (strength of the material), E and r (stiffness of cross section), A (area of cross section), KL (equivalent unbraced length), P (applied load), e (eccentricity of the applied load), and of the effects of residual stresses (if any) in the column section.

When dealing with the pile column, the parameters F_y, E, r, and A may be easily determined, since they are properties of the pile material and cross section. Evaluation of the KL, P, and e parameters and of the effects of residual stress is more difficult. In addition, Eqs. (6-5) through (6-8), which apply to the design of concrete sections, strictly speaking, are applicable only to sections containing reinforcement and not to those of plain concrete, such as are used for cast-in-place piling. For steel piles the factor $C_m/(1 - f_a/F'_e)$ must be estimated. Finally, piles are subject to damage during installation and to deterioration in place, to a greater degree than are columns in the superstructure, and the effects of these two possible incidents must be considered.

Methods for estimating the effects of these several variables and conditions are described in the following paragraphs. The methods necessarily are approximate and somewhat uncertain (which is one reason for the popularity of fixed values of allowable stress for piling), but they are sufficient for practical use.

a. Determining Equivalent Unbraced Length (KL) of Portions of Pile Embedded in the Ground

The pile in the ground is restrained laterally by the soil. This restraint tends to prevent buckling of the section and thereby to increase its load capacity as compared with the capacity of a like column of the same material, cross section, and length, installed in the superstructure of a building.

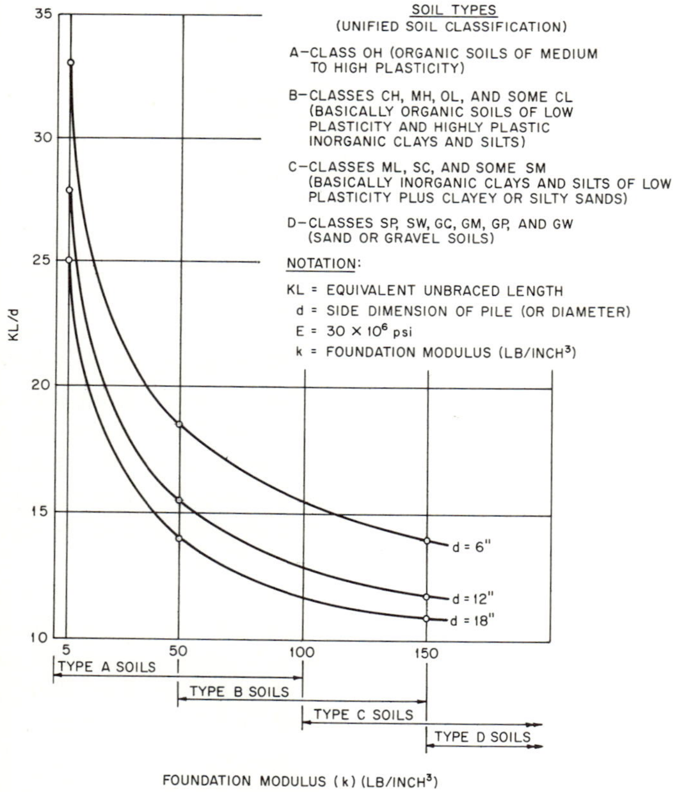

FIG. 6-3 Pile Column. Equivalent Unbraced Length (Steel Piles).

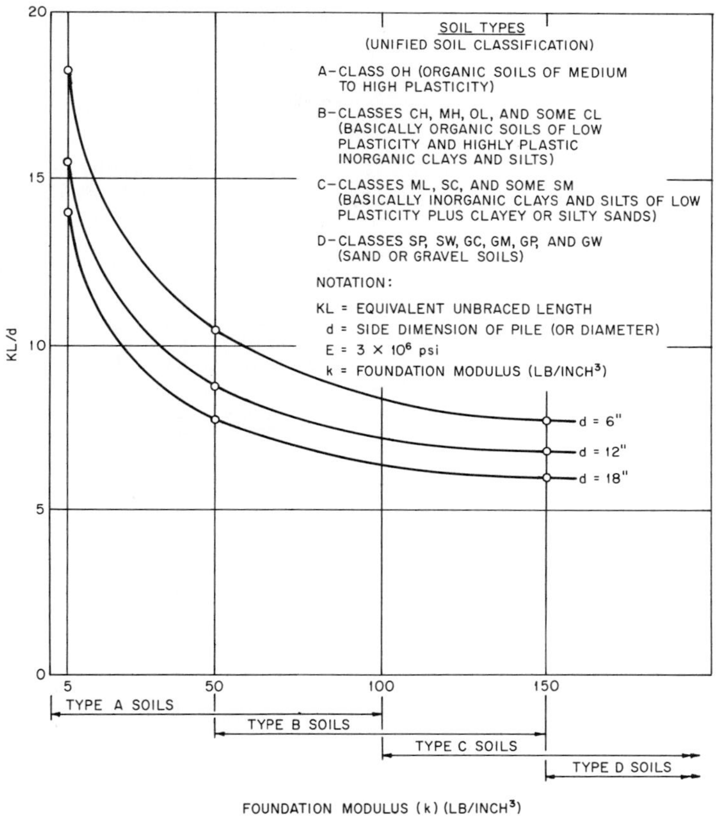

FIG. 6-4 Pile Column. Equivalent Unbraced Length (Concrete Piles).

Evaluation of the pile-soil interaction is a complex problem. For the interested reader, a mathematical derivation of expressions for stress and equivalent unbraced length is presented in Sec. 6E, Chap. Note 1. As may be noted, the expressions are cumbersome. However, the results are summarized in simplified form in Figs. 6-3 to 6-5.

Consider the implications of the data in Figs. 6-3 to 6-5, with respect to application of Eqs. (6-1) through (6-10).

1. *Timber Piles.* Equation (6-9) governs the structural design of timber piles. In Eq. (6-9), whenever the value of KL/d is less than the range given in Table 6-2, the value of f_a is equal to c* (the basic allowable stress on wood for loads applied parallel to the grain); i.e., the column acts as a short compression block without danger of buckling.

* In this regard, there is a problem in determining the c value applicable to timber piling, since piling normally is not stress graded. For suggested stress values, see Sec. 8A.2.

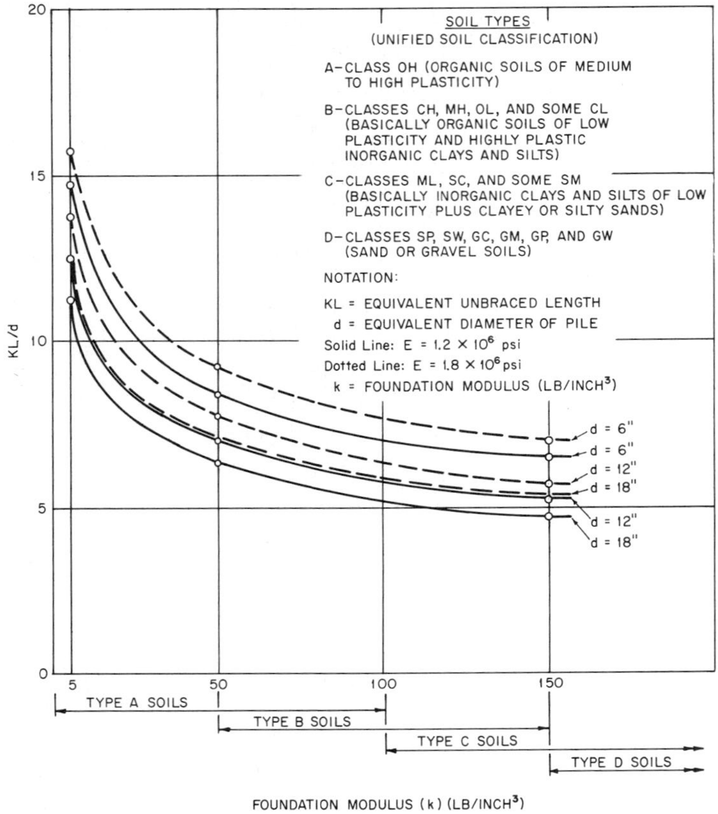

FIG. 6-5 Pile Column. Equivalent Unbraced Length (Timber Piles).

Comparing Table 6-2 with Fig. 6-5 it may be noted that for actual piles, except where the pile section is embedded in an unusually soft (or loose) organic clay (or silt), the dimensions are such that KL/d is less than the range given in Table 6-2 and so a timber pile *is*, in effect, a short column, except when embedded in unusually soft soil, as described.

TABLE 6-2 ***Values of KL/d for Round Timber Columns below Which the Allowable Compression Stress Is Equal to c***

Species of Wood	*KL/d*
Fir, Douglas	15–20
Hemlock, eastern	18–21
Hemlock, western	17–20
Larch	15–20
Pine, southern	11–21
Pine, Norway	18–21
Spruce, eastern	17–19

II. *Concrete Piles.* For a concrete column, the strength reduction for length would be derived from Eq. (6-8). The value of r is about $0.3d$. Substituting this equivalent in Eq. (6-8), it is found that the value of R (the reduction factor for unbraced length) is equal to 1.00 for values of h/d $(= KL/d)$ less than or equal to 2.6; that is, the column acts as a short column for ratios equal to or less than this value. For columns having a larger ratio of KL/d, the strength would be reduced.

Therefore, strictly speaking, concrete piles are *not* short columns unless they are sufficiently husky to qualify under the criterion $KL/d \leq 2.6$.

In this connection, it should be noted that the 1956 edition of "Building Code Requirements of the American Concrete Institute" [4] defined a short column as one having a KL/d ratio of 10, or less. Under this older definition, almost all concrete piles were short columns (see Fig. 6-4), and the use of that term in their design had meaning. That columns of this length/width ratio no longer qualify as short columns under the modern concrete code reflects the advances made in the state of our knowledge of column behavior, advances which have resulted in some amendment of the design standards for columns. These amendments simply have not been reflected in the design standards for piling.

III. *Steel Piles.* A similar situation prevails with steel piles. Reference 5 stipulates a reduction in allowable load for *all* ratios of L/r. In effect, there is no such thing as a short steel column.

For the KL/d ratios indicated in Fig. 6-3 (about 11 to 33) and assuming H piles for which $r = 0.25d$, a reduction in basic allowable stress of 9 to 57 percent applies. Thus, in theory, steel piles, also, are *not* short columns.

These conclusions, at least as regards concrete and steel piles, are not widely recognized. It is commonly considered that instability (in the form of buckling) of piling is not a problem. This appears to be largely due to the fact that there have been no troubles in this regard. Only Bjerrum [9] has reported actual instances of failure of piles by buckling, but these were unusually slender (KL/r about 100) sections embedded in the very soft clay soils of Norway.

The reason that there have not been any troubles, however, is simple and follows from a study of Figs. 6-3 through 6-5. So long as the designer uses only piles of conventional size (minimum 8 in. dimension), made of conventional materials (concrete strength not greater than 5,000 psi, structural steel having a specified minimum yield strength not exceeding 36,000 psi) and embeds the piles in soils of classes ML, SC, SM, and better, there is no problem. The pile indeed does act as a short compression block.

Even when the piles are embedded in CH, MH, OL, and CL soils, again if they are of conventional size and materials, as indicated, the reduction in allowable stress value for potential failure by instability is less than 20 percent, which is well within the usual safety factor for piling. Therefore the buckling problem can be, and usually is, neglected.

This does not mean, however, that the problem can be similarly neglected if unconventional sections (such as rails, rods, or sawn timbers, say 2 by 4s or 4 by 4s)* are to be used, if high-strength materials (say, 10,000-psi concrete or some of the high-yield steels) are employed, or if the piles are embedded in exceptionally soft soils. The reasons are:

1. High-strength materials mean more slender sections for a given load capacity. As the section becomes more slender, the KL/d and KL/r values increase.†

2. For a given material, the use of more slender sections (rod, rail, etc.) has the same effect.

3. Exceptionally soft soils (class OH, for example) may not provide adequate lateral restraint even for piles of conventional size and materials.

For these several cases, the buckling length (KL) must be evaluated. Figures 6-3 to 6-5 provide the necessary data for this purpose.

* The use of such sections is not entirely academic. The use of rod and rail piles in Norway is reported by Bjerrum [9]. The L. B. Foster Company currently (1968) is pioneering in the use of lightweight rolled-steel sections for support of light structures.

The use of light pile sections for loads of 5 to 10 tons is a potentially economic device. Unfortunately, its use is prevented by many building codes, which stipulate use of minimum 8-in. sections.

† *Examples:*

1. Compare KL/r for two steel H piles embedded in soil having a k value of 50 pci. One pile is of A7 steel (12BP53 section is required). The other pile is of A242 steel (8BP36 section is required).

12BP53

$$\frac{KL}{d} = 15.5 \text{ (Fig. 6-3)}$$
$$KL = 15.5 \times 12 = 185 \text{ in.}$$
$$r = 2.86$$
$$\frac{KL}{r} = 64$$

8BP36

$$\frac{KL}{d} = 17.5 \text{ (Fig. 6-3)}$$
$$KL = 17.5 \times 8 = 140 \text{ in.}$$
$$r = 1.95$$
$$\frac{KL}{r} = 72$$

2. Compare KL/r for two concrete piles embedded in soil having a k value of 50 pci. One pile is a 16-in.-diam section using 3,000 psi concrete. The other uses 10,000 psi concrete and is 9-in.-diam.

16-in.-diam Pile ($E = 3 \times 10^6$ *psi*)

$$\frac{KL}{d} = 8 \text{ (Fig. 6-4)}$$
$$KL = 8 \times 16 = 128$$
$$\frac{KL}{r} = \frac{128}{0.3 \times 16} = 26$$

9-in.-diam Pile ($E = 10 \times 10^6$ *psi*)

$$\frac{KL}{d} = 12 \text{ (Sec. E, Chap. Note 1)}$$
$$KL = 12 \times 9 = 108$$
$$\frac{KL}{r} = \frac{108}{0.3 \times 9} = 40$$

b. Determining Equivalent Unbraced Length of Portion of Pile Projecting above Ground

This problem probably is just as difficult as that of estimating the unbraced length of the embedded portion of the pile, but is handled, in practice, by the simple convention of considering the pile to be fixed against rotation at some arbitrary level (generally, 5 ft in firm ground and 10 ft in soft ground) below grade. These figures are arbitrary and appear to be conservative.

With the point of fixity so defined, the value of KL can be determined as for columns in the superstructure.

c. Determining the Applied Load P (Distribution of Axial Load along the Length of the Pile)

For piles of constant section embedded in uniform soil, insofar as structural design of the pile is concerned, the applied load P is the maximum reaction on the pile at the butt.

For piles of varying cross section (tapered, partially reinforced, or composite type) or for piles partly embedded in firm and partly in soft soils, however, it is mandatory to check the stresses at several cross sections along the length of the pile. This requires that the variation in axial load along the length be determined. No satisfactory general solution to this problem has been derived. Rules of thumb are used, instead. These rules provide (1) a determination of the hypothetical level along the pile length at which the pile still supports 100 percent of the pile load and (2) a determination of the load carried at the level of the tip. It is customary to estimate values of load at levels intermediate between these extremes on the basis of straight line interpolation.

I. *Level at which the pile still carries 100 percent of the applied load.* There are at least four criteria which are in use for this determination.

1. A level *two-thirds* of the embedded length of the pile measured up from the tip (building code of city of Philadelphia)
2. A level *one-third* of the embedded length of the pile measured up from the tip (see page 281 of Ref. 25 and page 7-13-21 of Ref. 35, for example)
3. The *top of satisfactory bearing material* in which the piles are embedded (building code of the city of Boston)*
4. The *top of the bearing stratum* (building code of the city of Los Angeles)*

These different criteria generally give different results and need to be considered in some detail.

* Depending on how "satisfactory bearing material" and "bearing stratum" are defined, criteria 3 and 4 may be the same, or different.

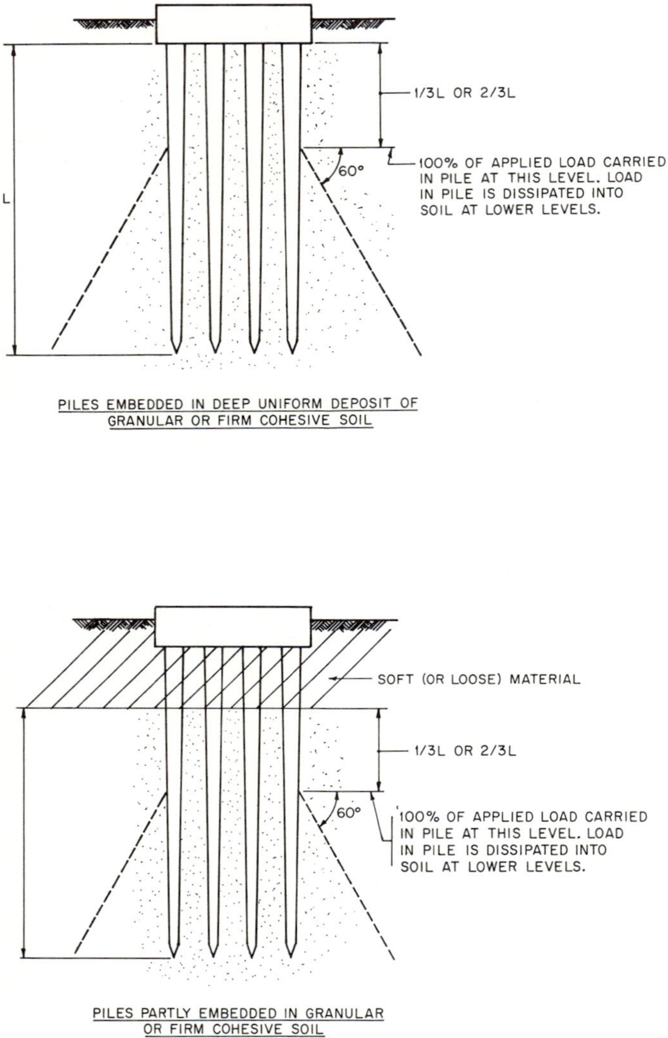

FIG. 6-6 Assumed Distribution of Load from Pile to Soil—Piles Embedded in Granular or Firm Cohesive Material.

Criteria 1 and 2 are widely used. Choice between them is a matter of individual judgment. However, they apply only to *relatively long* piles (length equal to two or more times the width of the group) of *uniform section* embedded in a *uniform deposit* of granular soil or in firm cohesive soils (see Fig. 6-6) and would not apply when the piles are embedded, wholly or partly, in strata which have not reached a condition of stability with respect to the weight of the overburden, such as is shown in Fig. 6-7. The reason for this limitation is illustrated by Fig. 6-7, which shows a tapered pile driven through 10 ft of newly placed fill,

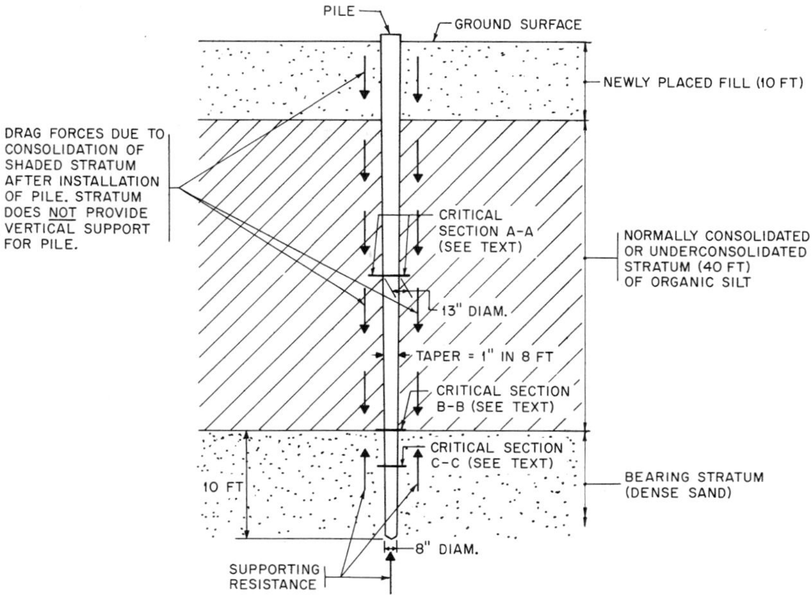

FIG. 6-7 A case in which the assumption of 100 percent of the pile load being supported at some arbitrary fraction of the embedded length above the tip (see Fig. 6-6) would not apply.

40 ft of a deposit of normally consolidated organic silt, and 10 ft farther into bearing in a dense sand. Were criterion 1 used, the full load on the pile would be assumed to be resisted at a level one-third of the pile length up from the tip or 20 ft below the ground surface, i.e., in the organic silt (critical section A-A). The load would be assumed to be dissipated into the soil below this level, and the lower sections of the pile would be assumed to carry lesser loads, perhaps with zero load assumed to be carried at the tip. Assuming an 8-in.-tip diam and a taper of 1 in. in 8 ft (a common value), the section available to resist this full load would have a diameter of 13 in.

As a matter of physical fact, the load actually would be redistributed down the pile because of consolidation of the silt under the weight of the newly placed fill until the full design load (and the drag forces, if they are sufficiently developed) is carried in the pile at the level of the top of the sand stratum (critical section B-B) or in the sand stratum (critical section C-C). The diameter of the pile at level B-B is about 9 in. The ratio of area at section A-A of 13-in. diam assumed resisting the pile load under criterion 1 and the area at section B-B actually resisting the load is 2.1:1.0. The overstress is 110 percent of the design stress. At level C-C, the overstress would be even greater.

Under criterion 2, a less severe, but nonetheless similar, condition of overstress would exist.

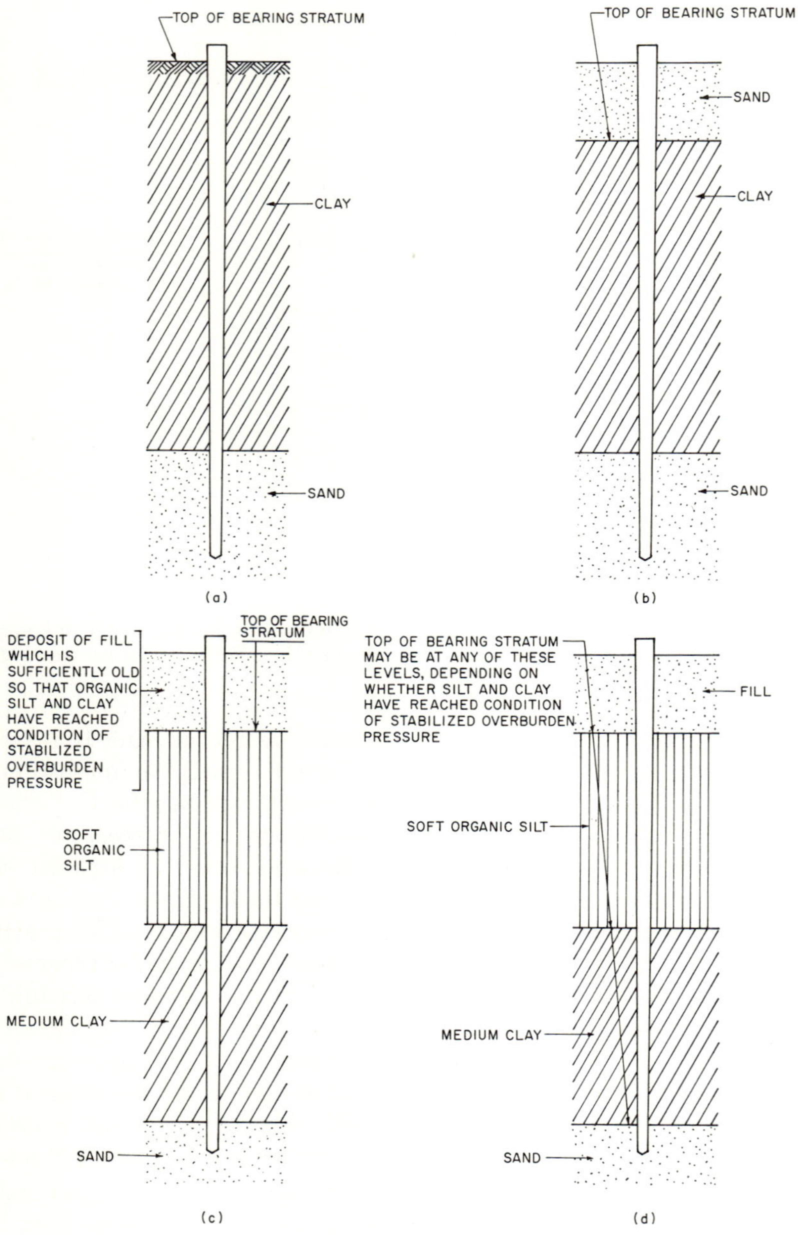

FIG. 6-8 Definition of a Bearing Stratum.

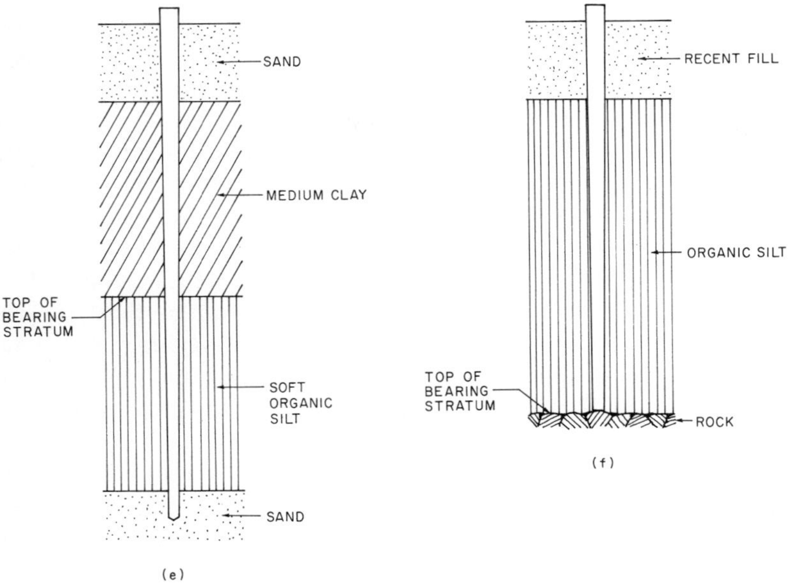

FIG. 6-8 *(Continued)*

The above limitation is commonly recognized. In fact criteria 1 and 2 were not originally intended to apply to the estimation of load in the pile section, but were intended for the less sensitive purpose of estimating soil pressures under the pile tips for purposes of settlement computations (see Ref. 25, for example). That they are widely used for estimating stress in the pile section simply is for lack of better criteria. Still, their use seems to work as the authors are not aware of any records of failure of tapered piles which are attributable to overstress in the pile. However, this lack of history of failure probably is due simply to the safety factor employed in the selection of allowable stress values. For concrete, it is in excess of 4 (based on ultimate strength). For steel, it is anywhere from 2 to 4 on the same basis.

Even when criteria 1 and 2 are used in the circumstances where they are intended to apply, the results tend to be very conservative.* Load tests (see Ref. 30, for example) consistently have shown that the primary

* As an example, consider a cast-in-place concrete pile, 40 ft long, embedded in a stratum of loose sand. Assuming a monotube section, type *F* shell, and f_c = 750 psi under criterion 1, the structural capacity would be 40 tons. For a Raymond step taper pile, the capacity would be 57 tons. Under the more realistic criterion that the load starts to distribute itself into the ground at the top of the bearing stratum, the distribution would begin at the butt, and the corresponding structural capacities would be 55 and 65 tons, increases of 37 and 15 percent, respectively, over the values for criterion 1.

load transfer, particularly for tapered piles, is in the extreme upper section of the bearing stratum and that, at working loads, relatively little of the load (even for ostensibly end-bearing piles) ever reaches the tip.

Because of these inconsistencies, the authors recommend the use of criteria 3 and 4 for estimating pile stresses, in preference to criteria 1 and 2, and presume to render criteria 3 and 4 equivalent to one another by defining the "bearing stratum" as *the lowest stratum penetrated by the pile which is not underlain by another stratum, also penetrated by the pile, of greater compressibility.* This definition is illustrated in Fig. 6-8 and intends, physically, to represent the bearing stratum as that one whose underlying strata will not settle out, leaving the overlying strata to hang up on, rather than support, the embedded pile, i.e., the uppermost stratum penetrated by the pile which can be relied upon to provide permanent and reliable support for the pile.

In practice, this definition is likely to be conservative, as it assumes that the pile is immovable with respect to the surrounding soil. Actually, some displacement of the pile occurs, and some resistance is developed along the full length, except in cases where a condition of stabilized overburden pressure has not been achieved so that an actual drag force is developed.

None of the above indicated criteria would apply when the pile installation utilizes a temporary drive casing, which is withdrawn. In such case, the development of resistance along the shaft may be impeded, or in the case where the concrete is forcibly extruded as the casing is withdrawn, it may be improved. Generally accepted rules for the estimation of load transfer between pile and soil for such piling have not been evolved, and resolution of the problem must be left to the reader's judgment. It is noted, however, that such piles generally are of constant section so that the problem of checking stresses at various sections need not be considered.

II. *Amount of Load Carried at the Tip.* Establishment of criteria for estimating the amount of load carried at the pile tip is a matter of even greater debate than the problem of establishing the level at which the pile carries full load. In theory, for a pure friction pile, the load at the tip is zero. For a purely end-bearing pile, it is the full load applied at the butt.

In practice, there seldom exists a pure friction or end-bearing pile, and some intermediate criteria need to be established. Some actual observations and sample expressions of opinion (and their source) are shown on Table 6-3. The reader is left to fit the individual case in hand to the cases described, as best he can.

d. Effect of Eccentricity of the Load Line (Moment in the Pile Section)

Virtually all piles (including those which, nominally, are considered to be axially loaded) are in fact subject to bending moment. The existence of moment is due to three basic causes: (1) distortion of the pile cap, (2) departure of the pile axis from a straight line, and (3) external shear forces.

I. *Deformation of the pile cap* occurs partly because the cap must span the column loads to the piles and, more importantly, because the piles in a group are likely to be of different length, dimension, or penetration resistance and so to have varying load-displacement characteristics. This deformation causes moment in the piles since they must bend to conform to the deformed configuration of the cap; i.e., the junction between pile and cap remains at a constant angle. In addition, pile location seldom is exact. This lack of exact positioning induces moment in the piling due to variable span conditions, just as in the superstructure.

The magnitude of these moments is not known. However, since the pile cap is rigid and the pile is flexible, the effect must be small and can be handled in a manner analogous to that used in the ACI code [4] for ultimate-strength design of columns. The procedure is to assume some minimum value of eccentricity, applicable to all sections. A value of 5 to 10 percent of the diameter (or side dimension) of the pile (corresponding to the ACI provision) is suggested. The lesser value would apply where pile installation is rigidly controlled and for less sensitive designs. The larger value would apply where less favorable conditions are contemplated. Specific value to be used is a matter for the reader's judgment.

II. *Distortion of the pile axis* is another source of bending in the pile section (see Fig. 6-43). Such distortion commonly is observed. It is caused by the presence of obstructions in the soil, variations in soil density, eccentric blow of the hammer, skewing the leads, or other factors which tend to cause the pile to walk off-line.

In cast-in-place, pipe, or caisson piles, the deviation of the pile axis can be seen by use of a droplight; it can be measured; and the stresses resulting therefrom can be calculated using instruments and methods of analysis as described in Refs. 15–17. If the stresses are excessive, the pile can be filled with high-strength concrete, reinforced, or the capacity reduced. This is a laborious matter, however, and so it is conventional to set certain limits on allowable distortions. If the distortions are within these limits, the stresses which result therefrom are neglected. If the limits are exceeded, the pile is rejected. If the cost of the piling is worth the delay and engineering involved, the alignment of the pile axis is measured, and the capacity is calculated as described in the above references. The usual limits on allowable distortion of the pile axis are specified as:

> Deviation of the pile axis from the vertical or from the specified batter shall not exceed 2 percent of the length of the pile. The alignment shall be such that some portion of the bottom of the pile can be seen by looking in from the top.

TABLE 6-3 ***Amount of Load Applied to Pile Butt Which Is Carried at Pile Tip According to Different Authorities***

Source	Type of pile and soil condition	Criteria or results
Building Code of City of New York (1963)	All piles more than 40 ft in length embedded in soil materials other than soft deposits such as peat, organic silt, or soft clays or loose silts and bearing on firm material such as rock, hardpan, or decomposed rock	75% of the load shall be assumed to be carried at the pile tip*
	All piles less than 40 ft in length embedded in soil materials other than soft deposits such as peat, organic silt, or soft clays or loose silts and bearing on firm material such as rock, hardpan, or decomposed rock	100% of the load shall be assumed to be carried at the pile tip*
Proposed building code of City of New York (1968)	Tapered piles embedded in granular soil materials	Not less than ⅓ of the full load on the pile to be assumed to act at the tip†
Sample computation in Ref. 31‡	Pile embedded in low strength clay soil	Negligible load carried at tip
Ref. 32‡	6-in. diameter pipe piles embedded 15 ft into stratum of soft, saturated clay	At failure, 10–15% of load carried at tip
Ref. 66‡	Vibro pile, 18-in. diameter, 28 ft long, driven in medium sand	Load at tip less than 30% of load applied at butt
	Same pile driven 37 ft into silt mixed with limestone and clay	Load at tip less than 30% of load applied at butt
Ref. 67‡	14BP73 sections driven through 50 ft of silt ($N = 20$) to 20–30 ft penetration into sand ($N = 50$)	Load at tip 30%, or less, of load at butt, except 50% at high load intensities
	Pipe piles driven through 50 ft of silt ($N = 20$) to 20–30 ft penetration into sand ($N = 50$)	Load at tip, 30%, or less, of load at butt
Ref. 30‡	14BP sections driven through 40 ft of sand ($N = 25$) to bearing on shale	About 50–70% of load resisted at tip
Ref. 68‡	Pipe step-taper pile, 10¾-in. tip, 17-in. butt, driven through 38 ft of sand and gravel ($N = 20–70$) to bearing in weathered shale	Less than 25% of load resisted at tip

FOOTNOTES TO TABLE 6-3

* The basis for these criteria is that for reasonably long piles (40 ft or more) embedded in firm or medium soils, even if the end bears on rock, some of the load (25% or more) is distributed to the soil above the rock. For shorter piles, the amount of elastic shortening is not enough to produce sufficient relative displacement between the pile shaft and the surrounding soil to develop the frictional or adhesive resistance of the surrounding soil.

† The basis of this provision is purely defensive. It is intended to prevent the use of freak piles having excessively slender tip sections. It has no theoretical basis.

‡ Caution: These cases are for single piles. In groups, a greater portion of the load at the butt would be transferred to the tip. The data indicated therefore should be used with a conservative factor of safety.

Of course, this procedure is only possible if the pile is inspectable; i.e., it is open for inspection.

For timber, steel H sections, precast concrete, and other noninspectable piles, similar distortion occurs, although the evidence of its existence is indirect (walking of the pile or inclination of the exposed portion of the shaft). Legendary accounts of H piles which turned around and popped out of the ground make stimulating conversation. Somewhat more reliable are the tales of "rocker systems" in which two piles formed a conjunctive (U) so that one rose while the other was driven down, and vice versa. In any event, for such piles, the amount of distortion of the pile axis (and its effect) cannot be evaluated quantitatively, and the stresses resulting therefrom are indeterminate. Therefore, there is a greater risk factor in the design of such piles than there is in the design of inspectable types. As would be done in similar circumstances in the design of the superstructure, it is suggested that the matter be handled by designing noninspectable piles for structural strength with a larger factor of safety than is used for the structural design of inspectable types. The ϕ' factors suggested in Chap. 8 provide for this effect.

The final source of eccentricity of the load line with respect to the pile axis is the presence of *lateral load* on the pile. This effect, in contrast to the effects of deformations of the pile cap and axis, is subject to rational analysis and, simply, is a matter of design. When lateral loads occur, batter piles should be provided, or if the lateral loads are small, the shear may be taken in the plumb piles (which are designed in accordance with Sec. 6C.1).

e. Effect of Residual Stresses in Pile Section

In addition to moment in the pile section, distortion of the pile axis during driving induces *residual stresses* in the pile materials. Expressions for determining these stresses are derived in Sec. 6E, Chap. Note 2. The expressions are complex, but the conclusions are relatively simple, namely, that residual stresses may be neglected for cast-in-place piling and, except for extreme cases, also may be neglected for piling made of material exhibiting a well-defined yield point (steel H and pipe sections) or when a vibratory hammer is used instead of an impact hammer.

For precast concrete or timber piles installed by use of an impact hammer, the effects of residual stress are more serious. Unfortunately, however, these stresses cannot be evaluated since they are due to offsets of the pile axis which cannot be measured. Therefore, again, they are a matter of calculated risk and may be considered by increasing the safety factor

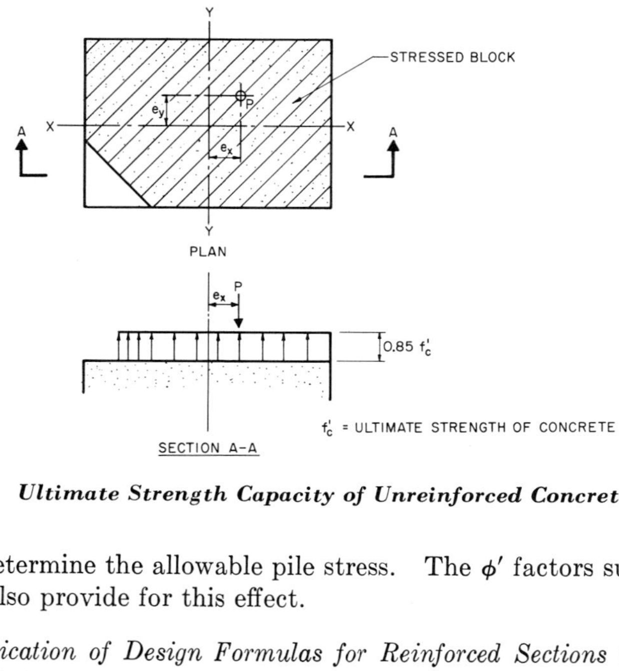

FIG. 6-9 Ultimate Strength Capacity of Unreinforced Concrete Section.

used to determine the allowable pile stress. The ϕ' factors suggested in Chap. 8 also provide for this effect.

f. Application of Design Formulas for Reinforced Sections to Unreinforced Sections

Strictly speaking, Eqs. (6-5) through (6-8) are applicable to the design of *reinforced* concrete columns only, and not to the design of unreinforced sections.

Precast piling generally is reinforced. Many cast-in-place piles are not, however, and in theory, Eqs. (6-5) through (6-8) would not be applicable to their design. Practice has been to neglect this discrepancy and to use Eqs. (6-5) and (6-6) (working stress design) for proportioning cast-in-place concrete pile sections. In such case, a value of p_g (the ratio of area of reinforcement to area of concrete section) equal to zero is used.

Practice with regard to the application of ultimate-strength design techniques to unreinforced sections is somewhat more nebulous since ultimate strength design of concrete columns includes, a priori, the assumption of eccentricity of the applied load, with the capacity of a given section calculated by proportioning between the capacities of an idealized axially loaded section (P_0) and a section containing balanced reinforcement (P_b). A P_b value does not exist for an unreinforced section. A possible approach would be to calculate the ultimate load capacity on the basis of a rectangular stress block whose centroid is under the line of action of the applied load (see Fig. 6-9).

g. Evaluating the Factor $C_m/(1 - f_a/F'_e)$

This factor reflects the additional distortion of the pile axis due to the presence of moment at the ends of the unbraced section and/or lateral forces between the brace points.

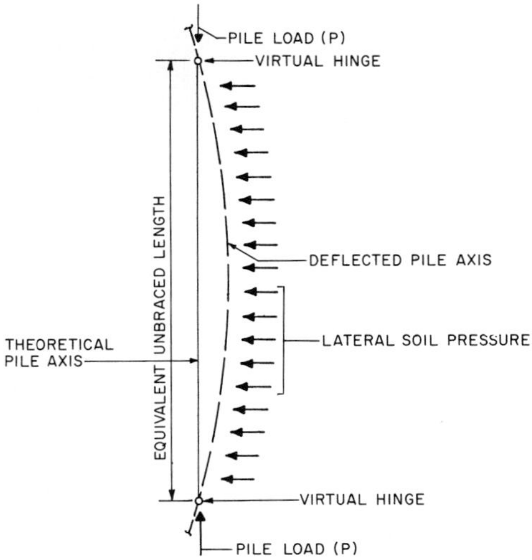

FIG. 6-10 Equivalent Unbraced Length for Pile. For situation indicated, factor $C_m/(1 - f_a/F'_e)$ may be taken as $+1.0$.

The equivalent unbraced pile length may be represented as in Fig. 6-10. For this case, the value of C_m is $+1.0$, and for short unbraced lengths, the fraction f_a/F'_e is small. Therefore, the factor $[C_m/(1 - f_a/F'_e)]$ may be taken as unity.

h. Evaluation and Effects of Damage and Deterioration

I. *Damage*

A. *Damage Due to Driving.* There have been many well-documented cases of serious damage (including fracture) to piles due to the stresses incident to driving. The problem of distortion (or fracture) due to driv-

FIG. 6-11 Crushing of Tips of Timber Piles Due to Overdriving. *(Courtesy Raymond Concrete Pile Company.)*

FIG. 6-12 Damage to Butt of Timber Pile Due to Hard Driving. *(Courtesy Raymond Concrete Pile Company.)*

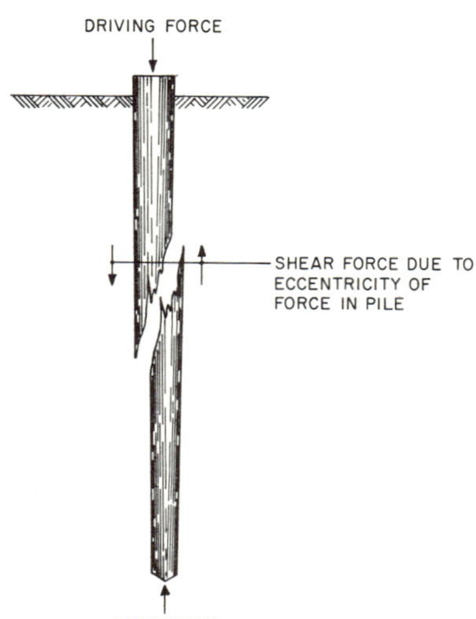

FIG. 6-13 Diagrammatic Representation of Broken Timber Pile.

FIG. 6-14 ***Damage to Steel H Pile Due to Overdriving.*** *(Courtesy Raymond Concrete Pile Company.)*

ing past obstructions has been noted in the previous section. Some other sources of difficulty and some examples are:

(1) *Timber Piles.* Figure 6-11 shows the condition of the tips of timber piles which were overdriven. Figure 6-12 shows the condition developed at the butt. Diagrammatic representation of a broken timber pile is shown in Fig. 6-13. Damage occurs from local crushing or fracture under hard driving, particularly when the piles are driven to sudden refusal (rock or hardpan) or when it is attempted to pierce a hard layer overlying a softer one. Obstructions, rock or rubble fills, or any other source of an abrupt increase in driving resistance can cause difficulty.

Steel shoes and banding of the piles are used to prevent damage due to these causes. Banding seems to be very helpful. Shoes do not seem to perform well, perhaps because the section just above the shoe is weak or, as some have reported, because the shoe cuts into and weakens the pile.

(2) *H Piles.* Figures 6-14 and 6-15 show the condition of the ends of steel H piles subject to hard driv-

FIG. 6-15 ***Damage to Steel H Pile Due to Overdriving.*** *(Courtesy Raymond Concrete Pile Company.)*

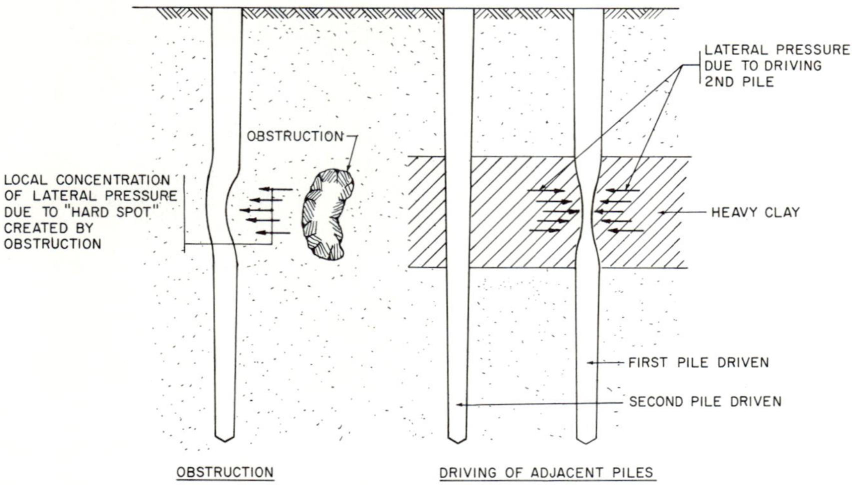

(a) CAUSE OF NECKED-DOWN OR BROKEN CAST-IN-PLACE CONCRETE PILES

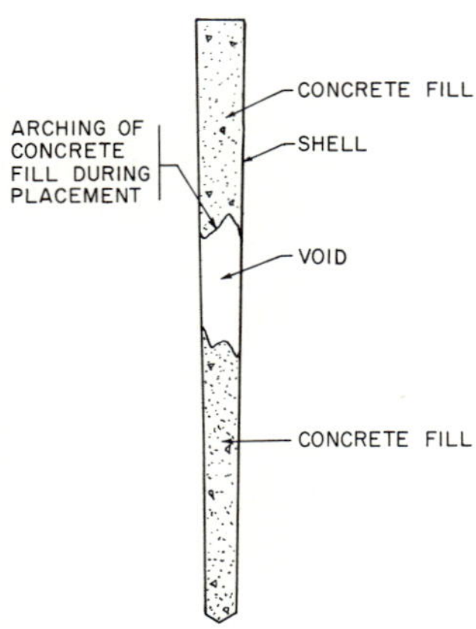

(b) CAUSE OF VOIDS IN CAST-IN-PLACE CONCRETE PILES

FIG. 6-16 Defects in Cast-in-place Concrete Piles.

ing (see also Ref. 20). Again the difficulty arises when there is hard driving (on rock or through obstructions or through rock or rubble fills).

(3) *Concrete Piles.* Concrete piles also are subject to damage due to driving. Precast sections are cracked and, sometimes, are fractured. Cast-in-place piles, particularly when they are cast without a shell or

when the shell is withdrawn, may be broken or necked down (Fig. 6-16*a*) by the lateral pressures incident to driving of adjacent piles, or seams of soil may be intruded into the concrete section [19, 21]. The continuity of the shaft also may be broken when pulling the shell.

Even when a permanent shell is used, the shell may be collapsed by the pressures incident to driving of adjacent piles, or if the concrete is still "green" when the adjacent piles are driven, its strength may be impaired by the vibrations incident to such driving.

When an open-end drive casing or an augered hole without casing is used, "blowing in" of the bottom of the hole can be a major problem, and it may be necessary to dewater or to use a bentonite slurry to keep the hole open until the concrete is placed.

B. *Other Defects*

(1) Cases of *voids* in cast-in-place concrete piles have been reported [18] (see Fig. 6-16*b*). Such voids are due to arching of the concrete fill as it is dropped down the narrow pile.

(2) *Splits and checks* often are developed in timber piles, particularly long piles, due to handling either from the truck to the stockpile or from the stockpile to the rig, more so if the handling is a little careless. The occurrence often is evidenced by a loud "pop" as the pile is hoisted. If a treated pile is being used and the split or check penetrates the treated shell (which is the usual case, as the treated shell likely is only an inch or two deep, and with some species, less), the effect is largely to invalidate the treatment by exposing the untreated core of the pile to attack by fungi or insects.

C. *Treatment of the Problem of Damage.* The problem of damage during installation is best handled by prevention. Types of pile which are prone to damage by hard driving or by the presence of obstructions should not be used when hard driving or obstructions are anticipated (alternatively, preboring or the use of a vibratory instead of an impact hammer may be considered), and they should be installed with loving care under the supervision of someone who knows what he is doing—not just battered into the ground while someone stands by and counts blows.

Sometimes it is necessary to use piling under conditions where a reasonably damage-free installation cannot be assured. For example, sometimes hard driving is encountered, unexpectedly, with timber or H sections already on the job. Then it may be economically advantageous to accept the probability of some damage and to reduce the allowable stress (increase the safety factor) as a compensating device. Unfortunately, with such strategy, the good piles are penalized for what *might* happen to the poor ones, but sometimes the overall economy still benefits.

In a similar vein, the use of timber, H piles, or precast piles at relatively low stress values may be economically attractive as compared with the alternative of reducing the risk factor by using pipe or other inspectable types. Of course, the risk factor in either situation must not be so great that the damage done to the piles overrides the increased safety factor and causes failure. The problem, though, is that it is exceedingly difficult to foretell what the risk factor might be and equally difficult to control the damage which occurs. Therefore, the wisdom is questionable in adopting the approach of increasing the safety factor by some arbitrary amount to cover possible contingent damage, rather than selecting a type of pile which is inspectable so that damage, if any, may be detected, or avoiding damage by selecting a type of pile suitable for site conditions.

II. *Deterioration.* After the pile has been installed, nature begins its slow but relentless attack on the material. The attack of fungi (rot) and insects on timber piles has been noted. Steel piles are subject to corrosion. All materials (timber, steel, and concrete) are subject to attack by chemicals in the soil, either chemicals generated within the soil or elements intruded from the surface.

The problem of causes and prevention of deterioration of structural materials is complex. For the interested reader, the matter is discussed in detail in Ref. 22. However, the best solution to the problem again is prevention, which means either using a type of pile which is not susceptible to the types of deterioration anticipated at the site or protecting against aggressive action by the use of coatings (concrete or epoxy, for example) or by preservative treatment (as for timber piles).

As with the problem of preventing damage during installation, this is not always economically feasible (although it probably always is physically possible), and again the alternate approach of increasing the safety factor may be used. The provision of sacrificial metal on steel piles is an example of this alternate approach. However, the problem of tailoring the factor of safety (allowable stresses) to the risk factor again is a baffling one. In the example cited, how much sacrificial metal should be provided? Should it be ⅛ in. (a common figure), or more, or less? If a bona fide problem of corrosion exists, is any reasonable amount of sacrificial metal likely to be adequate? Worse, since deterioration is a progressive phenomenon, the risk factor is a function of the service life of the structure. What service life should be assumed? What happens if the owner (present or future) of the building is unaware that he owns a "limited life" structure? Except in unusual circumstances, can a limited-life design be justified?

6A.4 Summary and Conclusions

It will be apparent from the text of the previous sections that the problem of selecting allowable stress values for piling is a complex one. As a matter of practicality, however, two relatively simple approaches are used. One is to assign fixed values of allowable stress to pile material. The other is to consider the pile as a short column.

Of the two, in modern practice, the use of fixed values of allowable stress is less favored because (1) it is inflexible and (2) it is ultraconservative. The reason for the latter simply is that the assigned values necessarily must allow for the many unknowns involved in the design, and unless complex rules and stress tables are established, designs involving few unknowns necessarily must be treated (and penalized) the same as less favorable designs in which many uncertainties are involved.

On the other hand, treating the pile as a short-column section, using the same values of allowable stress as are prescribed for short columns in the superstructure, results in the piles being underdesigned by comparison. For example, compare the concrete in a pile with that in a conventional building column. In the latter case, accuracy of straightness and alignment is assured within relatively narrow limits; a minimum percentage of longitudinal and tie or spiral reinforcement is specified to assure confinement of the concrete; forms are stripped and the concrete is inspected; and the concrete is or can be subject to maintenance inspection. In the case of the pile, the alignment and straightness are much less stringently controllable. In many cases there is no steel; the concrete may have been dumped 100 ft down a 10- or 12-in. tube, perhaps into a few inches of water at the bottom; maintenance inspection is impossible; the environment tends to be more aggressive; the member is subject to a highly indeterminate loading condition; and, except for precast sections, the concrete in place cannot be inspected, and placing conditions are more difficult. The allowable stress values for the concrete in these two cases should be different, with the pile being more conservatively proportioned.

This same conclusion applies to the case of timber or steel in pile sections versus the same material in the columns of the superstructure.

With the one approach (fixed values of allowable stress) being overly conservative, and the other (short column), by comparison, involving a degree of underdesign, a compromise solution is required. The authors propose the following.

6A.5 Proposed Basis for Selecting Allowable Stress Values for Design of Piling

a. Discussion

The concept of the pile as a column (with exception for tension piles and as previously noted) is valid and rational, provides a flexible and desirable basis for design, and is proposed for use.

As described in the previous sections, however, the conventional (and simple) short-column approach of the structural engineer is inapplicable;

a fuller, more rational application of column theory is required. The problems of applying this theory to the case of the pile column have been discussed.

The following procedure is suggested as a means for handling these complexities:

STEP 1: Select a type of pile and/or method of installation to suit conditions at the site and permit installation with reasonable assurance of *avoiding damage* to the pile. For example, if hard driving or obstructions are anticipated, do not use timber piles. Use an inspectable type, shelled, cast-in-place, or pipe pile, or use a vibratory hammer, where applicable. If a steel pile is to be used, select a heavy, stiff H section (or the vibratory hammer).* Similarly, if serious conditions of potential *deterioration* exist or are anticipated at the site, select a type of pile (or a coating) which will prevent such deterioration.

STEP 2: Determine the axial load acting on the pile.

STEP 3: Determine the lateral load acting on each pile. Assume the pile to be fixed at a point 5 ft below the pile cap in firm soils and 10 ft below the cap in loose soils. Calculate the moments consistent with this force and lever arm, and convert to an equivalent eccentricity of the load.

STEP 4: If the equivalent eccentricity is greater than or equal to one-tenth the diameter (or side dimension) of the pile, determine the corresponding moment. Otherwise use a minimum eccentricity of the load line equal to 5 to 10 percent (as previously discussed) of the diameter (or side dimension) of the pile as a representation of the *incidental moments* present in all piling. These moments are due to distortion of the pile cap and incidental distortion of the pile axis (not exceeding usual tolerances).

STEP 5: Select an appropriate special capacity-reduction factor (ϕ')† to reflect that *uncertainties and inaccuracies in construction* of piling are greater than those in construction of columns in the superstructure, that the axis of the pile unknowingly may be distorted beyond the incidental amount of usual tolerance,‡ and that the

* Other necessary or desirable precautions are discussed in Chap. 8.

† The notation ϕ' is intended to differentiate this factor from the ϕ factor used to calculate the design load (see Chap. 5).

‡ With regard to the matter of distortion of the pile axis, two cases apply: (1) For inspectable piles, no correction need be made. Any deviation from the specified tolerances in alignment (page 219) should be noted after the pile is installed and necessary adjustments made in the field by adding reinforcement, using higher-strength concrete, providing additional piles or other device, as applicable. If the alignment is within the specified tolerances, the minimum eccentricity values (Step 4)

type and magnitude of these unknown factors and distortions differ for different types of pile. Suggested values for the ϕ' factor are given in Chap. 8.

STEP 6: Calculate the critical design load and moment for which the pile section is to be proportioned. For working stress design use the P and M values from Steps 2 and 4, reduced for live-load reduction factor and for the probability factor representing the coincidence of wind or earthquake, and multiply by the ϕ' factor. For ultimate strength, multiply the reduced loads and moments by the load factors of Sec. 5A.2 and then by the ϕ' factor.*

STEP 7: Select a trial section.

STEP 8: Determine the value of KL/r, applicable to the *buckling* problem for the trial section. As noted, in most cases a value of zero may be used; i.e. the buckling problem may be neglected. Under the conditions in which failure by instability may not be neglected (as described on page 212), determine KL/r from Figs. 6-3 to 6-5.

STEP 9: Using the applicable KL/r value from Step 8 and the moment and load from Step 6, calculate the allowable pile load using Eqs. (6-1) to (6-10) and the allowable (or ultimate) stress values for the material in question, as for a conventional column.

STEP 10: Compare the allowable loads as finally determined (Step 9) with the design loads (Step 6). In the case of piles of variable sec-

---

adequately provide for the moments which occur. (2) For noninspectable piles, however, in all cases the calculated design capacity should be decreased to reflect the greater uncertainty of the installation. The amount of reduction depends on the sensitivity of the building, the type of soil, and the degree and competence of the inspection provided. The special ϕ' factors indicated in Chap. 8 include an allowance of 10 to 20 percent to compensate for this consideration.

The values of 10 to 20 percent derive from the equations given in Sec. 6E, Chap. Note 2. In this derivation, two classes of pile were considered: (1) those having a sweep which did not exceed the usual tolerance of 2 percent of the pile length and (2) those having sharper curvature.

The results indicate that if the sweep does not exceed the usual 2 percent tolerance, unless the pile is kinked, the eccentricity of the load line from the axis of the pile [even under conditions of extremely soft soil ($k = 10$ pci), and assuming entirely end-bearing resistance] does not exceed 10 percent of the kern distance. Accordingly, the capacity of the pile is about 90 percent of the capacity of the same member if considered as purely axially loaded.

For the sharply bent section, analysis was made of a pile having a sweep of 7.8 ft, or 8.6 percent of its length. The capacity was about 80 percent of that if axially loaded. Another extreme case is that noted in Ref. 15, fig. 8 (similar offset, but with a sharp kink in the pile axis). The reduction in capacity was 25 percent.

* The reader should note that, in general, the use of ultimate strength design procedures will provide economy as compared with the proportioning of the pile section under working stress criteria.

tion, check the capacity at several levels using the criteria on pages 213 through 218 to estimate the magnitude of the load carried in the pile section at each level.

Section 6B CAPACITY OF THE SOIL TO SUPPORT THE PILE

This problem is similar to that of determining the capacity of the soil to support a footing or pier, as described in Chap. 5. The same two factors must be considered: (1) safety against shear failure in the soil and (2) limitations on settlement. The pile capacity to be used for design is the lesser of the values dictated by the two requirements, divided by a suitable factor of safety.

6B.1 Safety against Shear Failure in the Soil

Calculation of the bearing capacity of a footing, as limited by considerations of potential shear failure in the soil, is a relatively straightforward proposition. The main problems are to determine the soil strength parameters ϕ and c and to decide which set of bearing-capacity factors to use (Terzaghi, Meyerhof, etc.). However, the bearing-capacity factors as proposed by the different authorities do not differ greatly, and the laboratory procedures for determining ϕ and c values are described in Chap. 5. Therefore the indicated procedures can be used for footings with reasonable assurance and the results are duplicable, within limits.

Calculation of the bearing capacity of piling (also as limited by considerations of potential shear failure in the soil) is nowhere near so simple. The primary reason is that no satisfactory means has been developed for determining the stresses in the soil which are imposed by the load as it is carried down the pile. Since allowable loads are established in design by the comparison of actual to allowable stress, it is apparent that allowable loads cannot be reliably determined if the stresses cannot be determined.

A second point is that the strength parameters of the soil are poorly defined. Much work has been done to establish strength criteria for pure soils such as clean sands and saturated clays, but the properties of the general run of soils, which are mixtures of different fractions, have not been extensively studied.

Still, pile foundations are designed and used, literally, by the hundreds of thousands, and usually with satisfactory results. If the stresses in, and the properties of, the soil are not known, how can this be done? The answer is that the designer assembles all the indicators of soil capacity (theoretical and empirical) available to him, evaluates each, and selects from among the differing results that value of capacity which his judgment suggests to be reasonable. There are four such indicators:

1. Computations based on *theory* (to the extent that theory has been developed); these are discussed in Sec. *a*
2. The *equivalent-footing analogy*, which is an empirical equivalent, and is discussed in Sec. *b*
3. *Penetration resistance* (Sec. *c* and Chap. 7)
4. *Load tests* (Sec. *d* and Chap. 9)

In practice, for a given case, each of the above four indicators will yield a different result (say, 30, 25, 28, and 62 tons).* The problem for the designer is to decide on which of these is the correct value or which value to interpolate. This is a matter of judgment, which must be based on a clear understanding of what limitations apply to each indicator. The following sections (and Chaps. 7 and 9) provide detailed discussion of each indicator and of the limitations and relative reliability of each.

a. Theoretical Determination

Under the application of load a pile is displaced with respect to the surrounding soil. For the usual case of axial downward load, the soil at and under the pile tip is compressed and reacts against the lower section of the pile (end bearing), and the shaft of the pile tends to drag down and compress the soil in which it is embedded (friction). The pile capacity, conventionally, is assumed to be the sum of the two resistances (end bearing and friction).

I. *Calculation.* The above described action is analogous to that of a deep pier foundation, as described in Chap. 5. In theory, the resistances of end bearing and friction and the total load capacity may be calculated from Eq. (5-4):

$$Q_{\text{ult}} = \overbrace{[\text{Eq. (5-1), (5-2), or (5-3)}] \times \text{area of base}}^{\text{end bearing}} + \overbrace{\Sigma(C_A + P_H \tan \delta)\, dA}^{\text{friction and adhesion}} \qquad (5\text{-}4)$$

The first term on the right side of Eq. (5-4) may be evaluated using the bearing-capacity factors indicated in Fig. 5-1, except that in loose to medium sandy soils, the term $\Sigma\gamma_d D_f N_q$ should not be taken as greater than that corresponding to a depth equal to about 10 times the pile diameter and about 20 times the pile diameter for dense soils.† The reason is that Eqs. (5-1) to (5-3) are derived for a shallow footing, wherein

* The indicated range is *not* exaggerated.

† The indicated values are derived from data by Vesic [56]. The actual relation is not so simple as indicated. The indicated values are intended to represent a simple and reasonable approximation of the more complex relation. For details see Refs. 56 and 57.

failure takes place along a failure surface extending up to the ground surface. The soil around the pile tip, however, is subject to a decrease in volume due to the stresses incident to the pile reaction; it becomes a local soft spot in the soil continuum, arching action develops over the soft spot, and so the vertical stresses acting on the horizontal plane adjacent to the bearing area have an upper limiting value.

In clay soils there also is an upper limit value, this time to the total end-bearing resistance. This value is about nine times the base area of the pile times the cohesion value of the soil ($q_u/2$).

Evaluation of the second term on the right side of Eq. (5-4) requires estimation of values for C_A and of $P_H \tan \delta$:

1. Approximate values of C_A may be taken from Table 4-2.* In clay soils, some designers derive the side shear resistance due to adhesion from shear-test data (see Chap. Note 4).

2. Values of $P_H \tan \delta$ may be approximated as described in Sec. 5A.1*b*, using $P_H = P_v K_n$ with K_n between, perhaps, 0.5 and 1.0,† and taking $\tan \delta$ from Tables 5-3 and 5-4. Alternatively, values of average total friction between pile and soil, as given in Table 6-4* may be used. As with the confining pressure contribution to end-bearing capacity in sand soils, there would be an upper limit to the normal pressure acting around the pile, again due to the arching effect. The limits appear to correspond, roughly, to those indicated for end-bearing considerations (i.e., the assumed hydrostatic pressure should not be taken greater than that corresponding to a depth equal to $10d$ for piles in loose sand and $20d$ for piles in dense material) [56].

This Eq. (5-4) represents the capacity of a single pile. It does *not* represent the capacity of a group of piles, a capacity which usually is less than the sum of the capacities of the individual piles constituting the group (see Refs. 23 to 26).‡ Therefore, the above derived results must be corrected for the group effect. Usually this is done by multiplying by one of the group reduction factors described in Sec. 6E, Chap. Note 3.

* The values in Tables 4-2 and 6-4 represent average values of resistance recorded in load tests (generally carried to failure).

† From Tschebotarioff [24]. Baker [28] also indicates a range of 0.4 to 0.7 (or a little greater) for straight-sided piles embedded in most granular soils, increased to perhaps 1.5 for tapered piles to reflect the wedging effect of the taper. Yang [65] suggests a range 0.5 to 0.7. Some investigators [43, 60] have indicated even higher values, up to 3.0.

‡ The reason, as indicated in Fig. 6-17, is that the extent of the soil under the pile tips stressed by a group of piles is different from that stressed by a single pile so that the stress intensity in the soil is different. Also, the type and behavior of the soil within the two different-sized bulbs of pressure may be different.

II. *Limitations of the Theoretical Approach.* If the various components of resistance (end bearing, and friction and/or adhesion) and the group reduction factor could be accurately evaluated, Eq. (5-4) would be an excellent way to determine the length of pile required to support a given

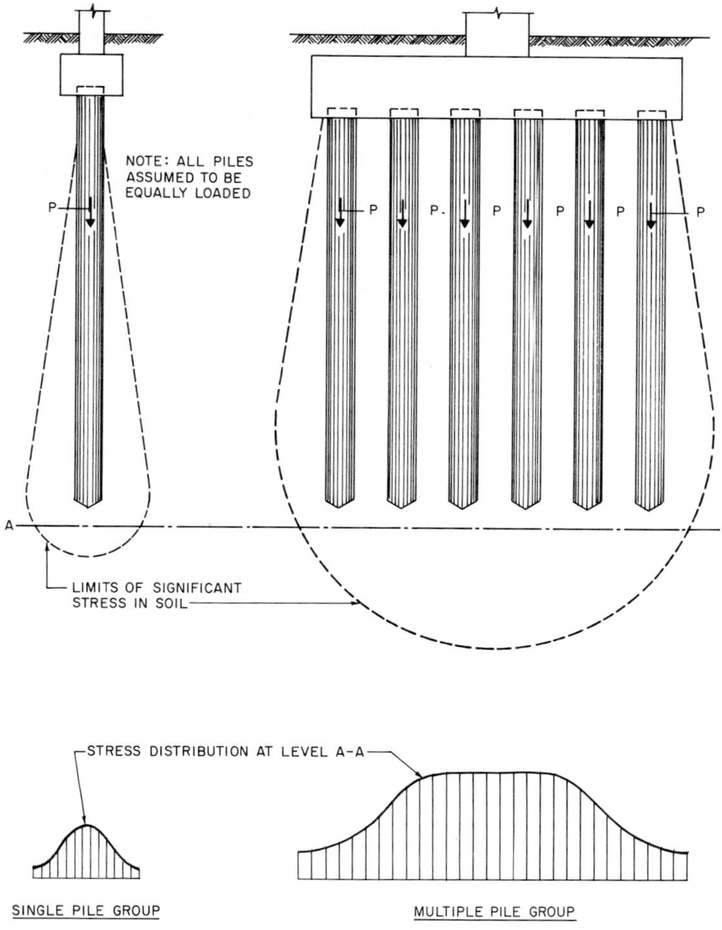

FIG. 6-17 Difference in Action of Single- and Multiple-pile Groups.

load, or the load which could be supported by a given type, size, and length of pile.

There are two enormous problems, however.

A. First, Eq. (5-4) gives the capacity of the soil to support load, but does not describe the settlements which will occur under that load. As a practical matter, the settlement consideration almost always controls for

TABLE 6-4 ***Friction between Pile and Soil****

Material	*Ordinary range of values, psf of bounding area of pile*†
Fine-grained soils:	
Very soft clay	125 ± 125
Mud	250 ± 200
Silt	300 ± 200
Soft clay	400 ± 200
Silty clay	600 ± 200
Sandy clay	600 ± 200
Medium clay	700 ± 200
Sandy silt	800 ± 200
Firm clay	900 ± 200
Dense silty clay	1,200 ± 300
Stiff clay	1,500 ± 400
Very stiff clay	3,000 ± 1,000
Hard clay	Over 4,000
Coarse-grained soils:	
Very loose sand and silt or clay	100 ± 50
Medium sand and silt or clay	500 ± 100
Dense sand and silt or clay	700 ± 100
Very dense sand and silt or clay	900 ± 100
Medium sand	1,200 ± 500
Sand and gravel	2,000 ± 1,000
Gravel	2,500 ± 1,000

* From Chellis [52], by permission.

† The ± figures indicate a range governed by the character of the soil. Not all soils falling in the same general category have equal properties.

buildings, and the capacity as limited by potential shear failure in the soil is of academic interest only.‡ The exceptions generally are limited to cases of friction piles in soft cohesive soils.

B. Second, it is exceedingly difficult to make an accurate evaluation of

‡ By way of illustration, consider a construction employing short (30 ft) piles, installed in a loose granular soil ($N = 15$) and supporting a building have a 20-ft column spacing. Using the 60° rule mentioned on page 240, the loaded area at the level of the tips of the piles has a diameter of 20 ft measured from the perimeter of each pile group. Substantially, then, the column loads are distributed to the soil uniformly over the entire plan area of the building. Assuming dry soil at 100 pcf and a minimum plan dimension of 40 ft for the building, then from Eq. (5-4), the ultimate bearing capacity is about 50 tsf which corresponds to a building about 500 stories high. The limitation on applied load, as governed by consideration of settlement, would be about 1.5 tsf basic, plus 1.5 tsf increase for embedment, plus 1.5 tsf increase for stabilized overburden pressure, or a total of about 4.5 tsf gross pressure, or only 9 percent of the load as limited by shear capacity.

the components of friction, etc. The reasons* are as follows:

(1) The *friction angle* and the *adhesive resistance* along the shaft vary with the type of pile, the pile material, the soil material, and the roughness and configuration of the pile surface.

For example, there are at least three common types of shelled, cast-in-place, concrete piles, including those having smooth, corrugated, and fluted walls. When driven into a given soil stratum, the amount of friction between pile and soil would be different for each type and would depend on the ratio of depth of corrugation or flute to pile diameter. The resistance for such a pile, in turn, would be different for the same pile installed without a shell or with a shell that is withdrawn.

Similarly, for timber piles, the adhesion achieved would depend on many factors, including such simple things as whether the pile was treated or untreated. (An untreated pile has a greater tendency to absorb moisture from the adjacent soil and so to develop a better adherence, at least in clay or silt soils.)

(2) The *coefficient of lateral earth pressure* also is poorly known. It varies with the size of the pile, the taper, the type and density of the soil, the amount of driving, the hammer, and the stress history of the soil. Values of 0.4 to 3.0 have been derived (see page 234). This range is beyond the limits of reasonable interpolation.

(3) *Development* of frictional resistance along the shaft of the pile requires displacement of the shaft relative to the surrounding soil. As indicated in Sec. 5A.1*b*, sometimes the displacement which is tolerable at design loads is sufficient to fully develop this resistance. More often it is not. In the general case, a condition of partial development is attained. Also the resistances of friction, adhesion, and end bearing generally are not developed simultaneously to the same degree. For example, consider a 10-in.-diam pipe pile, 20 ft long, driven closed-end through stiff clay (q_u = 4 tsf) to refusal on rock. In theory, the frictional resistance would be $\pi \times 0.83$ ft $\times$ 20 ft $\times$ 4,000 psf = 210,000 lb = 105 tons. Actually, since the base of the pile essentially is immovable and the elastic shortening (depending on the design stress) is only about $1/10$ in., there is virtually no relative displacement of the pile with respect to the surrounding soil; hence, virtually no strain or stress in the surrounding soil and no frictional resistance is developed.

(4) The frictional or end-bearing resistance tends to *dissipate with time* due to the effects of consolidation, vibration, temporary excess stresses in the soil induced by driving, and other causes.

(5) There are additional complications in that insertion of the piles generally *alters the character of the soil*. It may weaken the soil (as in some

* These reasons are essentially the same as those discussed in Chap. 5 with regard to the difficulties of estimating the capacity of deep pier foundations (see page 107).

clays) or strengthen it (as in loose sands). This tendency is cumulative as a group of piles is installed.

(6) The *taper of the pile*, if any, also is an important factor about which relatively little is known. There is little doubt that taper increases capacity in granular soil materials, perhaps 100 to 200 percent or more, but exactly how much is a matter of opinion.

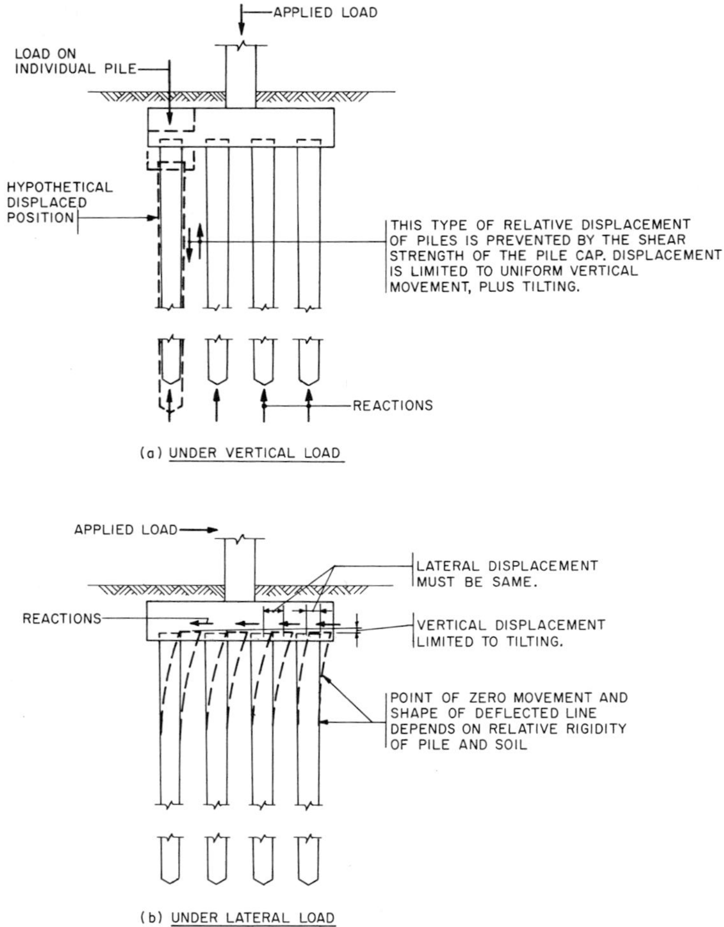

FIG. 6-18 Action of a Pile Group Due to Bridging Effect of Rigid Cap.

As a result of these uncertainties, the development of a rational, quantitative analysis of the capacity of the soil to support the pile without failure in shear (even if the distribution of the load were known) is far in the future. Equation (5-4) provides an indication only. The indication is not especially reliable except in certain limiting cases where the pile is clearly supported by end bearing, in which case capacity can be fairly well estimated from the first term of the equation.

b. Equivalent-footing Analogy

Figure 6-18 shows a group of piles, as conventionally installed. The piles are embedded in the ground, are loaded by a column or wall reaction, and are capped by a rigid cap which is designed to distribute the column or wall reaction to the several piles constituting the group.

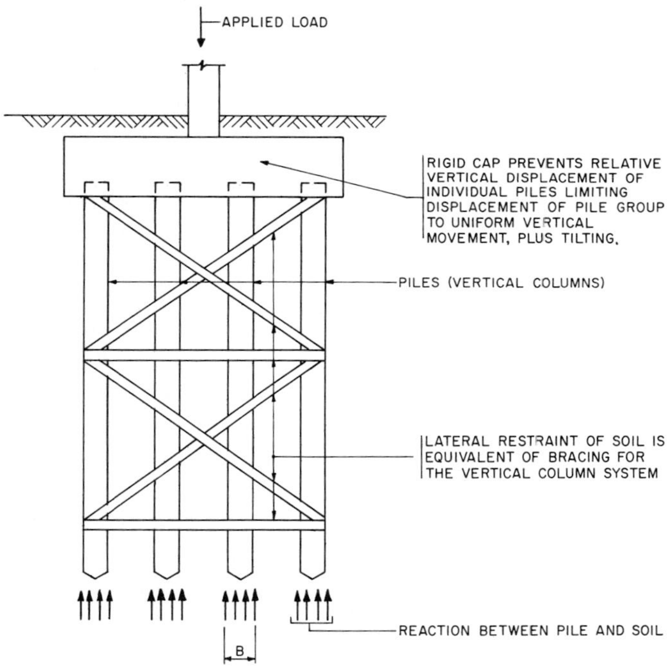

FIG. 6-19 Analogous Rigid Body.

So long as the cap does not fail in shear or flexure, relative axial displacement of the several piles in the group is prevented (Fig. 6-18*a*), and displacement of the pile group is limited to uniform vertical movement, plus tilting of the group, as a rigid body.

Under lateral load (Fig. 6-18*b*), the situation is the same. Again the presence of the rigid cap and the restraining effect of the soil confining the piles limit movement of the pile group to tilting, plus flexure, as indicated.

These actions correspond to those of a rigid body and the pile-soil continuum can be represented by an analogous structure, as shown in Fig. 6-19.

This structure is a familiar one. In it, the piles act as the vertical load-carrying members (the columns). The soil in which the piles are embedded provides lateral restraint (the bracing), and the soil below the pile tips functions as the foundation for the structural system above. Adequacy of the structure involves consideration of the adequacy of each component, the columns, the braces, and the foundation.

The columns (piles) will be adequate if the requirements of Sec. 6A are met. The braces (soil around the piles) will be adequate if buckling is prevented. This too will be accomplished if the requirements of Sec. 6A are met.

The capacity of the soil underlying the pile tips to function as a foundation for the pile system above is the only remaining consideration. It is evaluated by treating the soil at the base of each pile as a footing* for the pile.

The size of the footing, i.e., the lateral extent of the soil which reacts against the base of the pile (dimension B in Fig. 6-19), depends on the size and length of the pile, the type of soil, and the amount and distribution of the friction developed between the sides of the pile and the soil in which it is embedded. For a so-called end-bearing pile, little, or no, friction is developed, and the soil reaction below the pile is concentrated in a narrow area directly under the pile tip. For so-called friction piles, the reaction is spread over a wider area. Where the spread of the load is large, the analogous isolated footings for the individual pile-columns may overlap forming combined or mat footings.

The bearing capacity of footings, whether they be isolated, combined, or mat, is discussed in Chap. 5, and analysis of the capacity of the soil to support a given group of piling may be carried out in exactly the same manner as described there, provided that the size of the analogous footing can be determined.

Procedures for estimating this size for several different cases are as follows:†

I. *Friction Piles (Piles Which Receive Their Primary Support through Resistance along the Shaft)*

A. *Friction Piles in Granular Soils and in Firm or Stiff Cohesive Soils.* For friction piles installed in granular and in firm or stiff cohesive soils, the 60° rule applies.

(1) *Description.* The 60° rule for estimating the distribution of reaction below a shallow bearing area was described in Sec. 5C.2. A similar

* Hough [23] also has noted the analogy between the action of piles and pile groups under load and the action of footings, similarly loaded.

† For cases not described, or for composite cases, the reader is left to interpolate or extrapolate from the cases for which criteria are indicated.

rule of thumb may be used to approximate the distribution of load from a pile to the soil in which it is embedded. The procedure is described by Hough [23] and by Peck, Hanson, and Thornburn [25] and is stipulated in a number of building codes (Boston and Los Angeles, for example). It is illustrated in Fig. 6-20.

The procedure itself is simple. The problem is to determine the level (D' in Fig. 6-20) at which to start the assumed distribution. This problem is identical to that described in Sec. 6A.3*c* with regard to determining the level at which 100 percent of the applied load acts on the pile section, and the criteria described on pages 213 through 218 would apply.

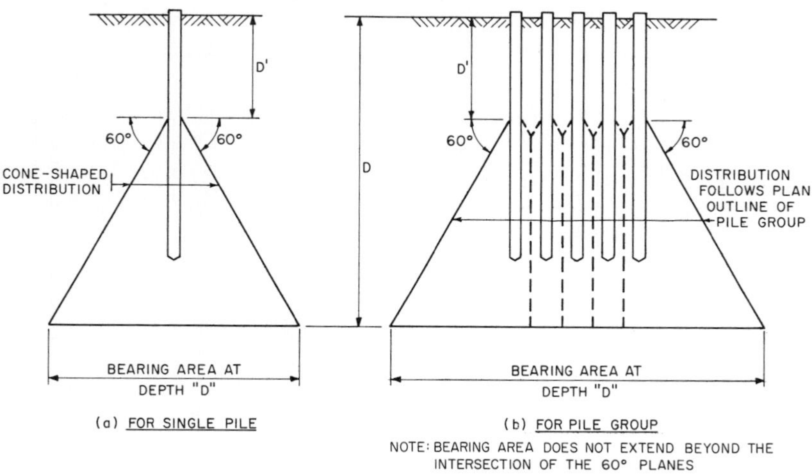

FIG. 6-20 Equivalent-footing Analogy—Friction Piles Embedded in Granular or Firm (or Stiff) Cohesive Soils.

(2) *Limits of Application.* Strictly speaking, the 60° rule is applicable only for friction-type piles embedded in substantially uniform granular soils, or in cohesive soils which are sufficiently firm or stiff to develop the indicated distribution within lines drawn 60° to the horizontal. When the piles are embedded in soft clay or loose silt soils, the criteria stipulated in Sec. B following, should be used. For piles embedded in medium cohesive deposits, either the 60° rule or the criteria of Sec. B might apply, and it is best to investigate the results of both.

Another exception would be where the pile installation utilizes a temporary casing which is withdrawn.

Finally, it is not known if the 60° rule applies to piles installed by use of a vibratory hammer. The grip of the soil along the length of the pile is likely to be less for such case than where the shaft is forcibly pounded into the ground. However, in granular soils, considerable compaction due to vibration will occur, and the rule seems a reasonable assumption.

B. *Friction Piles in Soft or Loose Cohesive Soils.* Soft or loose cohesive soils conventionally are assumed to be frictionless. Therefore, any distribution of the pile load to the soil would be dependent on how much soil is gripped by, and dragged down with, the pile. If the piles are long and closely spaced, the entire block of soil encompassed by the pile group will act as a unit. If the piles are short and widely spaced, each will punch a

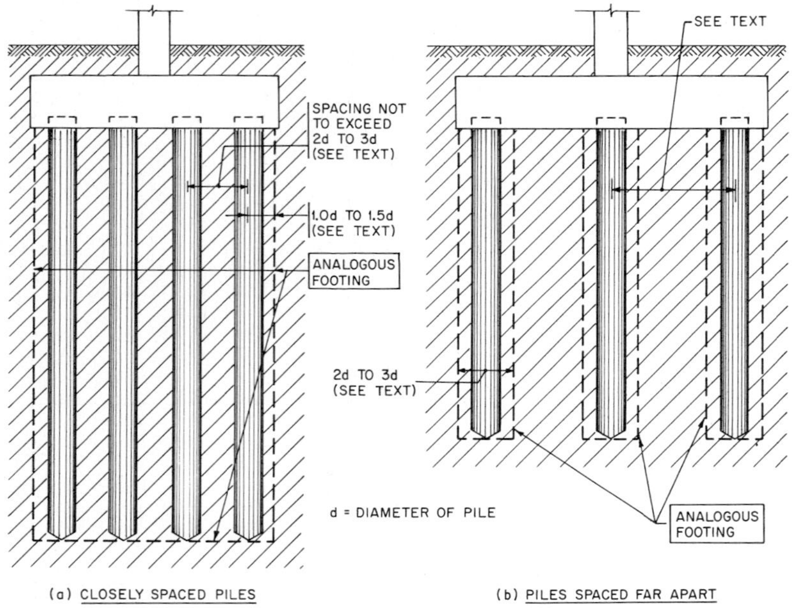

FIG. 6-21 Equivalent-footing Analogy—Friction Piles Embedded in Soft, or Loose, Cohesive Soils.

localized area of soil around the pile, rather than failure occurring as movement of the entire block encompassing the group (see Fig. 6-21).

Whitaker [34], investigating the bearing capacity of free-standing groups of model piles, found that for a given pile length and group size, there was a critical pile spacing. At spacings closer than critical, the failure was the block type. At larger spacings, some or all of the piles in the group failed individually. The critical spacing, as recorded in these tests, was about 2 to 3 times the pile diameter, being less for shorter piles and smaller groups. In practice, the action of the cap, which assures that the several piles in the group act as a unit, would tend to increase the critical spacing as compared to the values recorded by Whitaker.

Since conventional pile spacings are about 2 to 2½ pile diam, friction piles in soft or loose cohesive soils can be assumed to develop an equivalent footing consisting of a block of soil encompassing the group and extending 1 to 1½ pile diam beyond the center lines of the outer piles in the group.* The capacity of the block would be computed using Eq. (5-4), assuming that both end-bearing resistance and cohesive resistance along the sides of the block are simultaneously developed.

Alternatively, and only for purposes of computation of settlements (not of bearing capacity as limited by shear strength), Terzaghi and Peck [51] proposed that it be assumed that the load is distributed to the soil,

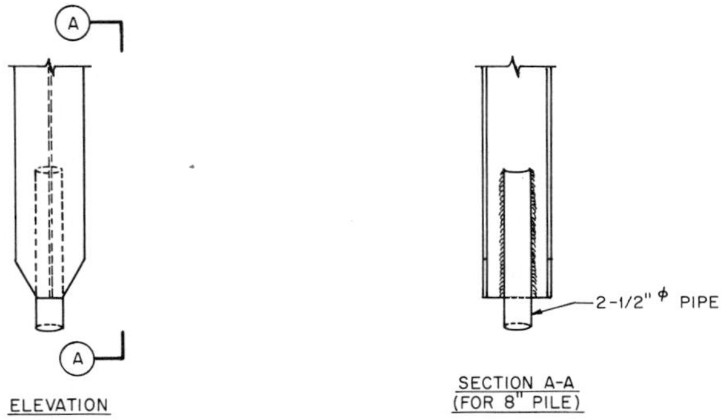

FIG. 6-22 End-bearing H Pile—Device for Socketing Pile into Rock Surface. *(After Bjerrum, Ref. 9.)*

as an equivalent-bearing area, at the level of the lower one-third point of the pile length similar to the analogy shown on Fig. 6-20.

II. *End-bearing Piles (Piles Which Receive Their Primary Support through Resistance at or near the Tip)*

A. *Piles Installed to End Bearing on Rock.* In theory, piles of this type would be classified as end bearing and all or nearly all of the load would be assumed to be carried at the pile tip. However, while a rule of thumb such as that described on page 218 is useful for estimating stresses in the pile cross section, it is excessively conservative with regard to estimating the concentration of stress in the soil at the pile tip.

If the pile penetrates through a really soupy soil material to refusal on a pronounced unconformity, on sound hard soils, then the full load on the pile probably is pretty much concentrated in direct bearing between the pile and the rock, and it is necessary to reinforce the tip or to adapt it for socketing into the rock, as in the Norwegian practice [9] (see Fig. 6-22).

In the more usual case, however, the surface of the rock is weathered or is overlain by some material of reasonable substance (sand, till, or a

* For tapered piles, the average diameter would be used for this computation.

medium to firm cohesive stratum). Experience indicates that, under conventional design loads (say 100 to 150 tons per pile for H piles and 200 to 250 tons per pile for pipe sections), if the piles are driven to refusal, the presence of even a few feet of such material is sufficient to distribute the pile load to an area of the rock surface of sufficient size to prevent overstressing that surface (see also Ref. 30).

B. *Piles Installed to End Bearing in Sand Strata* (Soil Classes SC, SM, SP or SW)

(1) *Discussion.* Meyerhof [33] has studied the problem of estimating the zone of compaction around displacement piles driven into cohesionless soils. His studies were directed at evaluating the compaction of soils near the shaft and base of displacement caissons of the Franki type, but Plantema and Nolet [59] found similar results for precast concrete piles driven into saturated fine to medium sand. Accordingly, the conclusions would seem to have application to the general case of displacement piles installed by impact hammers. These conclusions are:

1. For piles driven into a loose sand, the horizontal extent of the compacted zone (zone in which the ratio of major principal stresses is equal to the coefficient of passive earth pressure and therefore the zone in which the soil may be considered as virtually incompressible) along the shaft of the pile has an overall width of about five to six times the diameter of the shaft.

2. For piles driven into a deep stratum of loose sand the compacted zone extends below the pile base to a depth of about four to five times the base diameter.

These data may be used for design, as indicated in Fig. 6-23. The dimensions of the analogous footing, as indicated, are only for the limiting case of piles driven into a deep stratum of loose sand, however. For other cases, the analogous footing would be of smaller size.

For example, it will be apparent that a given compactive effort will be effective to a lesser depth in increasingly compact soils of infinite depth. Where the soils are of finite depth, the compactive effort will be less effective in soil of a given density if the soil being compacted is underlain by fine-grained material at sufficiently shallow depth that a sensible portion of the energy of compaction is absorbed by "quake" in the fine-grained strata.

By way of illustration, compare two piles, each driven to bearing at the same level in a shallow sand stratum extending 3 ft below the pile tips. In one case, the sand stratum is underlain by rock, in the other case by a deposit of peat. The sand overlying the rock would be well compacted under the hammer blows, whereas in the stratum underlain by peat, a portion of the energy of compaction would be dissipated in elastic or plastic deformation of the peat, providing only limited effective compaction in the sand.

In a specific case, the approach is that utilized under similar circumstances in previous chapters, namely, to establish the extremes and to interpolate between them, as follows:

CASE 1 *Pile Driven into Deep Stratum of Loose Sand:* The proposed Building Code of the City of New York (1968) defines this case as one in which a pile is driven into sandy material having an N value less than 30 and extending 20 ft or more below the tip of the pile (or, in the case of a pile having an enlarged base, 20 ft below the junction of the base and shaft).

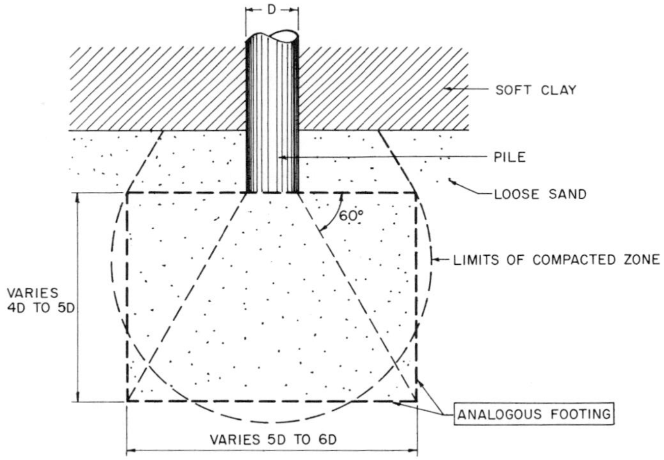

FIG. 6-23 Equivalent-footing Analogy—Piles Driven to End Bearing in a Deep Stratum of Loose Sand.

For such case, the analogous footing may be assumed to have dimensions as indicated in Fig. 6-23.

CASE 2 *Pile Driven into Shallow Stratum of Loose Sand Underlain by Hard Material:* Same criteria as for Case 1, but bottom of analogous footing not lower than top of hard material.

CASE 3 *Pile Driven into Shallow Stratum of Compact Material:* The above cited source defines this case as one in which a pile is driven into sandy material having an N value of 60 or greater and extending 10 ft below the tip of the pile (or, for piles with an enlarged base, 10 ft below the junction of base and shaft). It stipulates that the size of the analogous footing shall be assumed to be half the size designated for Case 1.

CASE 4 *Intermediate Conditions:* Linear interpolation between the indicated limits of N value and extent of bearing material below the pile tip (or junction of base and shaft) is suggested, giving equal weight to the two variables.

(2) *Limits of application*

(a) As indicated, the above criteria were derived for the case of Franki-type piles which, typically, are installed using a heavy impacting weight. Although there is limited confirmation of their use with other types of displacement piles [59], the general application of these criteria is assumed without experimental verification, and therefore they should be used with caution.

(b) The indicated criteria do not apply to nondisplacement piles (H sections and open-end pipe, for example). However, such piles are likely to penetrate deeply into any sand stratum into which they are driven so that the frictional component of resistance would be well developed and the 60° rule would apply.

(c) The indicated criteria do not necessarily apply when the piles are installed by use of a vibratory hammer, since the limits of the compacted area would be different than when the piles are installed by impact. Further study of this matter is required.

(d) The indicated criteria do not necessarily apply to piles installed to end bearing in gravel soils (soil class G). In such case, the compacted area around the pile tip is likely to be of lesser extent than indicated for piles installed in sand.

III. *Summary.* The foregoing rules provide means for estimating the size of the equivalent footing formed in the soil by the installation of a group of piles. After estimating this size, the pile group, which we do not know how to analyze, is transformed into a footing, whose action is better understood and which can be handled as described in Chap. 5. This does not necessarily make the analysis any more accurate, but it can be at least consistent and rational. The drawback is that the available rules apply only to

Friction piles installed in:

1. Uniform deposits of sand or gravel
2. Uniform deposits of firm or stiff cohesive soils
3. Uniform deposits of soft cohesive soils

End-bearing piles:

1. On rock
2. On sand

These constitute but a limited range of conditions. For example, the cases of friction piles installed in a layered continuum or in medium-cohesive soils, or end-bearing piles in gravel or hardpan are not covered. For these cases, the designer must interpolate results as best he can.

c. Penetration Resistance

Penetration resistance commonly denotes blows per foot recorded as the hammer drives the pile into the ground. It also denotes, albeit in less common use, rate of penetration (inches or feet per second or per minute)

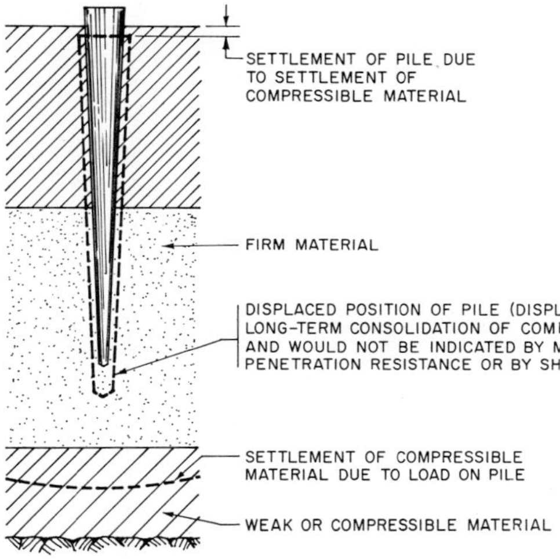

(a) PILES DRIVEN INTO FIRM STRATUM OVERLYING WEAK OR COMPRESSIBLE SUBSTRATUM

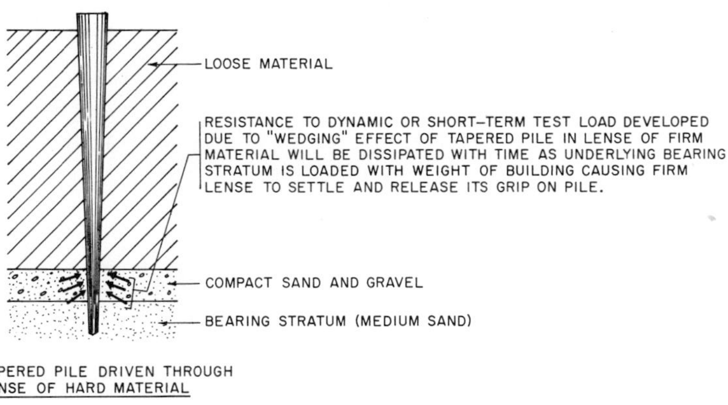

(b) TAPERED PILE DRIVEN THROUGH LENSE OF HARD MATERIAL

FIG. 6-24 Some Cases in Which Penetration Resistance and Load Tests Would Be Inapplicable as a Measure of Pile Capacity.

when installation is by use of a vibratory, rather than an impact, hammer. Whichever applies, penetration resistance is the most popular indicator of pile capacity, and almost all piles are driven to some resistance criterion.

The use of penetration resistance, as an indicator of pile capacity, is discussed in detail, in Chap. 7. For the reasons discussed therein, it is a

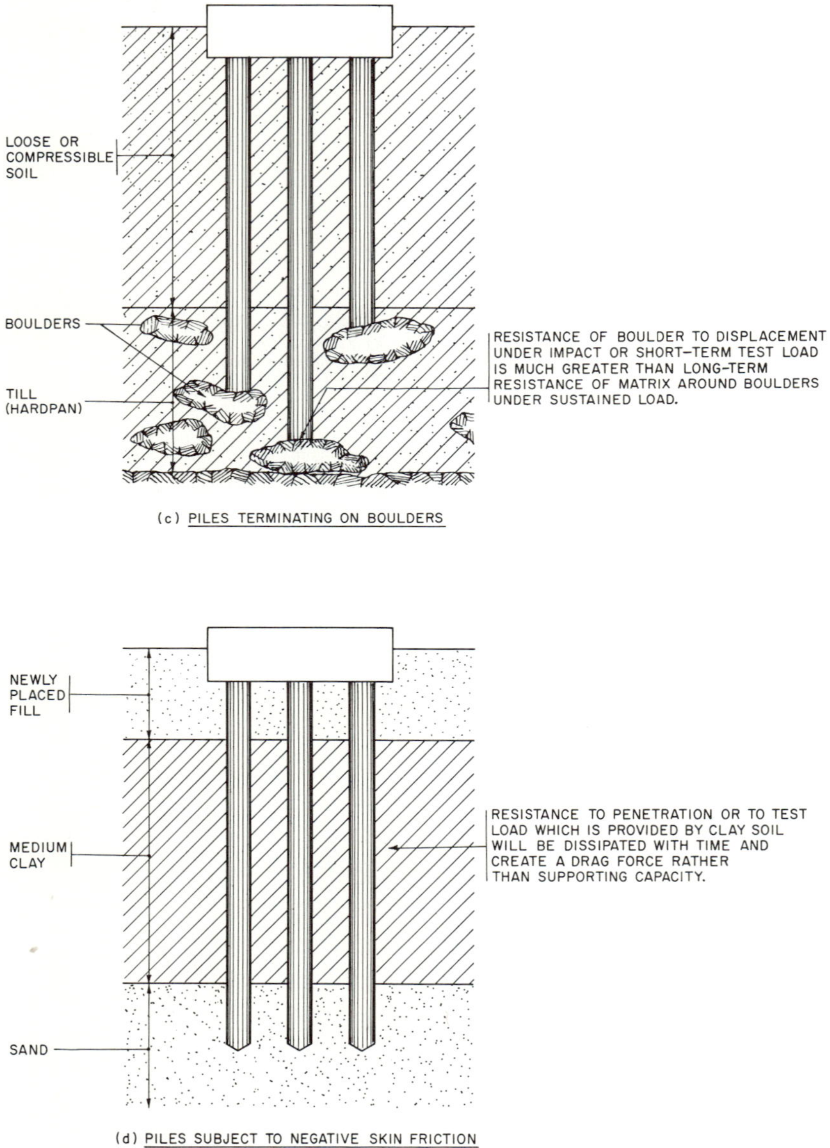

FIG. 6-24 *(Continued)*

grossly imperfect indicator and is far from universally applicable (see Fig. 6-24 and Sec. 7D, for example). Still, within its limitations, it is a valuable tool which is used as follows:

USE 1 *The Pile as a Probe:* The analogy between the behavior of a pile group and that of a pier or deep footing has been described. The implications of that analogy apply in the construction as well as in the design phase.

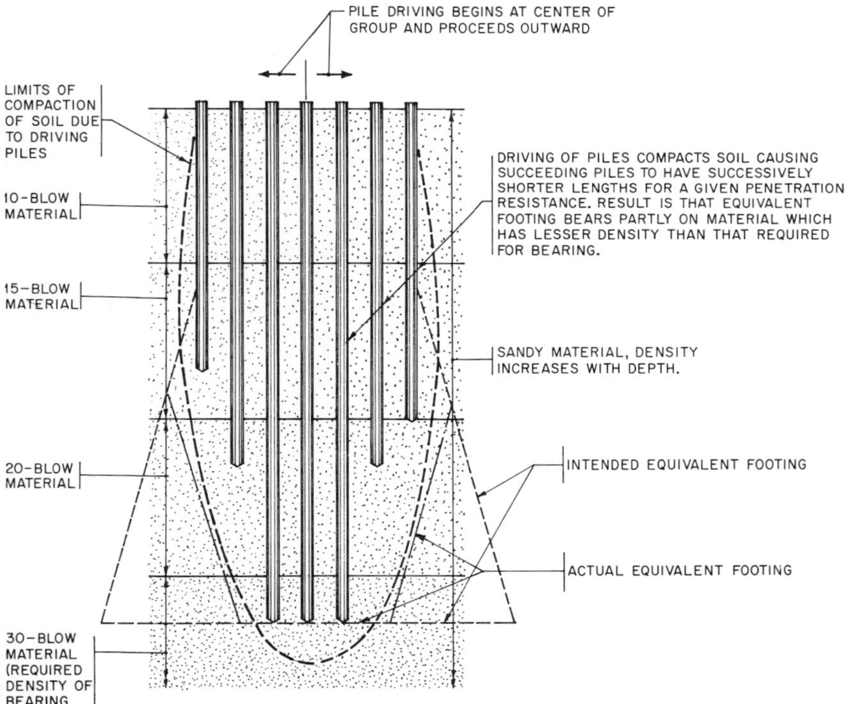

(e) COMPACTION DUE TO DRIVING GROUP OF PILES PRODUCES FALSE INDICATION OF RESISTANCE.

FIG. 6-24 *(Continued)*

When a footing is being constructed, the bearing area is uncovered, the excavation is dewatered, and the bottom is inspected to check that the soil at the subgrade level has the capacity assumed in the design and that there are no soft spots due to variations in soil type, density, or disturbance. Any excessively soft, loose, or disturbed ground is removed and replaced either with compact granular material or with lean concrete. For rock bearing areas, seams may be flushed out and grouted. Cavities are filled.

If the type of soil encountered is too soft or too loose to support the applied load or is of variable density, the bottom of the footing is lowered* below the elevation indicated on the plans, or benched until a satisfactory bearing stratum is reached. Every effort is made to assure that the bearing area in the soil consists of the right kind of material (i.e. the class and density, or consistency assumed by the designer) and is without significant flaws.

With a pile foundation, by definition, a visual check is impossible, but the need to verify the assumed class and density of the bearing stratum and to detect flaws therein remains. Penetration resistance may be used for this purpose.

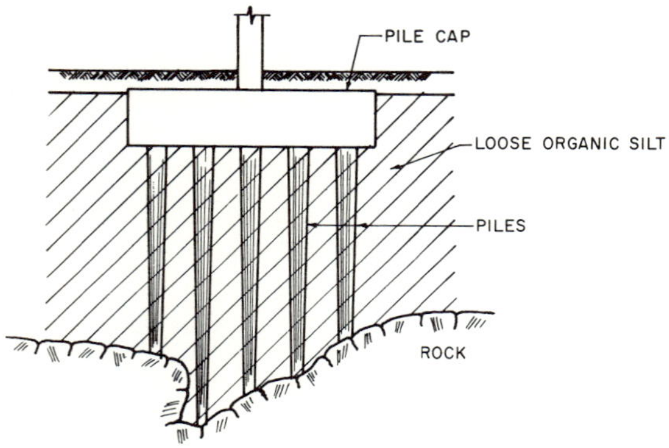

FIG. 6-25 The Pile as a Probe—Example 1.

As each pile is installed, it probes the soil being penetrated. The resistance to penetration is an indication of the type and density (or consistency) of the soil and may be used to detect the level at which the pile enters a given stratum and the relative density or soundness of that stratum. The accuracy of indication depends on the uniformity of the soil, the stratification at the site, the abruptness of the unconformities between the different strata, and the type and density of the soil. In a limited number of cases a good indication of stratification and density may be provided. For example:

EXAMPLE 1: Assume soil conditions at the site (Fig. 6-25) in question consist of loose materials overlying rock. The rock surface is irregular. The piles are heavily loaded, are to bear on the rock, and must not be terminated above the rock. In this case, penetration resistance provides a clear indication of when the pile has reached the level of satisfactory bearing material as there will be an abrupt increase in penetration resist-

* The footing level also may be raised if satisfactory bearing material is encountered at higher elevation than that indicated on the plans.

ance at that point. In fact because of the irregularity of the rock surface, it is the only clear indicator of this level.

EXAMPLE 2: A corresponding situation is illustrated in Fig. 6-26. The condition illustrated is common and will be familiar to seashore residents. This time the bearing stratum is a sand, rather than a rock, but the same requirement of penetrating to (and into) the bearing stratum applies. Again, only penetration resistance tells when this requirement has been satisfied.

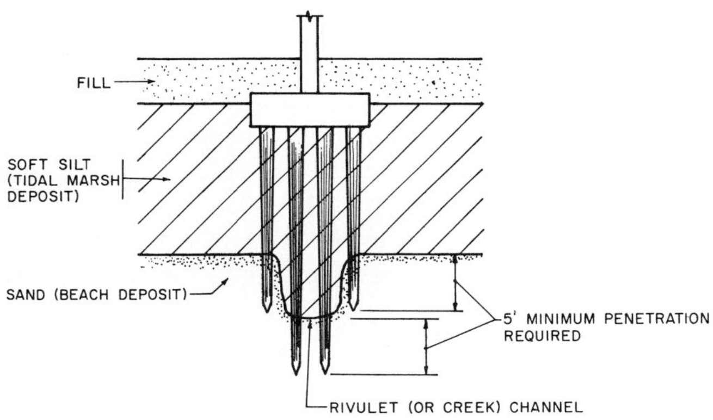

FIG. 6-26 The Pile as a Probe—Example 2.

The above examples represent somewhat idealized circumstances. In general, soil conditions are so mixed and variable that penetration resistance does not provide such a well-defined picture.

USE 2 *Driving Formulas:* The more common and popular driving formulas are described and discussed in Chap. 7. The *Engineering News* formula is an example:

$$P = \frac{2E}{S + L}$$

where P = pile capacity
E = energy of hammer
S = set per blow of the hammer
L = a constant

This formula and others of its genre are intended to provide a direct indication of capacity, related to the recorded penetration resistance, but in general the reliability* of the indication is exceedingly poor. In addition, there are many circumstances in which driving formulas are not

$$^* \text{Reliability} = \frac{\text{test load required to produce failure}}{\text{computed capacity}}$$

applicable at all (Fig. 6-24 and Sec. 7C). Therefore the use of penetration resistance for determination of capacity is not a simple matter of plugging values into a formula, but a complex problem of judgment and interpretation, which is discussed in Chap. 7.

USE 3 *As a Correlation with Experience and Local Practice:* Penetration resistance also may be used as a means of correlation with experience and local practice. This experience and practice provides the basis for the formulation of certain rules of thumb which are used in lieu of other criteria. An example would be where fixed values of penetration resistance are specified in lieu of driving formulas (see Table 7-4).

USE 4 *As a Correlation with the Behavior of Test Piles:* Perhaps the best application of penetration resistance as a measure of pile capacity is to relate the capacity of the many piles in a foundation to the capacity of those few piles which may be load tested. This procedure, which involves the estimation of an unknown pile capacity on the basis of similitude to a pile of known capacity, is discussed in the following paragraphs.

d. Load Tests (Principle of Similitude)

An example of the principle of similitude simply is to drive a pile and load-test it. If it carries the load satisfactorily and if the other nearby piles are of the same type and size and are driven in the same manner, presumedly they will carry the same load as the test pile and will behave in a similarly satisfactorymanner.

The principle is simple and logical, is widely used, and great faith is placed in the results. In fact, load tests generally are accepted as the "proof of the pudding."

Actually, for reasons discussed in Chap. 9, and as illustrated in Fig. 6-24, load tests are not a particularly reliable indicator of pile capacity, and the presumption that, because the load test pile behaves satisfactorily, so will the nearby piles, is a very large one.

However, if the load tests are properly conducted and are interpreted to consider the variables and factors described in Chap. 9, the similitude approach is a powerful tool. Moreover, there are situations, such as with augered, jacked, or screw piles, or with piles installed by vibratory hammer, in which there are no applicable direct correlations between capacity and penetration resistance, and the similitude approach is the only available check on calculated (Sec. 6B.1*a*) capacity.

6B.2 Settlement

The problem of settlement of pile foundations is treated essentially in the same manner as the problem of settlement of foundations on spread footings:

1. The anticipated settlement is computed, by whatever device may be applicable.
2. The amount of settlement which the building can tolerate is estimated.
3. The foundation is proportioned so that the anticipated settlement is less than the amount which can be tolerated.

a. Anticipated Settlement

No practical means of rational analysis of probable settlement of pile groups has been developed. This is largely because the problem of analysis of the quantitative distribution of load from the pile to the soil (i.e., how much is friction, how much end bearing, and how the friction or adhesion is distributed along the length of the pile) has not been solved (Sec. 6A). In addition, in the case of piles driven into clay soils, there is considerable uncertainty about the effects of the remolding of the soil which occurs as the piles are installed.*

As a result of these unknowns, estimates of settlement of pile foundations are made by recourse to empirical devices. Two such devices are in general use, as follows:

I. *The Equivalent-footing Analogy.* This analogy was described in Sec. 6B.1*b*. Its purpose is to establish the size of an equivalent footing created in the soil by the presence and installation of the piles. With this size established, the actual computation of settlement may be performed, as described in Sec. 5B.1, as for any other footing.

The equivalent-footing analogy is an arbitrary, empirical device. The basic concept, however, seems to be borne out by the limited amount of field and test data which actually have been analyzed with regard to settlement trends. Hanna [61], for example, basing his data on model tests of pile groups in sand, found that the settlement of the group increased as the number of piles in the group increased and that the settlement decreased as the pile spacing increased. Skempton [62] has found similar results for actual pile groups, and Whitaker [34], too, has found similar results, but for piles installed in clay soils. These results correspond to the action of footings. The increased settlement as the number of piles in the group increases corresponds to the increased settlement recorded

* It is widely recognized that the driving of piles into clay soils disturbs and remolds the soil, around the pile. This remolding, in most cases, increases the compressibility of the clay, at least temporarily. Accordingly, some engineers feel that the ultimate settlement of the foundation will be increased due to the development of a "drag" load created as the remolded clay consolidates and have suggested that friction piles in clay soils do more harm than good in reducing settlement. Other designers disagree with this contention, on the basis that the remolded zone is narrow and that its effects are dissipated before the building load is brought onto the foundation. The discussion remains unresolved.

with footings as the size increases while the bearing pressure is held constant. The decrease in settlement as the pile spacing is increased, also corresponds to the action of the footing since, in effect, the unit-bearing pressure is decreased as the pile spacing is opened up.

II. *Correlations with Load Test Data.* The second device in general use is to try to relate settlement of the foundation to settlement of test piles. In the general case, this approach has little merit, however, unless group load tests are utilized. The matter has been summarized succinctly by Peck [58], as follows:

Friction Piles

> There is absolutely no relationship between the load-settlement curve for the test pile and for the actual foundation containing many piles. The settlement predicted on the basis of the single pile is in error on the unsafe side, and there is no way of judging the magnitude of the error.

End-bearing Piles

> The results of a load test on a single pile, if those results consist of the relation between load and settlement of the pile head, cannot give us either the ultimate capacity of the point, which may be critical, or any conception of the time-settlement relationship for the foundation.

The authors must agree with the above conclusions, at least in the general case. The only exception known to them is the case of friction piles in sand, where both Skempton [63] and Meyerhof [64] have derived empirical relations between settlement of test piles and the pile group. The relation derived by Meyerhof is shown in Fig. 6-27.

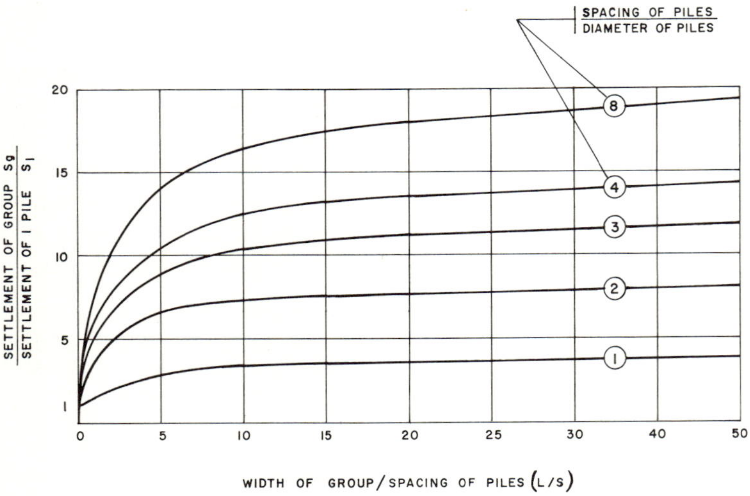

FIG. 6-27 Settlement of Pile Foundations in Sands. (*After Meyerhof, Ref. 33.*)

b. Allowable Settlement

The discussions and criteria of Sec. 5B.3 apply.

6B.3 Summary

a. Discussion

As indicated in the introduction to this section, evaluation of the capacity of the soil to support piling is analogous to determining the capacity of the soil to support footings. The same two considerations (safety against shear failure in the soil and limiting the settlement to tolerable values) apply and govern.

The introduction also described the basic difference in approach between the two cases; i.e., for footings, the physical constants to be used in solving the indicated equations for capacity and settlement are derived from reasonably well-defined and generally accepted procedures or standards, whereas in determining pile resistance, the physical parameters affecting capacity and settlement are so poorly defined that laboratory evaluation, even for assumed conditions, is so unreliable that it is seldom even attempted. Instead, reliance is placed on empirical indicators.

These indicators are (1) theoretical calculations based on Eq. (5-4), using empirically derived physical constants, (2) the equivalent footing analogy, (3) driving formulas or other correlations with penetration resistance, and (4) load tests.

The problem is that there are literally dozens of different kinds of piles, different shapes of piles, different sizes of piles, and different means and equipment for installing these dozens of different kinds, shapes, and sizes of piles. The human factor also is important, all in addition to the usual variables of type and stratification of the soil.

The result is that none of the available indicators of capacity is universally or especially reliable, and in a given case, each will generally indicate a different result. The variation may be small, or it may be several hundred percent.

The result, of course, is that the designer must do the best he can. This "best" consists of evaluating as many indicators as time and budget permit and then interpreting from among the varying results on the basis of judgment and experience. The previous text and the following chapters define the variables and influencing factors which must be weighed. The rest is up to the individual. The decision often is difficult. One hardly can go wrong by using the most conservative result derived from the listed indicators, and where experience is lacking or a sensitive structure is involved, this may be the prudent approach. In general, however, some compromise or mean value of the several indicators is used. The procedure roughly is as follows:

b. Procedure:

STEP 1: Assume a design capacity. Multiply by a safety factor* to get an installed capacity, and estimate the probable lengths of pile consistent with said installed capacity.

STEP 2: Lay out the size and configuration of several typical pile groups.

STEP 3: Check the safety of the typical groups against shear failure (under installed capacity) in the soil by use of Eq. (5-4) and/or by use of the equivalent-footing analogy.

STEP 4: If the results are not adequately safe, increase the assumed length of the piles and/or increase the size of the pile groups (reduce design capacity) until a satisfactory result is obtained.

STEP 5: Using these revised lengths, check the probable settlement of the typical groups and of the overall foundation (under design capacity) using the equivalent-footing analogy and the procedures of Sec. 5B.1.

STEP 6: If the results indicate excessive probable settlement, further increase the assumed lengths and/or increase the size of the pile groups (reduce capacity) until a satisfactory result is obtained.

At this point, the design is complete. The pile capacity and the minimum length of pile required to attain that capacity, both to assure safety against shear failure and to limit settlements to tolerable values, have been determined. It may be desirable to revise the design capacity and to vary the required lengths in order to achieve minimum total required footage of piling and minimum cost, but for any given capacity the pile grouping and the minimum required lengths are identified.†

The next problem is to select field procedures to be used to verify the consonance of actual soil conditions and properties and to compare these conditions with those assumed for purposes of design. Load tests and measurements of penetration resistance are the criteria used for said veri-

* Calculation of design loads using load factors as described in Sec. 5A.2 will provide a suitable safety factor.

† Note the reference to "minimum required lengths" and that no mention has been made, so far, of driving formulas or of load tests. The concept of minimum required pile lengths and that piles must be driven to and into an adequate penetration in an adequate bearing stratum *regardless of driving resistance* will be new to some designers (and field engineers). But piles merely are a means for transferring column or wall loads into the soil. The soil must be able to support that load, and the piles must be carried deep enough to reach soil which is capable of so doing or deep enough to spread the load over a large enough area (equivalent-footing analogy) that the bearing pressure at and below the pile tips is consistent with the capacity of the soil at the same levels. Driving according to formulas does not necessarily assure this. Neither do load tests. It is a matter of design, and under present practice the basis of design necessarily is Eq. (5-4) and/or the equivalent-footing analogy, with the latter used to evaluate probable settlements, and either one used to check shear capacity.

fication and the designer must:

STEP 7: Decide whether to drive to formula, or to length, or both. (See examples and discussion in Sec. 6B.3*c*.)
STEP 8: If it is decided to drive to formula, select which one (or ones) to use (see Chap. 7).
STEP 9: Decide whether or not to perform load tests (see Chap. 9).

The above nine steps constitute only the evaluation of the capacity of the soil to support the pile. In addition, the capacity of the pile as a structural member must be checked, as described in Sec. 6A. Numerous details such as selection of materials, spacing of piles, handling and driving stresses, and maximum upset values of capacity also must be considered. These are discussed in Chap. 8 and Sec. 6B.4.

Finally, all of the above discussion relates only to the capacity of the pile and soil to support axial, downward load. Lateral and uplift loads also occur. Capacity to resist such loads is discussed in Secs. 6C and 6D.

c. Some Examples

EXAMPLE 1: The following description is taken from Ref. 36 and relates to the pile foundation for the Oakland (California) Coliseum.

Soil conditions consist of relatively compressible organic silty clay overlying stiff clays and sandy clays with random layers of sand and gravel. Rock is at excessive depth.

Analysis indicated that pile loads of up to 80 tons each could be supported. This analysis was predicated on shear strength of the clay soil determined from unconfined compression tests. Based on considerations of economy, actual design loads varied from 40 to 80 tons per pile.

A series of load tests were performed during design (1) to verify the calculated capacity, (2) to allow a minimum factor of safety in the final design of the pile, (3) to develop pile-driving experience, and (4) to develop specifications for installation of the piling.

From the load test and analytic data, maximum and minimum required pile penetrations into the bearing strata were derived. The maximum penetration requirement was that which would develop a net ultimate capacity of twice the design load based on the assumption that the entire pile load was resisted by friction. The minimum penetration requirement was that which would develop the same load assuming that both friction and end bearing (as determined from load tests) were simultaneously developed.

A minimum penetration resistance (blow count) was established from the driving records of the test piles (principle of similitude).

The final driving criteria specified a minimum depth of penetration into the bearing stratum at which pile driving could be terminated if specified driving resistance were recorded; failing to record said resistance, driving was continued until the specified resistance was developed or until the maximum penetration requirement was met, whichever required the lesser penetration.

The contract specifications read:

> *Driving Refusal Criteria:* Drive all piles in a continuous operation to a minimum depth, noted as "Elevation A" on drawing schedules; all piles shall penetrate to at least this depth, regardless of the difficulty of driving beyond this depth; if blow counts as recorded on drawings as "Minimum-Acceptable Blow Count for Refusal" are reached before reaching "Elevation B" (on drawing schedule), cease driving; if this blow count is not realized, continue driving to "Elevation B." Achieving either the minimum blow count as the pile is driven between "Elevations A and B," or driving to "Elevation B," is the criterion for an acceptable pile.

Elevations A and B were given in the contract drawings for each zone of the site and for each type of pile, and the minimum acceptable blow count for refusal was given on the drawings for each type of pile.

The following points should be noted from this example:

1. Requirements for *minimum penetration into bearing stratum.*
2. Establishment of minimum required penetrations predicated on probable optimum conditions.
3. Establishment of maximum required penetrations predicated on most unfavorable probable conditions.
4. Use of principle of similitude (correlation between penetration resistance and capacity as indicated by load test) to interpolate within limits of minimum and maximum penetrations established by analysis and load test.
5. Piles were *not* driven to a formula, per se.
6. Piles were driven to minimum required penetrations, *regardless of penetration resistance.*
7. Piles could be terminated at maximum required penetrations, *regardless of penetration resistance.*

EXAMPLE 2: The following description is taken from Ref. 37 and relates to the use of timber piles in a clay soil.

Soil conditions consisted of strata of soft to stiff clay. Timber piles of 30 tons capacity were used.

Pile lengths were determined by calculation, taking the frictional resistance as the embedded surface area of the pile times the soil frictional or cohesive resistance (taken as one-half the unconfined compressive strength), and assuming the end-bearing resistance to be $3.7q_u$* times

* q_u = unconfined compression strength.

the area of the pile tip. Cohesive and end-bearing resistance were assumed to act simultaneously. A factor of safety of 2.0 to 2.5 was used.

Piles were driven to the indicated penetrations (ordered lengths) regardless of penetration resistance, which, in this case, was less than that required by the conventional (ENR) driving formula.

EXAMPLE 3: The following description also is taken from Ref. 37.

Soil conditions consisted of medium to dense granular soils overlying a compressible organic clay stratum, overlying dense silt and hardpan strata.

For reasons of limiting settlement, the piles had to be founded in the dense strata underlying the organic clay. Accordingly, a minimum tip elevation in the dense strata was established.

A load test was performed to establish a correlation between penetration resistance and capacity.

All piles were driven to the minimum tip elevations or to the indicated penetration resistance, whichever required the *greater* penetration.

6B.4 Pile Spacing

In addition to evaluating capacity and providing a sufficient number of piles, the designer must select the spacing between piles. All other things being equal, the smaller the spacing the more economical the design, as the size and reinforcement of the cap will be reduced.

The two considerations which govern the selection of pile spacing are (1) allowable load on the supporting soil and (2) feasibility of installing the piles at the given spacing.

The allowable load on the supporting soil is a matter of analysis as represented, for example, by the equivalent-footing analogy or other methods hereinbefore described. The spacing must be large enough so that the analogous footing formed in the soil has sufficient breadth to reduce the pressures imposed on the soil at and below the pile tips to tolerable values.

The matter of feasibility of installation is related to two problems: (1) installing the piles without their hitting each other and (2) installing the piles without displacing so much volume of soil as to cause excessive heave or excessive lateral pressure which might damage adjacent piles or adjacent buildings.

The first problem (installing piles so that they do not hit each other) occurs more often than might first appear, particularly with batter piles. It is susceptible to analysis. The spacing may be estimated by assuming that the deviation of the pile axis from the specified alignment will be some fraction of the maximum permissible value, and setting the piles sufficiently far apart so that, within the limits of this tolerance, they will not hit one another. The permissible tolerance usually is 2 percent of the pile length for batter, plus 3 in. out of position. A probability factor of, say, 50 percent is used to reflect the thought that adjacent piles are unlikely to toe-in together at the same time and to the same maximum deviation tolerance. Under this criterion, the pile spacing (center-to-center) becomes 3 in. plus 2 percent of the pile length, plus the pile diameter. For a 50-ft-long pile, 10 in. diam, the value is 3 + 12 + 10 in. = 25 in., which corresponds to the conventional values of 24 to 30 in. Narrower spacings may be used for shorter piles, and larger spacing should be used for longer piles.

TABLE 6-5 ***Minimum Pile Spacing According to Various Building Codes***

Source	*Minimum spacing (center to center)*
Building Code of City of St. Louis (1961)	Twice average diameter of round pile or 1¾ times the diagonal dimension of a rectangular pile or not less than 24 in. when driven into rock or not less than 30 in. when end bearing on materials other than rock
Building Code of City of Baltimore (1955)	Precast concrete piles: 2½ times the square root of the cross-sectional area of the butt
	Cast-in-place piles: twice average diameter of the embedded portion of the pile (or 30 in.)
	Cylindrical steel piles (concrete filled): diameter of the shell plus 10 in., but not less than 24 in.
	Note: When driven with closed ends, see spacing for cast-in-place piles
Los Angeles Building Code (1958)	2½ times the mean diameter of pile, but not less than 36 in.
National Building Code (1967)	Twice average diameter of round pile, but not less than 2 ft 6 in. (Piles not driven to rock)
	Greatest diameter of pile plus 12 in. (Piles driven to rock)
Southern Standard Building Code (1961)	Piles not driven to rock: twice average diameter of pile or twice diagonal dimension of a rectangular or rolled structural steel pile, but not less than 2 ft 6 in.
	Piles driven to rock: average diameter of a round pile plus 12 in. or diagonal dimension of a rectangular or rolled steel pile plus 12 in.
Building Code of City of Chicago (1963)	Twice diameter of round piles or twice diagonal dimension of rectangular or rolled steel pile, but not less than 30 in.
Building Code of City of Buffalo (1965)	Piles bearing on rock: twice diameter of round piles or 1.75 times the diagonal of the pile, but not less than 24 in.
	Other piles: same, but not less than 30 in.
National Building Code of Canada (1965)	End-bearing piles: butt diameter plus 12 in.
	Other piles: 2½ times average butt diameter or diagonal

The other problem, that of volumetric displacement, heave, and damage to adjacent piles, is not readily subject to analysis, and rule-of-thumb values are used. Terzaghi and Peck [51], for example, suggest a minimum spacing for friction piles of 2.5 times the diameter of the pile. Chellis [52] makes a similar recommendation and cites, as reasons for generous spacing, (1) problems of weaving of displacement piles when driving on close centers, (2) progressive tightening of the soil if spacing is too close (see Chap. 7), and (3) improved group action. Some typical values of spacing, for various different types of pile, are indicated in Table 6-5. These are code stipulations and represent minimum permissible values.

Section 6C RESISTANCE TO LATERAL LOADS

6C.1 Plumb Piles

A plumb pile resists lateral load by deflecting to mobilize the passive reaction of the surrounding soil. The amount and distribution of the reaction depends on the relative stiffness of the pile and the soil, and on the conditions of fixity of the ends of the pile.

A number of solutions to this problem have been derived on the basis of different assumptions of distribution of lateral load and/or the variation and mobilization of a "modulus of lateral soil pressure" or "modulus of subgrade reaction," corresponding to a modulus of elasticity for the soil. Two of the published solutions which determine the maximum total lateral load that can be resisted by a pile without overstressing it, and which give the lateral deflection at the ground level are:

a. General Solution

A general solution to the problem of determining moments and displacements of laterally loaded plumb piles, based on the Winkler model of a beam on elastic foundations with a variable modulus of lateral soil pressure to represent the reaction of the surrounding soil, is given by Matlock and Reese in Ref. 38. For purposes of design of the ordinary run-of-building foundations, it is sufficient to limit consideration to the case of an assumed linear increase in soil modulus E_s, with increased embedment x, that is, $E_s = kx$.

For this case, the coefficients of Fig. 6-28 apply, and moments and deflections may be computed as follows:

Procedure:

1. Assume a value of k, the constant of soil modulus variation, psi per inch of depth. See Fig. 6-3.

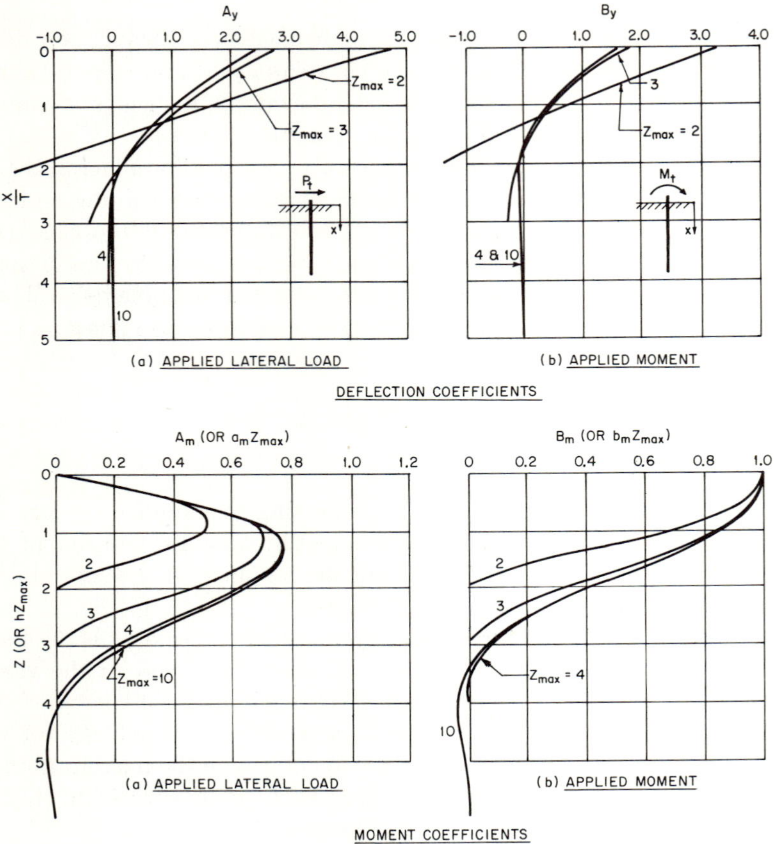

FIG. 6-28 Resistance to Lateral Loads—Plumb Piles—General Solution Coefficients. *(From Matlock and Reese, Ref. 38.)*

2. Compute

$$T = \left(\frac{EI}{k}\right)^{1/5}$$

where T = relative stiffness factor expressing the relation between the stiffness of the soil and flexural stiffness of the pile, in. of depth

E = the modulus of elasticity of pile material

I = the moment of inertia of the pile section

3. Compute

$$Z_{\text{max}} = \frac{L}{T}$$

where L is length of pile

4. For various values of x, enter chart (Fig. 6-28), and determine values of A_y or B_y (or A_m or B_m).
5. Calculate deflections as

$$y = A_y\left[\frac{P_tT^3}{EI}\right] + B_y\left[\frac{M_tT^2}{EI}\right]$$

6. Calculate moments as

$$M = A_m[P_tT] + B_m[M_t]$$

7. This basically is only a first approximation. It is possible, by twice differentiating the moment curve with respect to x, to obtain a plot of the soil reaction p versus depth x. A further plot of $E_s = p/y$ at each depth may be drawn therefrom, as a basis for further refinement of the assumed initial soil modulus variation $E_s = kx$. This leads to a trial and error solution toward final convergence, either modifying the constant of soil modulus variation k, or even considering a more complex variation according to a power function $E_s = kx^n$.

A factor of safety of at least 2.5 based on ultimate resistance should be incorporated in the design.

b. Simplified Solution

Broms [40] has developed a solution based on the assumption of two modes of failure: (1) shear in the soil and (2) bending of the pile (as governed by plastic yield resistance of the pile section). It is assumed that failure of the soil takes place in the case of short stiff piles and that plastic bending failure occurs with long piles. In the latter case, it is further assumed that the pile material has the capability of developing plastic hinges having sufficient rotational capacity to develop passive resistance of the soil along the pile and to cause redistribution of bending moment along the length. This may not be correct for unreinforced concrete piles or for timber piles, and the solutions should be treated with caution in such cases.

The subject solution is indicated in Figs. 6-29 and 6-30. Figure 6-29 relates to short piles. Figure 6-30 relates to long piles. Both cases should be checked in a given instance. As in the case of the general solution, a factor of safety of at least 2.5 should be applied to the calculated ultimate resistance.

Computations of deflection under working loads may be made using the coefficients of Fig. 6-31.

Solutions such as those indicated above (the General Solution and the Simplified Solution), while ostensibly simple and precise, actually represent rather gross approximations. There are two reasons.

The first is that no means has been derived which permits of accurate

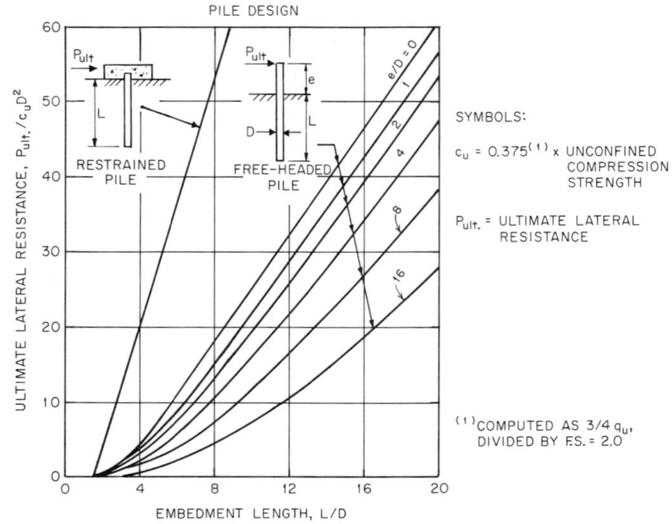

(a) ULTIMATE LATERAL RESISTANCE FOR COHESIVE SOILS RELATED TO EMBEDMENT LENGTH

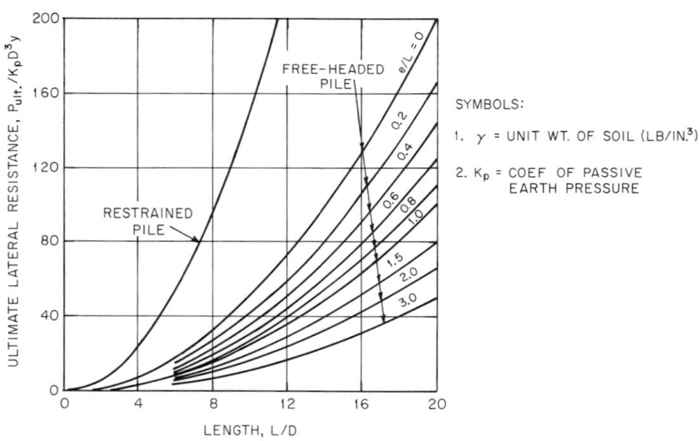

(b) ULTIMATE LATERAL RESISTANCE FOR COHESIONLESS SOILS RELATED TO EMBEDMENT LENGTH

FIG. 6-29 Coefficients for Calculating Ultimate Lateral Resistance of Short Plumb Piles. *(From Broms, Ref. 40.)*

evaluation of the necessary soil modulus or modulus of soil reaction. This is because the modulus is not a soil constant, but a variable that changes with a multitude of physical and geometrical factors, such as:

1. Structural rigidity (EI) of the pile
2. Size of pile

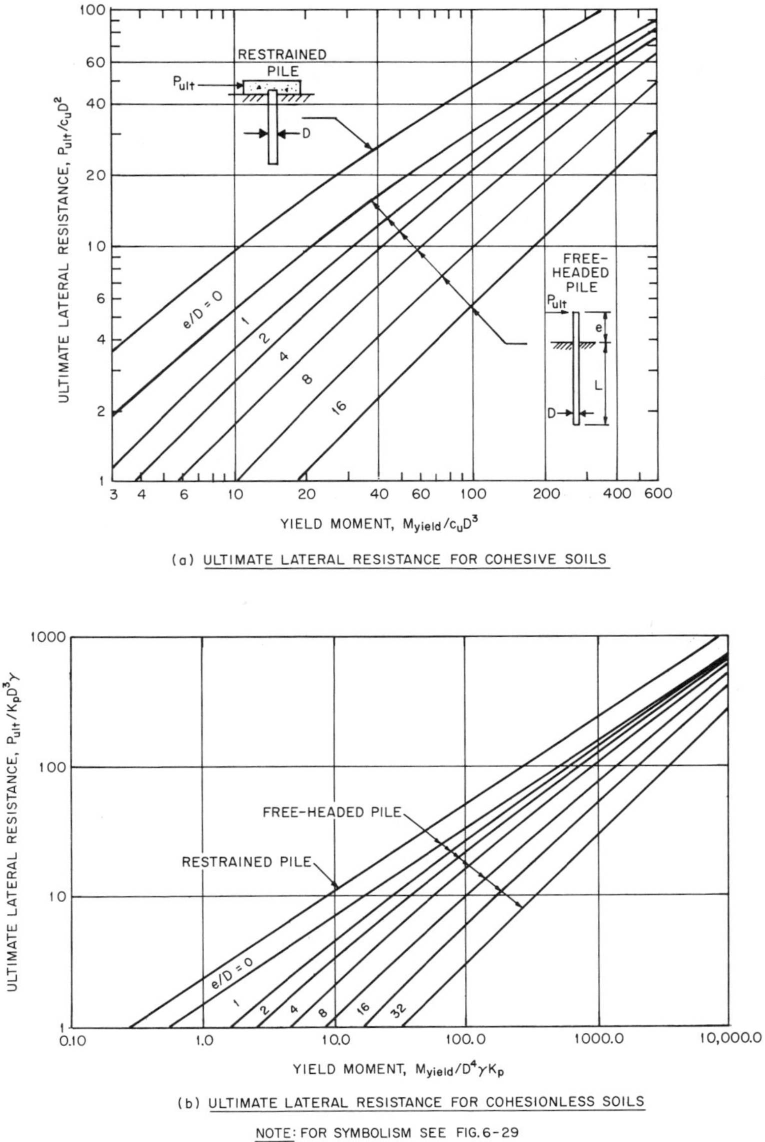

FIG. 6-30 Coefficients for Calculating Ultimate Lateral Resistance of Long Plumb Piles. *(From Broms, Ref. 40.)*

3. Type and density of the soil
4. Groundwater conditions
5. Stress history of the soil
6. Nature of applied load (static or dynamic)
7. Presence of adjacent piles

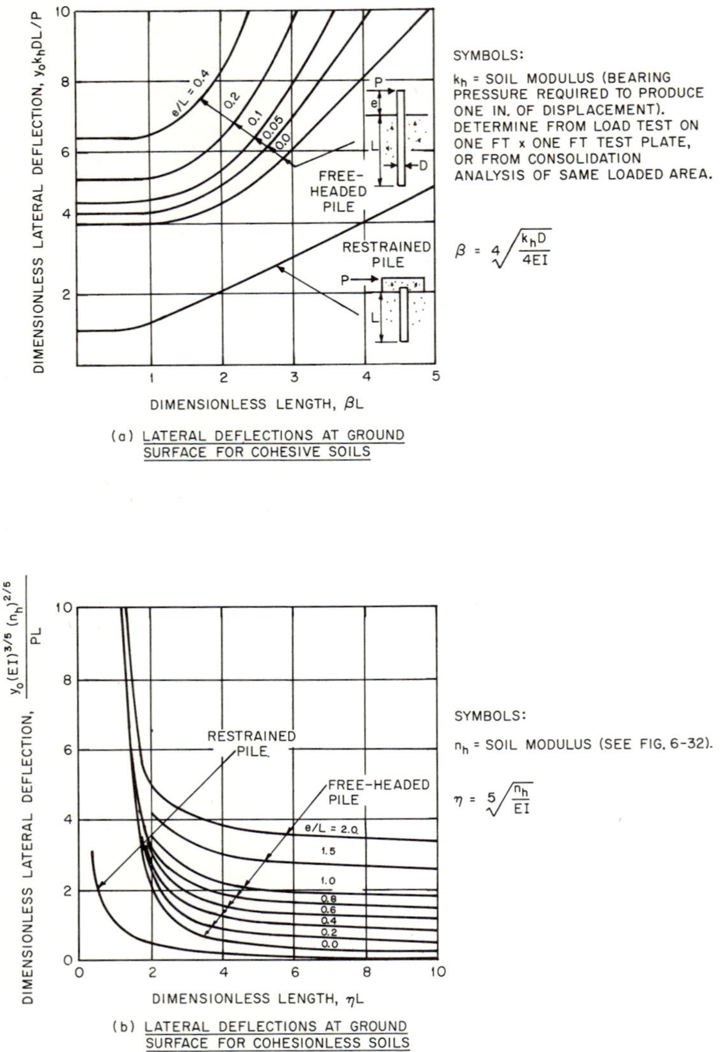

FIG. 6-31 Coefficients for Calculating Deflections of Plumb Piles under Lateral Load. (*From Broms, Ref. 40.*)

8. Presence of pile cap and resistance of soil under and abutting cap
9. Compaction of soil incident to driving adjacent piles

Because of the influence of variables such as those listed as (4), (7), (8), and (9), this modulus cannot be evaluated reliably even by actual load tests at the site. The best that can be rendered is a rough approximation such as that indicated in Fig. 6-32.

The second reason is that it is very difficult to establish precisely all the components which go to make up the probable magnitude of the lateral displacement which will occur and the amount of lateral displacement which can be tolerated.

The displacement which occurs consists of an elastic component, an inelastic component, and the effects of creep and shakedown.* The magnitude of the elastic displacement can be estimated using the indicated

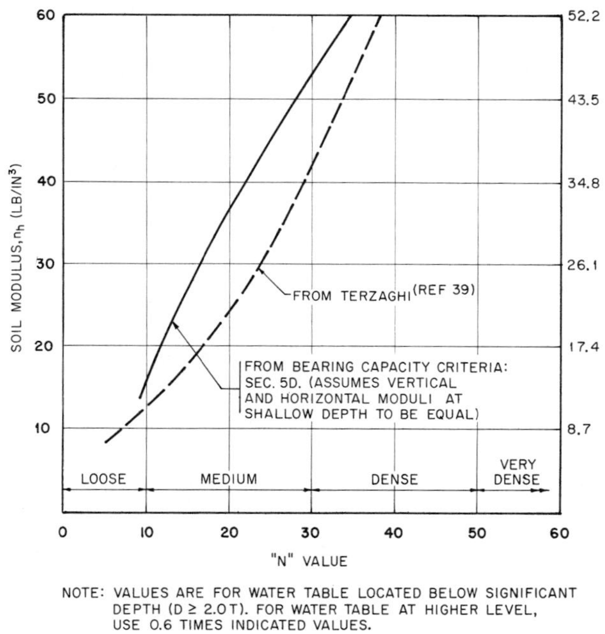

FIG. 6-32 Horizontal Soil Modulus (at Depth of 1 ft) for Granular Soils.

general and simplified solutions. The magnitude of the inelastic displacement and of the additional displacements due to creep and shakedown, however, are indeterminate, so that in the general case the total lateral movement of the foundation under the influence of a given lateral load can only be roughly estimated.

* Generally, the lateral loads which the foundation must resist are those resulting from wind and earthquake and are of an oscillatory nature. The soil under the pile cap likely is loose or soft (otherwise piles would not have been needed). Loose granular soil, particularly if saturated, tends to settle under vibratory load leaving a space around the pile (and under the cap), space which, in effect, reduces the degree of fixity of the top of the pile and creates a small cantilever extension of the pile above the level at which the pile is gripped by the soil. The effect of these conditions is to increase the lateral displacement of, and stress in, the pile under a given load.

The amount of lateral displacement which can be tolerated in the building also is difficult to establish and is the subject of much debate.*

SAMPLE COMPUTATION 6-1

Plumb Piles—Resistance to Lateral Loads

STATEMENT OF PROBLEM

Determine resistance to lateral load (about a strong axis) of 10BP42 pile (A36 steel), 60 ft long, embedded full length in a uniform medium sand ($N = 20$). The head of the pile is restrained by a rigid cap. Allowable lateral displacement is ¼ in.

1. SOLUTION BY GENERAL METHOD (Section 6C.1*a*)

STEP 1: *Capacity as Limited by Deflection*

a. E_s for soil = 30 tcf = 35 psi per inch (Fig. 6-32)

b. Allowable $y = 0.25$ in. at surface (x = zero)

c. $y = A_y \left[\dfrac{P_t T^3}{EI}\right]$

P_t = unknown

$EI = 30 \times 10^6 \times 210.8 = 6.324 \times 10^9$ lb-in.²

$$T = \left(\frac{EI}{k}\right)^{1/5} = \left(\frac{30 \times 10^6 \times 210.8}{35}\right)^{1/5} = 44.8 \text{ in.}$$

To determine A_y

$$Z_{\max} = \frac{L}{T} = \frac{720 \text{ in.}}{44.8 \text{ in.}} = 16.1$$

x = zero

$$\frac{x}{T} = O$$

From Fig. 6-28*a*, for $x/T = O$ and $Z_{\max} = 16.1$

$A_y = 2.3$ (approx.)

d. $$0.25 \text{ in.} = 2.3\left[\frac{P_t \times \overline{44.8}^3}{6.324 \times 10^9}\right] = \frac{2.07 \times 10^5}{6.324 \times 10^9} P_t$$

$$P_t = 0.25 \times \frac{6.324}{2.07} \times 10^4 = \boxed{7{,}600 \text{ lb}}$$

Controlling criterion

* The following criteria are cited:

1. A value of ⅜ in. arbitrarily is assigned by the New York City Building Code of 1947 as the permissible lateral displacement of the pile head when the proposed design lateral load is applied to the head of the pile as a test load.

2. In the Los Angeles building code, the lateral-load capacity is considered to be one-half the test load (applied laterally) which causes a deflection of ½ in. at the ground line, with 75 percent recovery.

3. Teng [46] indicates a value of ¼ in.

STEP 2: *Capacity as Limited by Moment*

a. $M = [A_m P_t T]$

b. $M_{\text{capacity}} = f_{\text{allowable}} S$

c. Compute $f_{\text{allowable}}$ from Sec. 6A
Assume result is 15,000 psi

d. $M_{\text{capacity}} = 15{,}000 \times 43.4 = 650{,}000$ in.-lb

e. $T = 44.8$ in. and $Z_{\max} = 16.1$ (step 1*c*)

f. max $A_m = 0.75$ (approx.), at depth $Z = 1.3T = 1.3 \times 44.8$ in. $=$ 4 ft

g. $$P_t = \frac{M}{A_m T} = \frac{650{,}000 \text{ in.-lb}}{0.75 \times 44.8 \text{ in.}} = 19{,}400 \text{ lb}$$

This is a first approximation. The validity of the initial assumption of $E_s = 35$ psi per inch now may be checked by following steps (*a*) through (*g*) using the value $P_t = 7{,}600$ lb and the apparent lateral resistance adjusted, as may appear to be required.

2. SOLUTION BY METHOD OF BROMS (Section 6C.1*b*)

STEP 1: *Check for Failure in the Soil*

$$\frac{L}{D} = \frac{60 \times 12 \text{ in.}}{9.72 \text{ in.}} = 74$$

From Fig. 6-29*b*, restrained head, it can be seen that, for this value of L/D, failure in the soil will not be the controlling criterion.

STEP 2: *Check for Bending Failure in Pile*

$$M_{\text{yield}} = 1.1 \times \overbrace{43.4 \text{ in.}^3}^{S} \times \overbrace{36{,}000 \text{ psi}}^{F_y} = 1.72 \times 10^6 \text{ in.-lb}$$

where 1.1 is the form factor for an H section.

Assume:

$$\gamma = 110 \text{ pcf} = 0.064 \text{ pci}$$

$$K_p = 3.0 \left(\text{coefficient of passive pressure} = \tan^2\left(45 + \frac{\phi}{2}\right)\right)$$

$$\frac{M_{\text{yield}}}{D^4 \gamma K_p} = \frac{1.72 \times 10^6}{(9.72)^4 \times 0.064 \times 3.0} = 1{,}000$$

From Fig. 6-30*b*, restrained head:

$$\frac{P_{\text{ult}}}{K_p D^3 \gamma} \approx 400$$

$$P_{\text{ult}} = 400 \times 3 \times (9.72)^3 \times 0.064$$
$$= 71{,}000 \text{ lb per pile}$$

STEP 3: *Check for Deflection*

$y_0 = 0.25$ in.

$\eta_h = 35$ pci (Fig. 6-32, corresponding to value of $E_s = 30$ tcf used in General Solution)

$(\eta_h)^{2/5} = 4.15$

$$\eta \text{ in.}^{-1} = \sqrt[5]{\frac{\eta_h}{EI}} = \sqrt[5]{\frac{35}{30 \times 10^6 \times 210.8}}$$

$$= \frac{1}{45}$$

$$L = 720 \text{ in.}$$

$$\eta L = \frac{720}{45} = 16$$

$$\frac{y_0(EI)^{3/5}(\eta_h)^{2/5}}{PL} \approx 0.1 \qquad \text{(Fig. 6-31}b\text{)}$$

$$y_0 = 0.25 \text{ in.}$$

$$P = \frac{0.25 \times 0.76 \times 10^6 \times 4.15}{720 \times 0.1} = \boxed{10{,}800 \text{ lb}}$$

Controlling criterion

This value compares with a value of 7,600 lb calculated by the general method. Deflection was the controlling criterion in both cases.

c. Prescription Values

As a result of these difficulties and uncertainties, many designers forego attempts to analyze lateral resistance of plumb piles and rely instead either on load tests (using the criteria indicated in the footnote on page 268) or on the assignment of arbitrary or empirical values such as those indicated in Table 6-6.

6C.2 Batter Piles

There are several methods in use for analysis of the loading and capacity of pile groups containing batter piles. These are:

a. Culmann's (Static) Method

In this method, it is assumed that the ends of the piles are hinged and so are subject only to axial load and that all piles battered in a given direction are equally loaded. The load in the individual piles is calculated from the requirements of static equilibrium as indicated in Fig. 6-33. The analysis can be made graphically (Culmann's method) or analytically.

Since there is no moment or transverse shear in the pile, the resistance of the pile is the resistance to axial load, and may be determined in accordance with Secs. 6A and 6B. This resistance should exceed the calculated load by a factor of safety as described in Sec. 6B.3.

TABLE 6-6 ***Allowable Lateral Loads on Plumb Piles According to Different Authorities***

Authority	Allowable lateral load on each plumb pile, tons			
New York City Building Code (1947)	1.0			
New York City Building Code (1968)	2.0			
McNulty, J. F. [41]		Med. sand	Fine sand	Med. clay
	*Free-end timber piles, 12.-in. diam.	0.75	0.75	0.75
	Fixed-end timber piles, 12.-in. diam.	2.50	2.25	2.00
	Free-end concrete piles, 16.-in. diam.	3.50	2.75	2.50
	Fixed-end concrete piles, 16.-in. diam.	3.50	2.75	2.50
Feagin [42]	1. Timber piles (allowable lateral displacement = $\frac{1}{4}$ in.)—4.50 2. Timber piles (allowable lateral displacement = $\frac{1}{2}$ in.)—7.00			
Teng [46]	Piles in deep bed of soft silts or clays—0.50			

* Safety factor of 3 applied to load required for $\frac{1}{4}$-in. lateral displacement.

b. Elastic Method

Several solutions to the problem of determining the forces on individual piles in a mixed group have been developed on the basis of elastic deformation of the pile [13, 43–45]. Usually it is assumed that the piles act elastically, the soil under the cap carries no load, the cap is infinitely rigid with respect to the piles, and the point of resistance of the pile is located, say, at the tip for end-bearing piles and perhaps one-third the length of the pile, measured up from the tip, for friction piles. Movement of the point of resistance of the pile is assumed to be negligible, and the effects of lateral restraint of the soil are ignored. The solution given in Ref. 13 will be given as an example.

Since the cap is assumed to be infinitely rigid with respect to the piles,

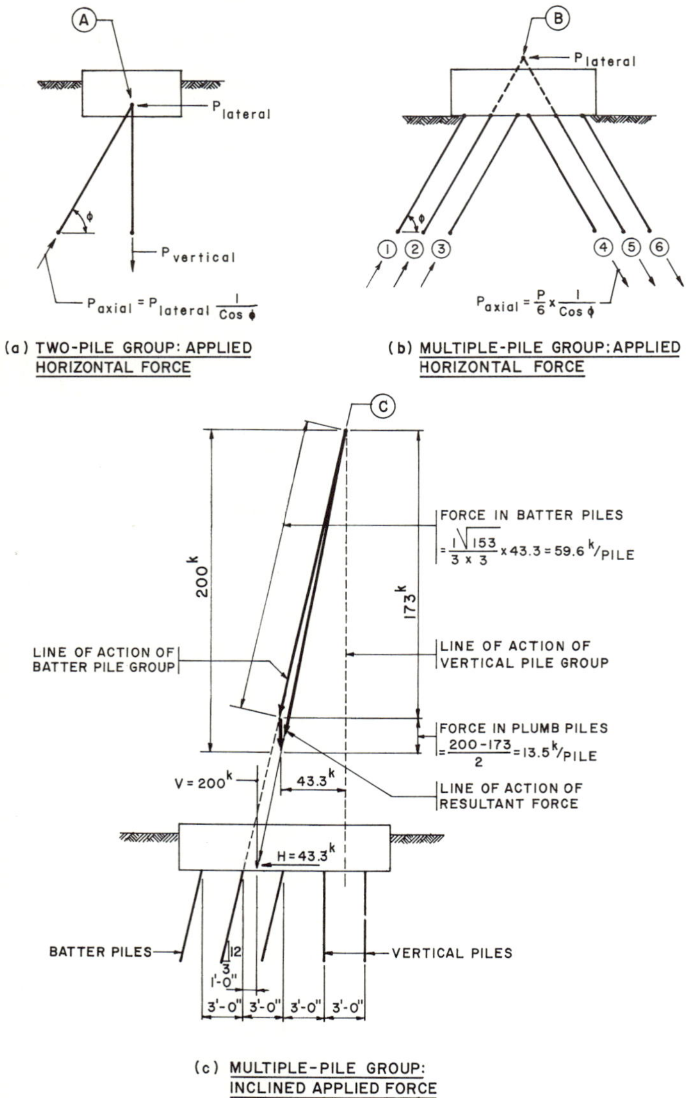

FIG. 6-33 Forces in Piles of Batter Pile Group—Static Method. Line of action of applied force concurrent with lines of action of piles.

its change of position resulting from application of load is equivalent to rotation with respect to a point called the *center of rotation* (see Fig. 6-34).

The procedure, assuming that all piles are the same type and section (the usual case), is as follows:

STEP 1: Locate center of rotation. This is done by taking two arbitrary points along the line of action of the resultant force acting on the

FORCES IN PILES:		
PILE NO.	p	FORCE
1 =	+18"	$+17.3^k$
2 =	-25"	-24.0^k
3 =	+3"	$+2.9^k$
4 =	+31"	$+29.8^k$
5 =	-24"	-23.1^k

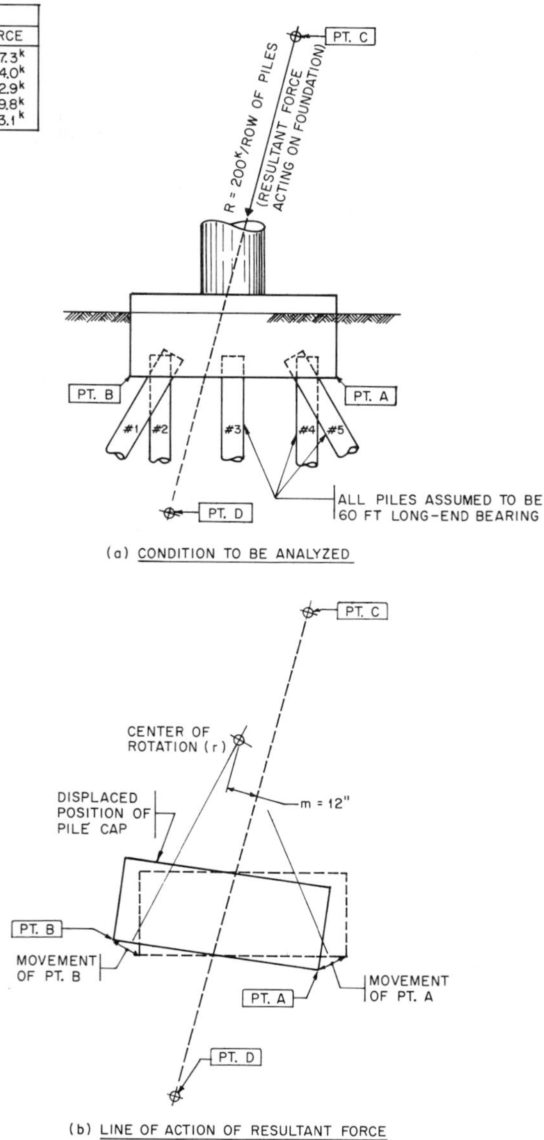

FIG. 6-34 Elastic Analysis of Mixed-pile Group.

foundation (points C and D in Fig. 6-34*b*) and determining the forces which would have points C and D as a center of rotation. This procedure is indicated in Figs. 6-34*c* and 6-34*d*. The force polygons are constructed as follows:

1. Determine the perpendicular distances from points C and D

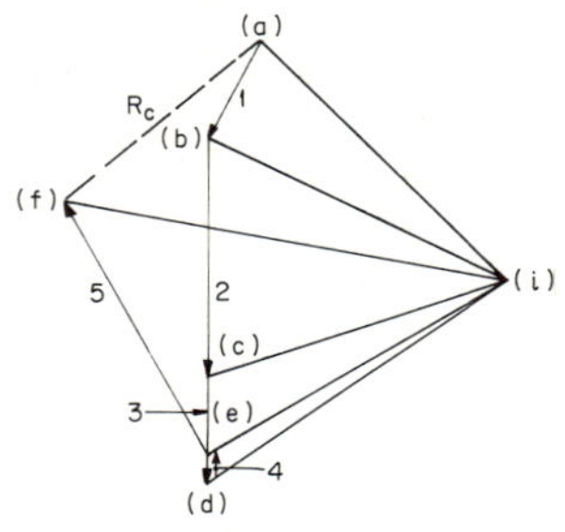

(c) FORCE POLYGON FOR PT. C

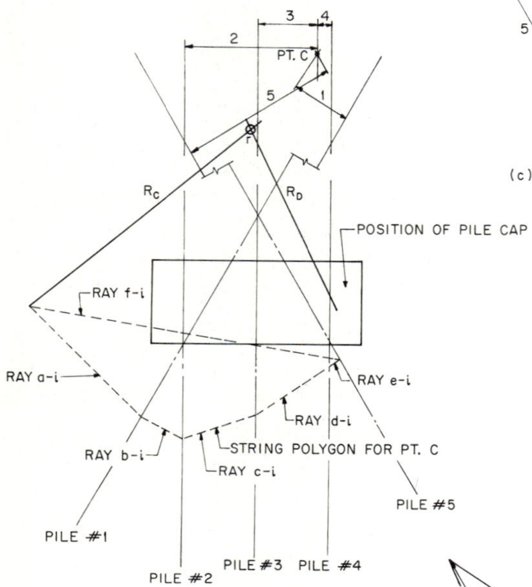

(e) STRING POLYGON FOR PT. C

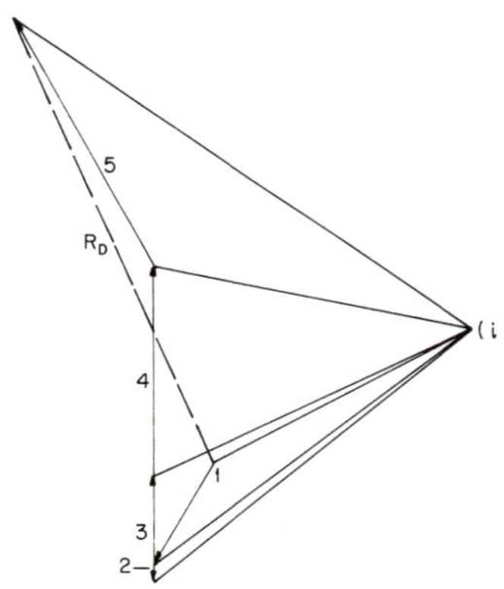

(d) FORCE POLYGON FOR PT. D

FIG. 6-34 *(Continued)*

to the axes (projected) of the piles (for the point *C* these are distances 1 through 5 in Fig. 6-34*e*).

2. Distances 1 through 5 are laid off parallel to the respective piles as a force polygon (Fig. 6-34*c*).

3. Through an arbitrary pole *i*, project rays *ai*, *bi*, etc.

4. Construct string polygon (Fig. 6-34*e*) starting at any arbitrary point along the axis of pile 1.

5. Repeat process for point D (string polygon not shown).
6. Project resultants R_C and R_D to point r (center of rotation).

STEP 2: Determine the forces in the several piles from

$$P = \frac{p}{l}\left[\frac{Rm}{\Sigma p^2/l}\right] \tag{6-11}$$

where P = load in a given pile, lb
R = resultant applied load, lb
m = perpendicular distance from center of rotation to line of action of resultant applied load
p = distance from centerline of pile to center of rotation; p shall be considered positive if a shortening of the pile will increase the angle of rotation, otherwise negative
l = length of pile measured from the assumed hinge at the top to the point of resistance, in.

The lateral displacement of the foundation can be calculated from the elastic axial deformation of the piles by use of a Williot diagram.

c. *Other Methods*

Other, more sophisticated methods have been developed for the analysis of mixed pile groups, methods which consider the lateral restraint of the soil and fixed-end conditions of the piles.

The British Code of Civil Engineering Practice [47], for example, notes the use of dummy piles to simulate fixed-end conditions of the actual piling.

Hrennikoff's [48] procedure (which is illustrated in Fig. 6-35 and Sample Computation 6-2) takes into consideration the elastic resistance of the surrounding soil and the rotation and translation of the pile. The procedure described by Francis [44] also takes account of these influences. Both methods are relatively laborious, however, and involve the same problems of evaluating applicable soil modulus and tolerable lateral displacement of the foundation, described with regard to evaluating the lateral resistance of plumb piles.

These sophisticated solutions are, perhaps, more properly applicable in the analysis of earth-retaining structures, piers, breasting dolphins, trestles, and like constructions than in the design of routine building foundations where their use possibly has more value in terms of the satisfaction of the designer than in the practical significance of the computed pile loads.

1. DETERMINE THE ELASTIC CONSTANTS FOR A SINGLE PILE:

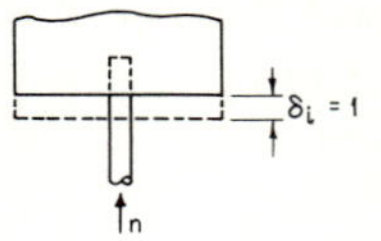

a. DETERMINE FORCE n CORRESPONDING TO UNIT VERTICAL DISPLACEMENT.

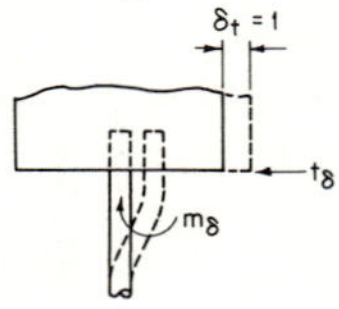

b. DETERMINE RESISTANCE t_δ AND END MOMENT m_δ DUE TO UNIT TRANSVERSE DISPLACEMENT. FROM THEORY OF ELASTICITY:

$$t_\delta = \frac{k}{\beta}, \quad m_\delta = \frac{k}{2\beta^2} \quad \text{PILE HEAD FIXED}$$

$$t_\delta = \frac{k}{2\beta}, \quad m_\delta = 0 \quad \text{PILE HEAD PINNED}$$

WHERE:

k IS THE COEFFICIENT OF HORIZONTAL SUBGRADE REACTION

$$\beta = \sqrt[4]{\frac{k}{4EI}}$$

E = MODULUS OF ELASTICITY OF PILE MATERIAL

I = MOMENT OF INERTIA OF PILE CROSS-SECTION

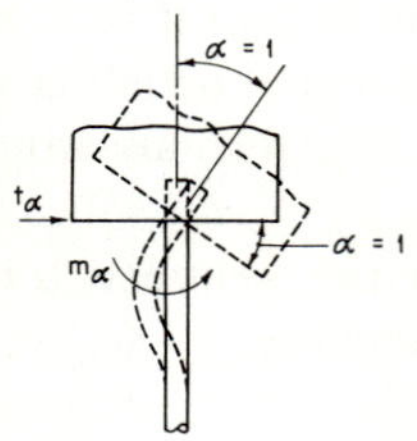

c. DETERMINE RESISTANCE t_α AND END MOMENT m_α FROM THEORY OF ELASTICITY:

$$m_\alpha = \frac{k}{2\beta^3} \qquad m_\alpha = t_\alpha = 0 \text{ FOR FREE PILE HEAD}$$

$$t_\alpha = m_\delta$$

2. DETERMINE THE ELASTIC CONSTANTS FX, FY, M OF THE FOUNDATION AS A WHOLE FOR UNIT DISPLACEMENTS Δx, Δy, AND UNIT ANGLE CHANGE α (ASSUMING PILE FOOTING INFINITELY RIGID).

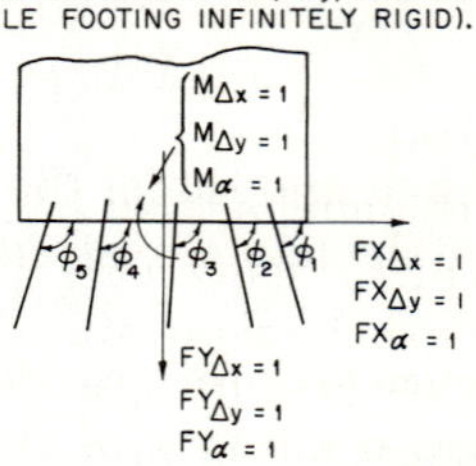

FROM THEORY OF STRUCTURES:

$$FY_{\Delta x = 1} = FX_{\Delta y = 1}$$

$$M_{\Delta x = 1} = FX_{\alpha = 1}$$

$$M_{\Delta y = 1} = FY_{\alpha = 1}$$

FROM GEOMETRY AND STATICS

$$FX_{\Delta x = 1} = -\Sigma(n \text{ COS}^2 \phi + t_\delta \text{ SIN}^2 \phi)$$

$$FY_{\Delta x = 1} = -\tfrac{1}{2}(n - t_\delta)\ \Sigma(\text{SIN } 2\phi)$$

$$M_{\Delta x = 1} = -\tfrac{1}{2}(n - t_\delta)\ \Sigma(X \text{ SIN } 2\phi) + m_\delta\ \Sigma(\text{SIN } \phi)$$

WHERE X = COORDINATE OF PILE HEAD FROM AN ARBITRARILY CHOSEN ORIGIN. PILE HEADS TO THE LEFT OF CENTER HAVE NEGATIVE X.

$$FY_{\Delta y = 1} = -\Sigma(n \text{ SIN}^2 \phi + t_\delta \text{ COS}^2 \phi)$$

$$M_{\Delta y = 1} = -\Sigma(n \text{ SIN}^2 \phi + t_\delta \text{ COS}^2 \phi)\ X - m_\delta\ \Sigma(\text{COS } \phi)$$

$$M_{\alpha = 1} = -\Sigma\left[(n \text{ SIN}^2 \phi + t_\delta \text{ COS}^2 \phi)\ X^2\right] - 2m_\delta\ \Sigma(X \text{ COS } \phi) - Nm_\alpha$$

WHERE N = TOTAL NUMBER OF PILES IN THE FOUNDATION

FIG. 6-35 ***Analysis of Mixed-pile Groups—Method of Hrennikoff.*** *(Part I—Procedure.)* *(After Teng, Ref. 46.)*

3. SOLVE THE FOLLOWING EQUATIONS FOR Δx, Δy AND α

$$FX_{\Delta x=1}\ \Delta x + FX_{\Delta y=1}\ \Delta y + FX_{\alpha=1}\ \alpha + H = 0$$

$$FX_{\Delta y=1}\ \Delta x + FY_{\Delta y=1}\ \Delta y + FY_{\alpha=1}\ \alpha + V = 0$$

$$FX_{\alpha=1}\ \Delta x + FY_{\alpha=1}\ \Delta y + M_{\alpha=1}\ \alpha + M = 0$$

WHERE Δx, Δy, AND α ARE THE PILE FOUNDATION DISPLACEMENTS UNDER THE ACTION OF A FORCE H ALONG THE X-AXIS, V ALONG THE Y-AXIS AND A MOMENT M ABOUT THE CENTER OF THE PILE GROUP

4. DETERMINE THE PILE DISPLACEMENTS

FROM GEOMETRY FOR THE INDIVIDUAL PILE:

δ_l = LONGITUDINAL DISPLACEMENT (COMPRESSION)

$= \Delta x \cos\phi + \Delta y \sin\phi + \alpha X \sin\phi$

WHERE X = COORDINATE OF PILE HEAD

δ_t = TRANSVERSE DISPLACEMENT, TO THE RIGHT

$= \Delta x \sin\phi - \Delta y \cos\phi - \alpha X \cos\phi$

5. DETERMINE THE PILE FORCES

P_l = LONGITUDINAL FORCE (COMPRESSION) $= n\delta_l$

P_t = TRANSVERSE FORCE (ACTING ON THE PILE FOOTING TO THE RIGHT) $= -t_\delta \delta_t + M_\delta \alpha$

P_m = MOMENT (ACTING ON THE FOOTING CLOCKWISE) $= M_\delta \delta_t - M_\alpha \alpha$

FIG. 6-35 *(Continued)*

SAMPLE COMPUTATION 6-2
Analysis of Mixed Pile Groups—Method of Hrennikoff *(Part II—Sample Computations)*

STATEMENT OF PROBLEM

Determine forces in indicated pile bent due to loads, as shown. Assume all piles 10BP42, 60 ft long. Assume K (coefficient of horizontal subgrade reaction) = 35 psi per inch = 350 psi for pile width of 10 in. Pile heads assumed to be pinned.

Note: Bent and forces indicated are same as for sample computation shown in Fig. 6-34.

PROCEDURE (Steps are those enumerated in Fig. 6-35.

STEP 1: *Determine Elastic Constants for Individual Piles*

a. Vertical force (n) for unit displacement (from load test) = 90 kips

b. For horizontal displacement:

$$\beta = \sqrt[4]{\frac{k}{4EI}} = \sqrt[4]{\frac{350}{4 \times 30 \times 10^6 \times 210.8}} = \frac{1}{92} = 0.0109$$

$$t_\delta = \frac{k}{2\beta} = \frac{350}{0.0218} = 16{,}000 \text{ lb} = 16 \text{ kips}$$

$$M_\delta = \text{zero}$$

c. For rotation:

$$M_\alpha = t_\alpha = \text{zero}$$

STEP 2: *Determine Elastic Constants for Pile Group*

$\phi = 60°$	$\phi = 90°$	$\phi = 120°$
$\sin \phi = 0.866$	1.00	0.866
$\cos \phi = 0.500$	Zero	−0.500
$\sin 2\phi = 0.866$	Zero	−0.866
$\sin^2 \phi = 0.750$	1.00	0.750
$\cos^2 \phi = 0.250$	Zero	0.250

$$FX_{\Delta x=1} = -(90 \times 0.25 \times 2 + 90 \times \text{zero} \times 3 + 16 \times 0.75 \times 2 + 16 \times 1.00 \times 3) = -117 \text{ kips}$$

$$FX_{\Delta y=1} = -\tfrac{1}{2}(90 - 16)(0.866 \times 1 \times 2 + \text{zero} \times 3) = -64 \text{ kips}$$

$$M_{\Delta x=1} = -\tfrac{1}{2}(90 - 16)(30 \text{ in.} \times 0.866 - 30 \text{ in.} \times 0.866) + \text{zero} = \text{zero}$$

$$FY_{\Delta y=1} = -(90 \times 0.75 \times 2 + 90 \times 1.00 \times 3 + 16 \times 0.25 \times 2 + 16 \times \text{zero} \times 3) = -413 \text{ kips}$$

$$M_{\Delta y=1} = -(90 \times 0.75 \times 2 + 90 \times 1.00 \times 2 + 16 \times 0.25 \times 2 + 16 \times \text{zero} \times 2)(30 - 30 \text{ in.}) = \text{zero}$$

$$M_{\alpha=1} = -(90 \times 0.75 \times 2 + 90 \times 1.00 \times 2 + 16 \times 0.25 \times 2 + 16 \times \text{zero} \times 2)30 \times 30 \text{ in.} = -290{,}000 \text{ in.}^2\text{-kips}$$

STEP 3: *Solve Simultaneous Equations*

$$-117\Delta x - 64\Delta y - \text{zero} + 55 = 0$$
$$-64\Delta x - 413\Delta y - \text{zero} + 195 = 0$$
$$\text{Zero} + \text{zero} - 290{,}000\alpha + 1{,}750 = 0$$
$$\Delta x = 0.24 \text{ in.}$$
$$\Delta y = 0.44 \text{ in.}$$
$$\alpha = 0.006 \text{ radians}$$

STEP 4: *Solve for Pile Displacements*

Pile no.	$\Delta x \cos \phi + \Delta y \sin \phi + \alpha X \sin \phi = \delta_l$
1	$0.24 \times (+0.5) + 0.44 \times 0.866 + 0.006 \times 30 \times 0.866 = 0.65$ in.
2	$0.24 \times 0 + 0.44 \times 1.0 + 0.006 \times 30 \times 1.0 = 0.62$ in.
3	$0.24 \times 0 + 0.44 \times 1.0 + \text{zero} = 0.44$ in.
4	$0.24 \times 0 + 0.44 \times 1.0 + 0.006 \times (-30) \times 1.0 = 0.26$ in.
5	$0.24 \times (-5) + 0.44 \times 0.866 + 0.006 \times (-30) \times 0.866 = 0.11$ in.

STEP 5: *Solve for Pile Loads*

Pile no.	Load $= n\delta_l$
1	$90 \times 0.65 = 60$ kips
2	$90 \times 0.62 = 56$ kips
3	$90 \times 0.44 = 40$ kips
4	$90 \times 0.26 = 24$ kips
5	$90 \times 0.11 = 10$ kips

n = Vertical force for unit displacement (step 1*a*)
δ_l = Displacement

6C.3 Bracing of Piles, and Other Matters Relating to the Resistance of Piles to Lateral Loads

a. Bracing of Plumb Piles

In Sec. 6A.2, it was noted that, because of differences in pile length, eccentricity of the centroid of the pile group with respect to the column reaction, deformation of the pile cap, and other factors, some bending moment exists in virtually all piles. For static equilibrium, this bending moment necessitates the development of lateral resistance in the abutting soil.

The amount of lateral resistance which can be developed, depending on the size and type of pile and the type of soil, is in the order of 1 to 4 tons per pile (see Table 6-6).

The resistance which would be required for, say, a 60-ton pile where the applied load is at the limit of usual tolerances,* assuming a factor of safety of 2.0, would be 4½ to 7½ tons for a pile embedded in a loose granular soil† and 1½ to 5 tons‡ for a pile embedded in soft or loose cohesive soil. (For piles of less than 60 tons capacity, the required force would be less.) Thus, for high-capacity piles and, to a lesser extent for piles of lower capacity, eccentricities are developed in the pile section, due to normal vagaries of installation and length, which require the development of lateral resistance which is a substantial part of, or greater than, the lateral resistance which is available in the soil. As a result, most

* Usually 3 in. allowable out-of-position and 2 percent allowable deviation from plumb or the specified batter.

† Figure 6-30*b*. Diameter of pile (typically) would be 8 to 14 in. Assume γ = 0.06 pci, $K_p = 3.0$, $c_u = 1.5$ psi. The applied moment is 60 tons $\times$ 3 in. = 360,000 in-lb. For factor of safety = 2.0, ultimate design moment M_y would equal 720,000 in-lb. Therefore, $\frac{M}{D^4\gamma K_p} = 100$ to 1,000. Entering Fig. 6-30*b*, for a restrained pile, $\frac{P_{\text{ult}}}{K_p D^3 \gamma}$ would equal 60 to 200. The required P_{ult} = 9,000 to 15,000 lb (lateral), which exceeds the available resistance.

‡ Repeating the above procedure, M/C_uD^3 would equal 12 to 1,000. Entering Fig. 6-30*a*, read $P_{\text{ult}}/C_uD^2 = 10$ to $100^{(+)}$. Calculate required P_{ult} = 3,000 to 10,000 lb (lateral), which is about equal to the available capacity.

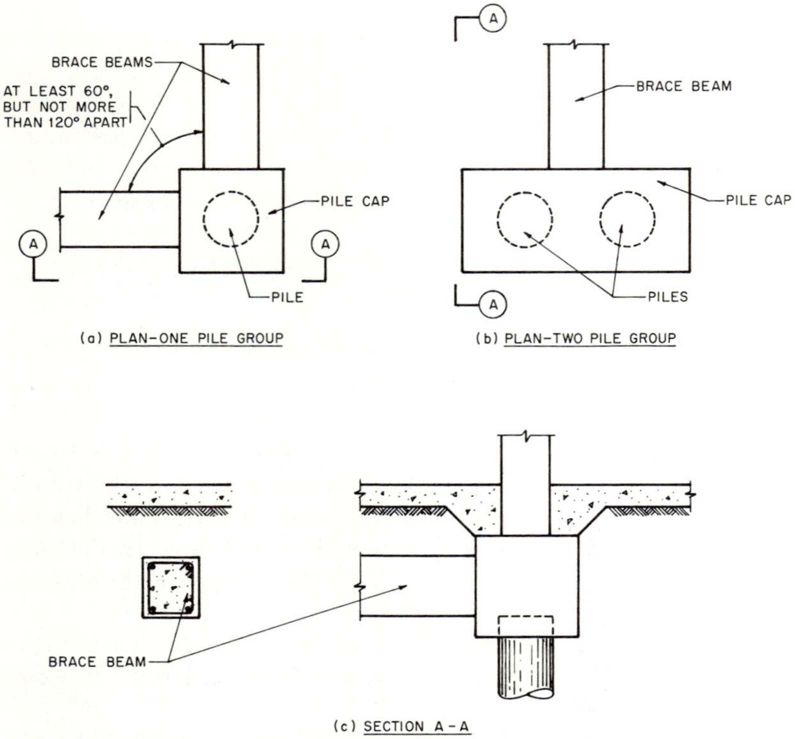

FIG. 6-36 Bracing of One- and Two-pile Groups.

building regulations require that one- and two-pile groups be braced to provide supplementary lateral resistance. Single-pile groups are required to be braced in two directions. Two-pile groups (or lines of piles) need be braced only in one direction (see Fig. 6-36). For larger groups (three piles, or more) confined by a rigid cap, the eccentricity is assumed to cause an increase (or decrease) in the vertical pile load, rather than developing lateral movement (see Fig. 6-37).

The brace should be designed for an axial compression equal to 3 to 5 percent* of the axial capacity of the piles being braced, plus a lateral

* These figures are derived from the following:

1. For the previously indicated example of a 60-ton pile and a 3-in. eccentricity of the load line, the lateral force required to prevent lateral movement is ½ to 3 tons, i.e., between 1½ and 6 percent of the axial load. These are upper limit values, however, because (1) some lateral displacement can be tolerated so that the pile need not be fully fixed and (2) the moment resistance of the brace beam counters the applied moment due to eccentricity.

2. The British "Civil Engineering Code of Practice" [47] prescribes a 5 percent value.

3. The resistance required to brace an axially loaded compression member [49] is about 2 to 4 percent of the axial load.

load equal to the overlying weight of the soil, slab-on-grade, and live load on said slab-on-grade.* Design of the concrete brace should conform to ACI-318 [4].

The braces must transfer their horizontal reaction to some abutment. A larger pile group (three piles, or more) usually is considered as an ample

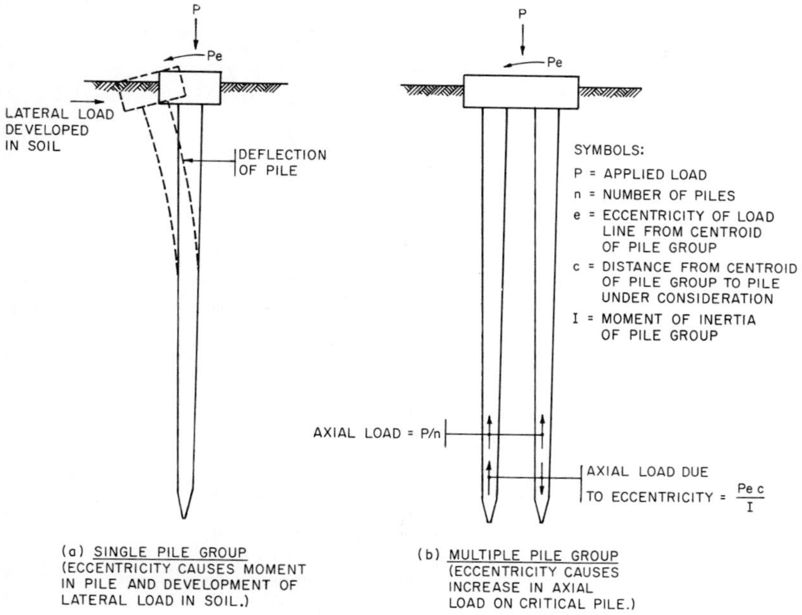

FIG. 6-37 Effect of Eccentricity of Load Line on Pile Load.

abutment. Alternatively, batter piles, an anchor wall, or a deadman anchor may be used in lieu of brace beams.†

b. Bracing of Batter Piles

Batter piles are not used singly, but pairs must be braced transverse to the line of batter, as for groups of two of plumb piles.

In addition, sometimes it is overlooked that, in order to develop a horizontal reaction in a batter pile, there must be a horizontal action (force); i.e., the sum of the horizontal forces must balance. For example,

* The lateral load requirement is derived from the probability that the soil near the ground surface will be compressible (otherwise piles would not be required) and so will tend to settle away from under the brace.

† The use of brace beams also may be omitted when the piles (or pile caps) are braced by the adjacent ground or basement floor slab, provided that the piles (or caps) are adequately tied to the slab and that the distance from the bottom of the piles cap to the slab level is not so large that the flexibility of the columns or walls invalidates the bracing action of the slab.

in the pile arrangement indicated in Fig. 6-38*a*, the batter piles are ineffective because there is no counterbalance for the horizontal component. Similarly, the horizontal component must necessarily be accompanied by a vertical component so that the arrangement indicated in Fig. 6-38*b* also

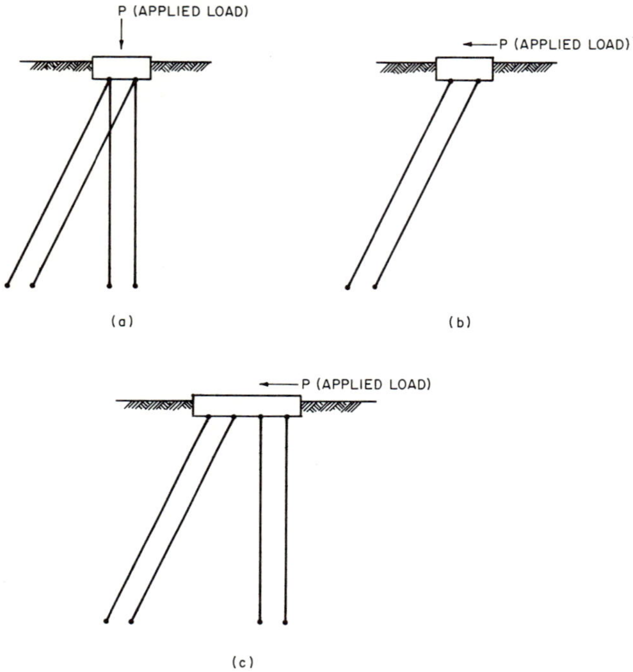

FIG. 6-38 Bracing of Batter Piles.

is ineffective and that in Fig. 6-38*c* works only if there is adequate pullout capacity in the plumb piles.

c. Short Piles

There are many occasions when it is desirable to use short piles (less than 10 ft long). Under some circumstances (such as when a shallow depth of running ground is encountered directly overlying rock or when piles are being driven to rock and suddenly a few of the piles in a partly driven group fetch up short because of irregularities in the rock surface) piles less than 10 ft long may be economically desirable in lieu of the use of footings at the lower level.

The problem is that many engineers (and some building codes) do not consider anything less than 10 ft long to be a pile. The apparent reason for this is the problem of lateral stability discussed above. Since it is commonly considered that conventional building piles are fixed in the ground at a depth of 5 to 10 ft, a pile 10 ft or greater in length would have substantial resistance to overturning and so would be acceptable. Shorter piles could be rotated or shaken out of the ground with comparative ease and would be unacceptable.

However, the required lateral resistance can be achieved by other means. For example, if the pile is fixed into the cap and the cap is braced so that the pile-cap-brace continuum acts as a rigid frame and if provision is made to prevent sliding of the foundation at the interface between the pile tips and the bearing stratum,* there is no reason why very short piles cannot be used.

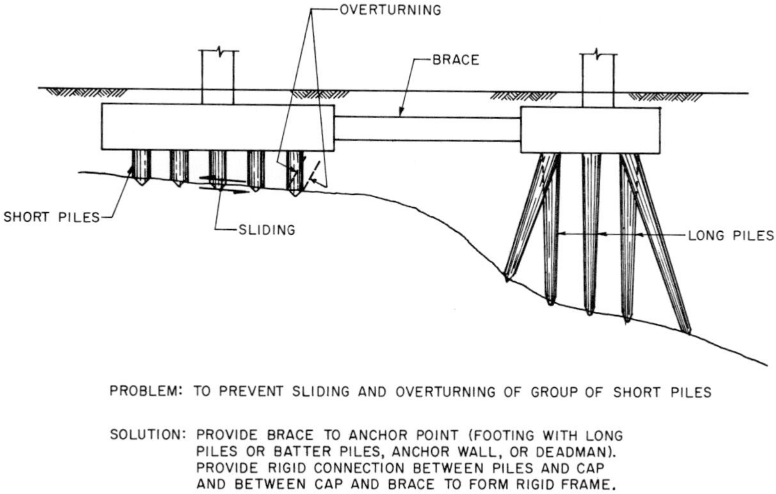

FIG. 6-39 Bracing of Short Piles.

d. Consonance of Lateral Resistance Components

When there is a large lateral load to be resisted, the required number of piles may be controlled by lateral, rather than by vertical, load. In such a case, it often occurs to the designer to utilize the passive pressure of the soil abutting the foundation (within the limits indicated in Sec. 4H) to resist part of the lateral force and to use piles to resist the rest. Unfortunately, this seldom works, and the reader is cautioned against the practice. The reason is that the passive resistance cannot be developed within the limits of lateral displacement which can be tolerated in the structure.†

* See Fig. 6-39.

† As a rule of thumb, the lateral displacement required to develop full passive pressure on a vertical face presentment is about 1 to 2 percent of the embedded height of the face (measured from the ground surface).

A similar situation prevails with regard to simultaneous development of the lateral resistances of plumb and batter piles in the same pile group. The plumb piles, generally, are unable to develop significant lateral load capacity within the limits of lateral displacement, which occur in the batter piles, from shortening and displacement in the axial direction.*

e. Splicing of Piles

The common prohibition against splicing piles within the upper 10-ft section also relates to the problem of lateral stability and is intended to

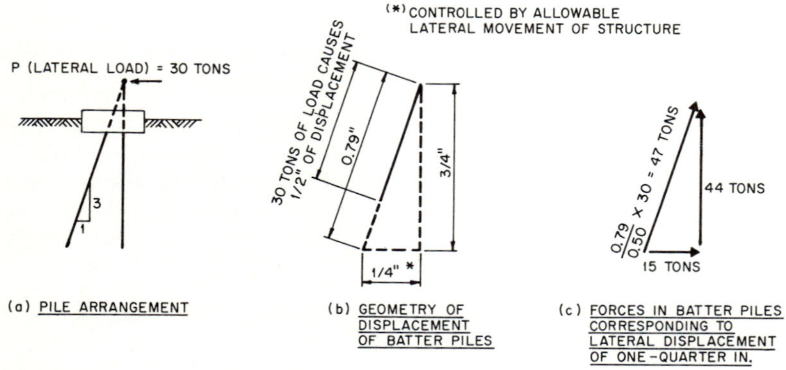

FIG. 6-40 Consonance of Lateral Resistance Components.

prevent the possible formation of a hinge mechanism within the portion of the piles which has relatively poor lateral support.

f. Special Requirement for Bracing of Pile Groups Where Earthquake Is a Factor in the Design

In this case, there is a special need to prevent the individual pile caps from moving relative to each other under the influence of the earthquake vibrations. To satisfy this requirement, ties are provided to interconnect all the pile caps. The ties should be designed to resist (in tension or compression) a horizontal force equal to 10 percent of the larger pile cap loading [54]. The ties also must be capable of resisting the lateral loads (weight of overlying soil, slab, and live load), as previously indicated.

* For example, assume that there is an arrangement of plumb and batter piles as shown in Fig. 6-40 and that the piles would undergo an axial displacement of ½ in. under the design load of 30 tons. The horizontal load corresponding to ¼ in. of lateral displacement would be 15 tons (see Fig. 6-40) for the batter piles and only 1 to 2 tons for the plumb piles. Therefore the batter piles carry virtually all (90 to 95 percent) of the lateral load.

Section 6D RESISTANCE TO UPLIFT FORCES

Resistance to uplift forces is provided by the frictional and cohesive resistances developed along the pile shaft.

These resistances are represented by the second term of Eq. (5-4). As previously indicated, evaluation of this term (whether by theoretical or laboratory procedures) is unreliable. Accordingly, recourse again usually is made to empirical procedures such as load tests or rules of thumb, as follows.

6D.1 Load Tests

A pullout load test is an excellent device for approximating resistance to uplift.* The total pullout resistance of a pile group is not the sum of the resistances of the individual piles, however, and extrapolation from load test results to pullout capacity of a group must be made. One method is indicated in Fig. 6-41.

In lieu of performing load tests, the computations indicated in Fig. 6-41 may be made on the basis of assumed failure planes or values of cohesion, either as dictated by experience, or using known, conservative values.

CAUTIONS:

1. Uplift resistance in sand or gravel (cohesionless) soils tends to be relatively low, and the angle $\phi/2$ (Fig. 6-41) may be substantially less than the value of 30° conventionally assumed when calculating resistance to downward loads.

2. Piles installed in cohesionless soils and subject to a vibratory load are best assumed to have zero resistance to uplift, unless the bottom of the pile is belled or other provisions are made to develop a positive mechanical anchorage. The loss of resistance due to vibration would not be reflected in a static load test.

3. Piles subject to lateral as well as uplift loads tend to dissipate part of the uplift resistance because of lateral motion which may leave an open annular space around the upper part of the pile. This effect, also, would not be reflected in a static load test.

6D.2 Rules of Thumb

Rules of thumb relating pullout capacity of a pile to its capacity to resist downward axial load are in common use. One such rule is that for

* A point to be remembered is that the frictional resistance to uplift on a pile is not necessarily the same as the resistance to that same load on the same pile when it acts to press the pile into the soil. Aside from the obvious effects of taper, the uplift load relieves the vertical pressure in the soil and so tends to relieve the lateral (normal) pressure between the soil and the pile. This decrease in normal pressure causes a corresponding decrease in frictional resistance. It is for this reason that apparent values of frictional resistance derived from conventional load tests (see Chap. 9) do not necessarily apply in making an estimate of uplift capacity.

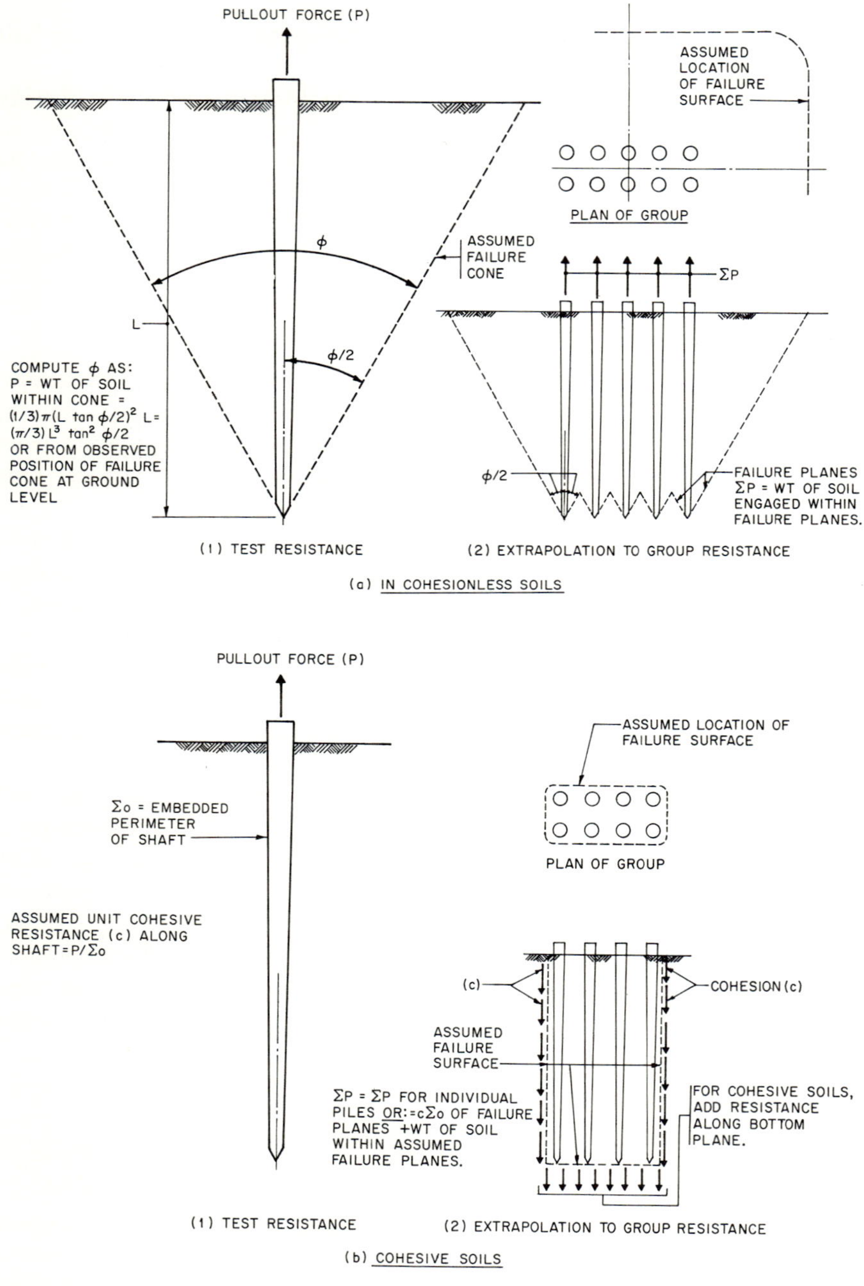

FIG. 6-41 Pullout Capacity Determined by Test.

friction piles pullout capacity is assumed to equal one-third to one-half the capacity to resist compressive load.

A guide rule such as this does not apply to short piles, to end-bearing piles, to piles where creep is a factor, or to a pile that is subject to an oscillatory load which would shake it out of the ground. Within its

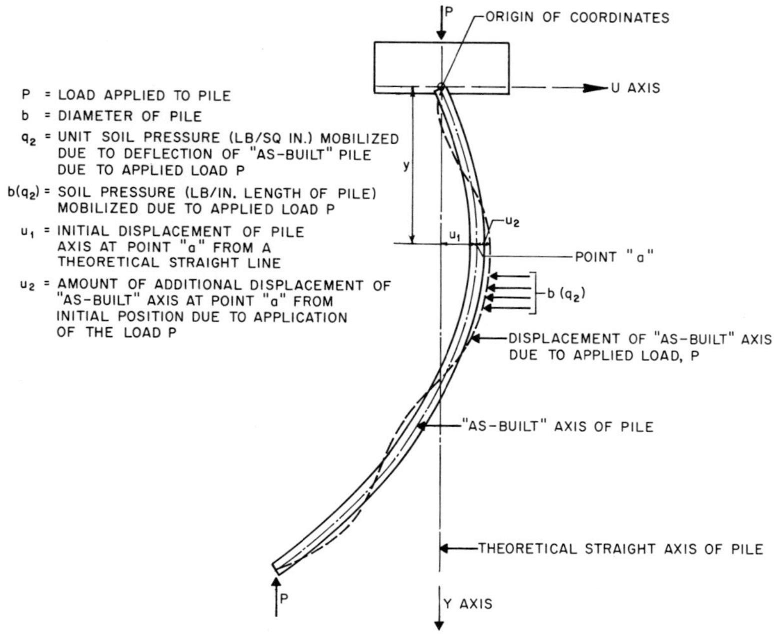

FIG. 6-42 Diagrammatic Representation of Pile and Notation.

limited range of application, however, the rule seems to give reasonable results.

Section 6E CHAPTER NOTES

1 Equivalent Unbraced Length of the Pile Column

The condition and the applicable symbolism to be considered are shown in Fig. 6-42. Let u_1, q_1, M_1, and S_1 be, respectively, the lateral displacement, the lateral pressure, the bending moment, and the shear force on the pile prior to the application of the design load P, and assume a purely end-bearing pile in which the entire applied load P is transmitted to the point of the pile.

Let u_2, q_2, M_2, and S_2 be, respectively, the *additional* lateral displacement, lateral pressure, bending moment, and shear force on the pile due to adding the load P.

After application of the load P, the pile is in equilibrium under the action of two systems of forces, the system (P,H_p) and the system (q_2,H_2) (see Fig. 6-43). The first system has the relation

$$H_p = \frac{Pc}{L} \tag{A1}$$

and the moment M_p in the pile section at any point, at y distance from

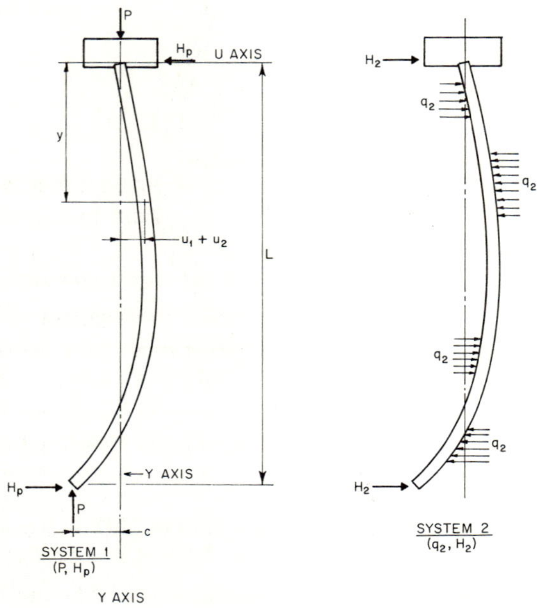

FIG. 6-43 Force Systems Acting on Pile.

the top, due to the first system of forces may be expressed as

$$M_p = P(u_1 + u_2) - H_p y = P\left(u_1 + u_2 - \frac{cy}{L}\right) \tag{A2}$$

The moment M_q due to the second system of forces may be defined by the fundamental expression

$$\frac{d^2M_q}{dy^2} = q_2 b \tag{A3}$$

The total moment at any point of the pile is

$$M_2 = M_p + M_q \tag{A4}$$

The moment M_2 is related to the second force system by the fundamental expression

$$M_2 = -E_2 I_2 \frac{d^2 u_2}{dy^2} \tag{A5}$$

NOTE: E_2 and I_2 refer to the final material and section properties of the pile, which (because of filling with concrete, as in the case of a pipe pile) may differ from E_1 and I_1 at the time of driving the pile. Thus

$$\frac{d^2 M_2}{dy^2} = \frac{d^2 M_p}{dy^2} + \frac{d^2 M_q}{dy^2} = P\left(\frac{d^2 u_1}{dy^2} + \frac{d^2 u_2}{dy^2}\right) + q_2 b = -E_2 I_2 \frac{d^4 u_2}{dy^4}$$

or

$$\frac{d^4 u_2}{dy^4} + \frac{P}{E_2 I_2}\frac{d^2 u_2}{dy^2} + \frac{q_2 b}{E_2 I_2} = \frac{-P}{E_2 I_2}\frac{d^2 u_1}{dy^2} \tag{A6}$$

It will be seen that the right-hand term of Eq. (A6) is defined by the initially bent shape of the pile and represents the rate of change of curvature of the pile axis before applying the load P.

Equation (A6) is the general differential equation used to derive expressions relating the equivalent unbraced length of the pile to the properties of the pile and the soil. Several solutions have been derived.

a. Conventional Solution

This solution assumes the initial shape of the pile to take the form of a Fourier series:

$$u_1 = \sum_n c_n \sin \frac{n\pi y}{L} \qquad n = 1, 2, 3 \ldots \tag{A7}$$

where c_n is a constant length.

If one assumes that there is a linear correlation between q_2 and u_2, namely,

$$q_2 = \kappa u_2 \tag{A8}$$

where κ is the so-called foundation modulus (pounds per cubic inch), then Eq. (A6) has the following solution satisfying the boundary conditions of no displacement or moment at top and bottom:

$$u_2 = \sum_n d_n \sin \frac{n\pi y}{L} \qquad n = 1, 2, 3 \ldots \tag{A9}$$

where $d_n = \dfrac{c_n P}{\kappa b L^2/n^2\pi^2 + (n^2\pi^2 E_2 I_2 - P)/L^2}$

Then from Eq. (A5),

$$M = P \sum_n \frac{c_n}{1 + \kappa L^2 b/n^2\pi^2 P_n - P/P_n} \tag{A10}$$

where $P_n = n^2\pi^2E_2I_2/L^2$ = Euler's buckling load for the nth mode. Equations (A9) and (A10) determine the initial (buckling) load P_{cr} as the value of P when the denominator becomes zero, namely,

$$P_{cr} = P_n + \frac{\kappa L^2 b}{n^2\pi^2} \tag{A11}$$

The critical or lowest buckling mode is that obtained by differentiating with respect to n and setting equal to zero, i.e.,

$$P_{cr} = 2\sqrt{\kappa bEI} \tag{A12}$$

The equivalent effective length (KL) is obtained by equating P_{cr} from Eq. (A12) to the Euler load $\pi^2EI/(KL)^2$ to obtain

$$KL = 0.70\pi\sqrt[4]{\frac{E_2I_2}{\kappa b}} \tag{A13}$$

or $$KL = 0.70\pi\sqrt[4]{\frac{E_2I_2}{c}}$$

where KL = equivalent unbraced length, in.
c = foundation modulus, psi
E_2I_2 = final stiffness of the pile, lb-in.^{2}

This solution corresponds to that presented by Cummings [8] and is cited by Bjerrum [9] (see also Timoshenko [10]).

For a round pile, I may be expressed as $0.05d^4$, where d is the pile diameter in inches, and c may be expressed as κd, where κ is the foundation modulus expressed in pounds per cubic inch.

Then

$$KL = 0.7\pi\sqrt[4]{\frac{E(0.05)d^4}{\kappa d}}$$

or $$\frac{KL}{d} = 1.04\sqrt[4]{\frac{E}{\kappa d}} \tag{A14}$$

From Eqs. (A13) and (A14), critical values of KL/d may be plotted as in Figs. 6-3 to 6-5 (Sec. 6A.1) for steel, concrete, and timber sections.

It should be noted that beyond the elastic range, E decreases with increasing stress according to the tangent modulus and that the value of P_{cr} obviously cannot exceed that value of P which corresponds to yielding or crushing of the cross section.

Unfortunately, this solution has some deficiencies. First, it does not take account of the initial (residual) bending moments in the pile prior to the application of the load. Second, the shape presupposes the existence of a lateral shear at the top of the pile during and following driving, which usually is contrary to the physical facts. Last, the assumption of a constant foundation modulus is incorrect.

These deficiencies are discussed in some of the more extended papers on the subject, e.g., Walter [11].

b. Solution of Marcus [12]

The solution of Marcus represents an initially deflected shape of the pile as shown in Fig. 6-42 and does not involve the supposition of lateral shear at the top of the pile. As with the conventional solution, however,

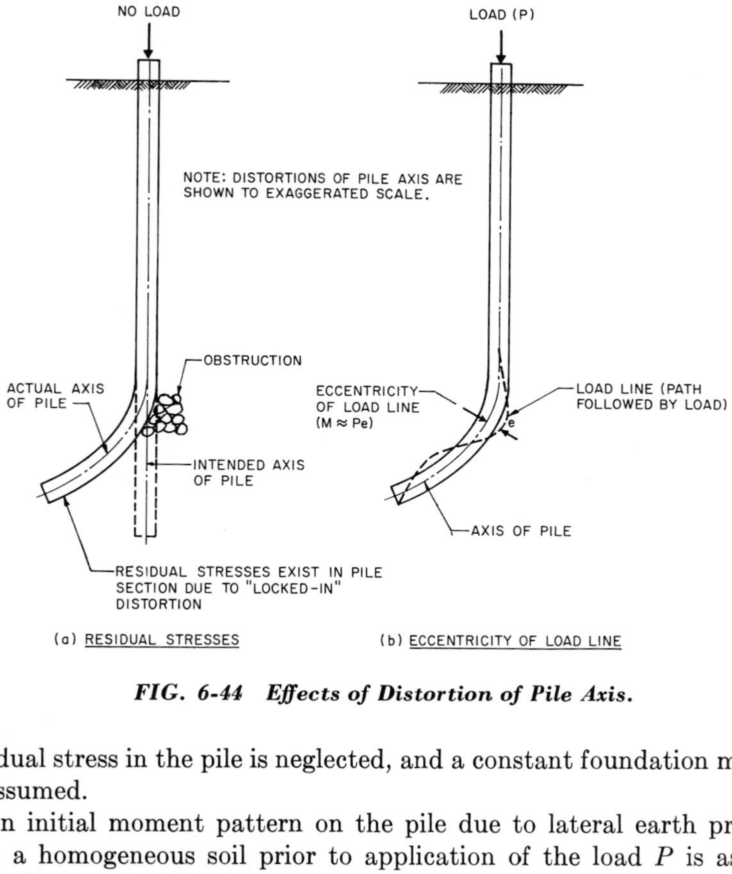

FIG. 6-44 Effects of Distortion of Pile Axis.

residual stress in the pile is neglected, and a constant foundation modulus is assumed.

An initial moment pattern on the pile due to lateral earth pressures q in a homogeneous soil prior to application of the load P is assumed according to the following equation.

$$M_1 = M_q = M_c(\sin ky - ky \cos ky) \tag{A15}$$

where M_c is a constant bending moment and k a constant parameter. This assumption corresponds roughly to the deflected shape usually observed in practice (Figs. 6-42 and 6-44). The boundary conditions at the ends ($M = V = q = 0$ at $y = 0$, and $M = 0$ at $y = L$) in the initially unloaded pile are satisfied.

Assuming for convenience that Eq. (A8) is valid after the load P is applied, the introduction of Eq. (A15) in Eq. (A6) yields

$$\frac{d^4u_2}{dy^4} + \frac{P}{E_2I_2}\frac{d^2u_2}{dy^2} + \frac{\kappa b u_2}{E_2I_2} = \frac{P}{E_1I_1}\frac{M_c}{E_1I_1}(\sin ky - ky \cos ky) \tag{A16}$$

A particular solution of this equation is

$$u_2' = f_1 \sin ky - f_2\, ky \cos ky \tag{A17}$$

where f_1 and f_2 are constant lengths to be determined from the equations

$$\begin{aligned} f_1\left(k^4 - \frac{P}{E_2I_2}k^2 + \frac{b\kappa}{E_2I_2}\right) - 2f_2k^2\left(2k^2 - \frac{P}{E_2I_2}\right) &= \frac{P}{E_2I_2}\frac{M_c}{E_1I_1} \\ f_2\left(k^4 - \frac{P}{E_2I_2}k^2 + \frac{b\kappa}{E_2I_2}\right) &= \frac{P}{E_2I_2}\frac{M_c}{E_1I_1} \end{aligned} \tag{A18}$$

or with the notation

$$k^4 - \frac{P}{E_2I_2}k^2 + \frac{b\kappa}{E_2I_2} = \frac{1}{a^4} \tag{A19}$$

$$\begin{aligned} f_2 &= \frac{Pa^4M_c}{E_2I_2E_1I_1} \\ f_1 &= \frac{Pa^4M_c}{E_2I_2E_1I_1}\left[1 + 2k^2L^2\left(2k^2L^2 - \frac{PL^2}{E_2I_2}\right)\frac{a^4}{L^4}\right] \end{aligned} \tag{A20}$$

A second value u_2'' is determined by the homogeneous equation

$$\frac{d^4u_2''}{dy^4} + \frac{P}{E_2I_2}\frac{d^2u_2''}{dy^2} + \frac{\kappa b u_2''}{E_2I_2} = 0 \tag{A21}$$

the solution of which is

$$\begin{aligned} u_2'' = {} & C_1 \sin(m_1y) \operatorname{Cosh}(m_2y) + C_2 \cos(m_1y) \operatorname{Sinh}(m_2y) \\ & + D_1 \sin(m_1y) \operatorname{Sinh}(m_2y) + D_2 \cos(m_1y) \operatorname{Cosh}(m_2y) \end{aligned} \tag{A22}$$

where C_1, C_2, D_1, and D_2 are constant lengths and m_1 and m_2 are parameters determined by the conditions

$$m_1^2 = \sqrt{\frac{\kappa b}{4E_2I_2}} + \frac{P}{4E_2I_2} \qquad m_2^2 = \sqrt{\frac{\kappa b}{4E_2I_2}} - \frac{P}{4E_2I_2} \tag{A23}$$

The values of the constants C_1, C_2, D_1, and D_2 are determined from the boundary conditions

$$u_2 = u_2' + u_2'' = 0 \qquad \text{for } y = 0 \text{ and } y = L$$

$$\text{and } M_2 = -E_2I_2\left(\frac{d^2u_2'}{dy^2} + \frac{d^2u_2''}{dy^2}\right) = 0 \qquad \text{for } y = 0 \text{ and } y = L$$

for $y = 0$, $D_1 = D_2 = 0$. For $y = L$, it is required that

$$C_1 \sin (m_1L) \operatorname{Cosh} (m_2L) + C_2 \cos (m_1L) \operatorname{Sinh} (m_2L) = -f_1 \sin kL + f_2kL \cos kL \quad \text{(A24)}$$

and

$$C_1[2m_1m_2 \cos (m_1L) \operatorname{Sinh} (m_2L) + (m_2^2 - m_1^2) \sin (m_1L) \operatorname{Cosh} (m_2L)] + C_2[m_2^2 - m_1^2 \cos (m_1L) \operatorname{Sinh} (m_2L) - 2m_1m_2 \sin (m_1L) \operatorname{Cosh} (m_2L)] = k^2[f_1 \sin kL - f_2(2 \sin kL) + kL \cos (kL)] \quad \text{(A25)}$$

Equations (A24) and (A25) determine C_1 and C_2 in terms of m_1, m_2, P, M_c, and k, and can be used to determine the values of u_2'' and M_2. To the bending stresses determined from M_2, of course, must be added the axial stress P/A and the initial residual bending stress due to M_1.

It is of interest to note that this treatment yields critical values for P in Eqs. (A18), wherein if P has the value

$$P_{cr} = \frac{E_2I_2}{L^2} k^2L^2 + \frac{\kappa bL^2}{k^2L^2} \quad \text{(A26)}$$

f_1 and f_2 become infinitely great. This value differs from Eq. (A11) for $n = 1$ only insofar as k^2L^2 is substituted for $n^2\pi^2$.

Also in Eq. (A23), the parameter m_2 is a real quantity, provided P is equal to or less than the value defined by

$$P_{cr} = 2\sqrt{\kappa bE_2I_2} \quad \text{(A27)}$$

which is identical with Eq. (A12).

Further, if the determinant of Eqs. (A24) and (A25) becomes zero, a third critical condition derives therefrom; namely, $m_2 = 0$, $m_1 = \sqrt[4]{\kappa b/E_2I_2}$ and again $P_{cr} = 2\sqrt{\kappa bE_2I_2}$ as in Eq. (A27).

Equations (A16) through (A25), plus the values $m_2 = 0$ and

$$m_1 = \sqrt[4]{\kappa b/E_2I_2}$$

define the solution to the stress condition in the pile. Such solution is, of course, limited to consideration of an end-bearing pile, but it is well known that the effect of a given deformation of the pile axis is less critical for a friction pile than for one loaded in end bearing, and so the solution represents a conservative upper limit to the stresses existent in the member.

2 Residual Stresses Due to the Initial Deformation of the Pile Axis

In addition to the axial and bending stresses caused by applying the external load P to the pile (see Refs. 15–17), there are residual stresses locked into the pile section because of its initial deformation. Three cases must be considered.

1. The first is that of a cast-in-place concrete pile constructed using a thin shell casing. In such case, the concrete fill is initially unstressed

(being placed after the pile shell is in place), and the shell is not normally counted on to carry stress. Accordingly, the thin shell cast-in-place pile may be considered as initially unstressed.

2. The second case is that of a steel pipe or H pile. Here, the steel section has residual stress due to its initial curvature. Two subcases must be considered.

a. The first is where the pile is kinked or otherwise very sharply curved, locally. Such cases occur, and measurements made of the curvature of actual bent piles show that the stresses at points of sharp curvature generally equal the yield stress of the material. However, it must be considered that these are *residual* stresses and not stresses due to an applied, external load. It is well known that the two types of stress differ in their effects and that, for materials which yield plastically, a residual stress pattern in a highly indeterminate structure has no significant effect on the ultimate load capacity of the member, provided that failure due to instability is prevented. In fact, this is the basis for current theories and practice of plastic design.

The instability problem must be considered, of course, since the presence of residual stress in a compression member tends to depress the critical buckling load [10]. The effect, however, does not appear to be significant for relatively low ratios of L/r (say 40, or less) [14] and therefore may be neglected insofar as piling is concerned.

b. The second subcase is where the pile is smoothly and uniformly curved, without any sharp kinks, i.e., points of high local stress concentration. Equation (A15) may be used to evaluate the magnitude of initial (residual) stresses for such condition. The values of shear forces S and lateral pressure q_1 are*

$$Sq = \frac{dM_q}{dy} = kM_c ky \sin ky \tag{A28}$$

$$bq = +\frac{d^2M_q}{dy^2} = k^2M_c(\sin ky + ky \cos ky) \tag{A29}$$

Note that $M_q = S_q = q = 0$ for $y = 0$. The condition $M_q = 0$ for $y = L$ is satisfied provided

$$\sin kL - kL \cos kL = 0 \tag{A30}$$

This equation has a series of roots, k_1, k_2, k_3 . . . . The deflection u_q associated with the bending moment $M_1 = M_q$ is determined by the equation

$$E_1I_1 \frac{d^2u_q}{dy^2} = -M_q = M_c(ky \cos ky - \sin ky) \tag{A31}$$

* The following analysis is by Dr. Henri Marcus.

One solution of this equation is

$$E_1 I_1 u_q = \frac{1}{k^2} M_c(3 \sin ky - ky \cos ky - 2ky) \tag{A32}$$

$$\text{with } E_1 I_1 \frac{du_q}{dy} = \frac{1}{k} M_c[ky \sin ky - 2(1 - \cos ky)] \tag{A33}$$

The boundary values for $y = 0$, $M_q = du_q/dy = 0$, are satisfied. If $u_q = +c$ at $y = 1$, then

$$M_c = \frac{-E_1 I_1 ck^2}{2kL + kL \cos kL - 3 \sin kL} = \frac{E_1 I_1 ck^2}{2(kL - \sin kL)} \tag{A34}$$

This equation determines the moment M_c in terms of the displacement of the tip of the pile relative to the top of the pile which, in turn, may be related to the residual moment M_q by Eq. (A31). As indicated in the following example, the residual stresses are considerable. However, as described above, the effects of residual stresses on column capacity are different from the effects of stresses due to applied loads and, for highly indeterminate members like piles, may be neglected.

EXAMPLE: Consider an unfilled steel pipe pile, 12 in. OD, with a thickness $t = 0.20$ in., a moment of inertia $I_1 = 136$ in.^{4}, and a length $L = 100$ ft ($kL = 4.494$). The tip is displaced horizontally relative to the top by $c = 24$ in., i.e., 2 percent of the pile length. From Eq. (A34),

$$M_c = \frac{(30 \times 10^6) \times (136)(24) \times (4.494)^2}{2 \times (4.494 + 0.976) \times (1{,}200)^2} = -125{,}520 \text{ in.-lb}$$

A point of maximum curvature occurs approximately at the point $y = 5L/6$. For this point, Eq. (A31) gives

$$M_q = +125{,}520(3.75 \cos 3.75 - \sin 3.75)$$
$$= 125{,}520(-3.07 + 0.57) = 313{,}000 \text{ in.-lb}$$
$$f = \frac{Mc}{I} = \frac{313{,}000 \times 6}{136} = 13{,}800 \text{ psi}$$

3. The third case is that of timber or precast concrete piles. Here, the material is initially stressed and has no well-defined yield point. Accordingly, the problem of residual stress is much more important than with steel piling.

Theoretically, the stresses should be evaluated, per Eq. (A31). This is not practical, however, as the offsets of the pile cannot be measured. Therefore, the effects of distortion of the pile axis, for precast and timber piles, are a matter of conjecture which must be considered in establishing the required safety factor.

3 Group Reduction Factors

a. The Method of Feld [27]

In this method, the value of load capacity of an individual pile in a group of friction piles is the value of an isolated single pile, less one-sixteenth for each adjacent diagonal or row pile (see Fig. 6-45).

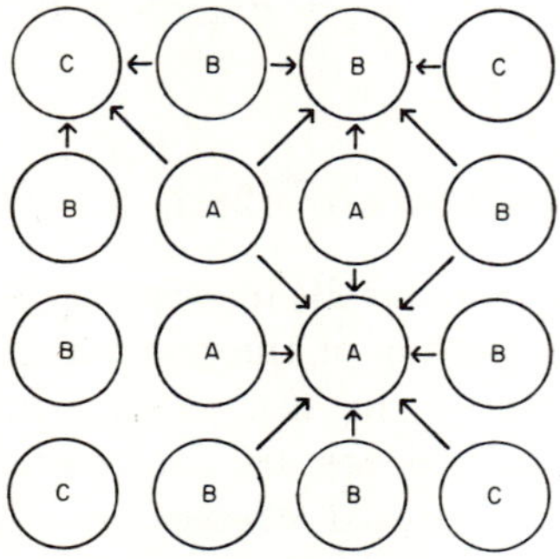

PLAN OF PILE GROUP

CAPACITY:

(A) PILES: 4 @ $\left(\frac{16}{16} - \frac{8}{16}\right) = \frac{32}{16}$

(B) PILES: 8 @ $\left(\frac{16}{16} - \frac{5}{16}\right) = \frac{88}{16}$

(C) PILES: 4 @ $\left(\frac{16}{16} - \frac{3}{16}\right) = \frac{52}{16}$

$\frac{172}{16}$ = 10.76 PILES

FIG. 6-45 Group-reduction Formula of Feld.

b. The Equation of Converse-Labarre

This equation, which is specified in the Uniform Building Code and in AASHO specifications, is

$$N = \text{efficiency} = 1 - \theta\left[\frac{(n-1)m + (m-1)n}{90mn}\right]$$

where n = number of piles per row
m = number of rows
θ = arc tan d/s (degrees)
d = pile diameter, ft
s = pile spacing (center to center), ft

The pile capacity, per pile in the group, is taken as N times the capacity of an isolated pile of the same construction.

c. The Los Angeles Group Action Formula

This formula is similar to that of Converse-Labarre, but using radians instead of degrees. The formula is

$$\text{Efficiency} = 1 - \frac{d}{\pi smn}[m(n-1) + n(m-1) + \sqrt{2}\,(m-1)(n-1)]$$

where n = number of piles per row
m = number of rows
s = pile spacing (center to center), ft
d = pile diameter, ft

d. The Seiler-Keeney Formula [Ref. 29]

The formula is

$$\text{Efficiency} = \left[1 - \frac{11s}{7(s^2-1)}\frac{(m+n-2)}{(m+n-1)}\right] + \frac{0.3}{m+n}$$

where the symbols are as for the Los Angeles formula.

The above formulations all have an empirical or semiempirical basis.

4 Side Shear along Pile Surface Embedded in Clay Soil

The side shear along a pile surface embedded in clay soil may be approximated from the empirical data of Table 4-2. However, the range of values indicated in this table for a given soil is large, and the necessary interpolation involves considerable uncertainty. Therefore, many designers prefer to predicate the assumed side shear resistance on the basis of laboratory shear tests performed on material recovered from borings taken at the site.* Assumptions are made (1) that the shear strength so determined is fully developed and (2) that no frictional resistance develops, i.e., that the side shear is a function only of the shear strength of the soil and is independent of the normal pressure.

* The use of soil shear strength values in lieu of values of adhesion between soil and pile appears to be based on the common observation that piles withdrawn from clay soils show a veneer of soil adhering to the pile, indicating that the adhesive resistance between pile and soil exceeds the shear strength of the soil. There are occasions, however, particularly when driving into very sensitive or hard clays when this is not true. In the case of the sensitive clay, the remolding incident to driving reduces the strength of the soil in a limited area adjacent to the pile. This has been discussed in connection with the computation of drag forces (Sec. 4D). In the case of the hard clay, the pile tip may open a hole slightly larger than the pile shaft, an opening which remains open, obviating the grip of the soil along the shaft. This may occur, for example, with pipe or cast-in-place shell piles, when the closure plate is a little larger than the diameter of the smooth shaft [28] or with H piles, where the tip is reinforced by the addition of plates [55].

Criteria which are used for estimating the shear strength of the soil include (1) one-half the unconfined compressive strength $c = q_u/2$ [24, 36, 37], (2) yield strength as indicated by the ring shear test [50], which amounts to about one-quarter the unconfined compressive strength, and (3) $0.35q_u$ [53]. The range of values, therefore, is $0.25q_u$ to $0.50q_u$, that is, 100 percent which makes for a considerable variation in results.*

The proponents of the use of ring-shear data point out that the unconfined compression test represents soil strength under short-term loading conditions, whereas the critical condition, in place, would be the long-term yield strength, which is best measured by the ring-shear apparatus.

On the other hand, the unconfined compressive strength is widely used and apparently with success.†

The recommended value $0.35q_u$ appears in Special Report 36 of the Highway Research Board and is predicated on load test data. Subsequent studies (Special Report 67 of the Highway Research Board) indicates that this value may be too high for some types of pile.

No conclusions in this matter are offered. All three (and other) values are used by competent designers. The reader must decide for himself, perhaps using the more conservative approach for sensitive buildings and the less conservative criterion for routine constructions.

REFERENCES

1. Longworth, J.: Some Aspects of Pile Foundations, "Timber Pile Foundations," American Wood Preservers Association, 1967.
2. Comparative Loading Tests on Piles of Three Different Kinds of Materials, *ibid.*
3. Foundation Piling: A Survey of Practice, Report 4, Building Research Advisory Board, National Academy of Sciences, National Research Council, Washington, D.C., 1962.
4. "Building Code Requirements for Reinforced Concrete," ACI 318, American Concrete Institute, 1963.
5. "Specification for the Design, Fabrication, and Erection of Structural Steel for Buildings," American Institute of Steel Construction, 1963.
6. "Standard Specifications for Highway Bridges," The American Association of State Highway Officials.
7. "National Design Specification for Stress Grade Lumber and Its Fastenings," National Lumber Manufacturer's Association, 1962.
8. Cummings, A. E.: The Stability of Foundation Piles against Buckling under Axial Load, *Proceedings of the Highway Research Board,* vol. 38, pt. II, pp. 112ff, December, 1938.

* There also is considerable debate as to what value to use for q_u, i.e., yield stress or peak stress. The use of yield stress is popular as a device for avoiding possible progressive failure.

† Use of the unconfined compressive strength would not be applicable when the piles are to be embedded in particularly sensitive or unusually preconsolidated clays. Where such soils are to constitute the bearing stratum, some expert advice should be sought.

9. Bjerrum, Laurik: Norwegian Experience with Steel Pile Foundations to Rock, *Journal of the Boston Society of Civil Engineers,* pp. 155–198, 1957.
10. Timoshenko, S., and J. M. Gere: "Theory of Elastic Stability," McGraw-Hill Book Company, New York, 1961.
11. Walter, H.: Das Knickproblem bei Spitzenpfahlen, deren Schaft ganz oder teilweise in nachgiebigem Boden steht, *Bautechnik-Archiv,* no. 6, pp. 40–66, 1951.
12. Marcus, Henri: Unpublished work derived pertinent to investigations conducted in connection with development of provisions of New York City building code relating to piling.
13. Andersen, Paul: "Substructure Analysis and Design," The Ronald Press Company, New York, 1956.
14. Nagaraja Rao, N. R., and Lambert Tall: Columns Reinforced under Load, *Lehigh University, Fritz Laboratory Report* 286, April, 1962.
15. Johnson, S. M.: Determining the Capacity of Bent Piles, *Journal of the Soil Mechanics and Foundations Division, Proceedings of the American Society of Civil Engineers,* vol. 88, pp. 65–76, December, 1962.
16. Parsons, J. D., and S. D. Wilson: Safe Loads on Dog-leg Piles, *Transactions of the American Society of Civil Engineers,* vol. 121, pp. 695–716, 1954.
17. Broms, Bengt B.: Allowable Bearing Capacity of Initially Bent Piles, *Journal of the Soil Mechanics and Foundations Division, Proceedings of the American Society of Civil Engineers,* vol. 89, pp. 73ff, September, 1963.
18. Defective Piles Halt Airport Work, *Engineering News-Record,* Nov. 9, 1961, p. 22.
19. Feld, Jacob: Failure of Extrusion Pile Foundation, *Proceedings, Technical Conference on Design of Foundations for the Control of Settlements,* American Society of Civil Engineers, June, 1964.
20. *Engineering News-Record,* July 21, 1949, p. 53.
21. Hobbs, N. B.: Unusual Necking of Cast-in-situ Concrete Piles, *Proceedings, Fourth International Conference on Soil Mechanics and Foundation Engineering,* London, vol. II, p. 40, 1957.
22. Johnson, S. M.: "Deterioration, Maintenance, and Repair of Structures," McGraw-Hill Book Company, New York, 1965.
23. Hough, B. K.: "Basic Soils Engineering," sec. 11-8, p. 331, The Ronald Press Company, New York, 1957.
24. Tschebotarioff, G. P.: "Soil Mechanics, Foundations, and Earth Structures," McGraw-Hill Book Company, New York, 1957.
25. Peck, R. B., W. E. Hanson, and T. H. Thornburn: "Foundation Engineering," pp. 257–265, John Wiley & Sons, Inc., New York, 1953.
26. Kyrnine, D. P.: "Soil Mechanics," pp. 216–229, McGraw-Hill Book Company, New York, 1947.
27. Feld, J.: discussion of "Friction Pile Foundations," by F. M. Masters, *Transactions of the American Society of Civil Engineers,* vol. 108, 1943.
28. Baker, C. N., Jr.: How to Determine Pile Depth of Embedment, *Wood Preserving News,* January, 1967.
29. Seiler, J. F., and W. D. Keeney: The Efficiency of Piles in Groups, *Wood Preserving News,* vol. 22, no. 11, November, 1944.
30. D'Appolonia, E., and J. P. Romualdi: Load Transfer in End-bearing Steel H-Piles, *Journal of the Soil Mechanics and Foundations Division, Proceedings of the American Society of Civil Engineers,* vol. 89, no. SM3, March, 1963.
31. Novick, D.: How to Select the Most Satisfactory and Economical Pile Foundations, *Wood Preserving News,* vol. 44, no. 12, December, 1966.
32. Seed, H. B., and L. C. Reese: The Action of Soft Clay along Friction Piles, *Transactions of the American Society of Civil Engineers,* vol. 122, 1957.

33. Meyerhof, G. G.: Compaction of Sands and Bearing Capacity of Piles, *Journal of the Soil Mechanics and Foundations Division, Proceedings of the American Society of Civil Engineers,* vol. 85, no. SM6, December, 1959.
34. Whitaker, T.: Experiments with Model Piles in Groups, *Geotechnique,* vol. 7, no. 4, 1957.
35. Design Manual DM7, "Foundations," U.S. Navy, Bureau of Yards and Docks, Washington, D.C.
36. Hallenbeck, J. J., and R. E. Johnston: Pile Foundation for Oakland Coliseum, *Civil Engineering,* January, 1967, pp. 57–61.
37. Novick, D.: How to Select the Most Satisfactory and Economical Pile Foundations, *Wood Preserving News,* vol. 44, no. 12, pp. 5–8, December, 1966.
38. Matlock, H., and L. C. Reese: Generalized Solutions for Laterally Loaded Piles, *Journal of the Soil Mechanics and Foundations Division, Proceedings of the American Society of Civil Engineers,* vol. 86, no. SM5, October, 1960.
39. Terzaghi, K.: Evaluation of Coefficients of Subgrade Reaction, *Geotechnique,* vol. 5, no. 4, 1955.
40. Broms, Bengt B.: Design of Laterally Loaded Piles, *Journal of the Soil Mechanics and Foundations Division, Proceedings of the American Society of Civil Engineers,* vol. 91, no. SM3, May, 1965.
41. McNulty, J. F.: Thrust Loading on Piles, *Journal of the Soil Mechanics and Foundations Division, Proceedings of the American Society of Civil Engineers,* vol. 86, no. SM2, April, 1956; also discussion, no. SM4, October, 1956.
42. Feagin, L. B.: Lateral Pile Load Tests, *Transactions of the American Society of Civil Engineers,* vol. 102, 1937.
43. Henry, F. D. C.: "The Design and Construction of Engineering Foundations," McGraw-Hill Book Company, New York, 1956.
44. Francis, A. J.: Analysis of Pile Groups with Flexural Resistance, *Journal of the Soil Mechanics and Foundations Division, Proceedings of the American Society of Civil Engineers,* vol. 90, no. SM3, May, 1964.
45. Westergaard, H. M.: The Resistance of Pile Groups, *Engineering Construction,* May 22, 1918.
46. Teng, W. C.: "Foundation Design," p. 227, Prentice-Hall, Inc., Englewood Cliffs, N.J., 1962.
47. Civil Engineering Code of Practice No. 2, "Earth Retaining Structures," pp. 151–164, The Institute of Structural Engineers, 11 Upper Belgrave St., London, SW 1, 1951.
48. Hrennikoff, A.: Analysis of Pile Foundations with Batter Piles, *Transactions of the American Society of Civil Engineers,* vol. 115, 1950.
49. Winter, G.: Lateral Bracing of Columns and Beams, *Journal of the Structural Division, Proceedings of the American Society of Civil Engineers,* vol. 84, no. ST2, 1958.
50. Housel, W. S.: Internal Stability of Granular Materials, *Proceedings of the American Society for Testing and Materials,* vol. 36, 1936.
51. Terzaghi, K., and R. B. Peck: "Soil Mechanics in Engineering Practice," 2d ed., John Wiley & Sons, Inc., New York, 1967.
52. Chellis, R. D.: "Pile Foundations," McGraw-Hill Book Company, New York, 1961.
53. Comparative Behavior of Friction Piles, *Highway Research Board Special Report* 36.
54. From working draft of proposed USASI Standard A58, 1967.
55. Tschebotarioff, G. P.: Personal communication.

56. Vesic, A. S.: Investigations of Bearing Capacity of Piles in Sand, *Conference on Deep Foundations,* Mexico City, 1964.
57. Berezanter, V. G., V. S. Khristoforov, and V. N. Golubkov: Load Bearing Capacity and Deformation of Piled Foundations, *Proceedings, Fifth International Conference on Soil Mechanics and Foundation Engineering,* vol. 2, 1961.
58. Peck, R. B.: Lecture on pile foundations delivered before metropolitan section, ASCE, 1966.
59. Plantima, G., and C. A. Nolet: Influence of Pile Driving on Sounding Resistance in a Deep Sand Layer, *Proceedings, Fourth International Conference on Soil Mechanics and Foundation Engineering,* 1957.
60. Ireland, H. O.: Pulling Tests on Piles in Sand, *Proceedings, Fourth International Conference on Soil Mechanics and Foundation Engineering,* 1957.
61. Hanna, T. H.: Model Studies of Foundation Groups in Sand, *Geotechnique,* vol. 13, 1963.
62. Skempton, A. W.: Discussion, Session 5: Piles and Piled Foundations, *Proceedings, Third International Conference on Soil Mechanics and Foundation Engineering,* 1953.
63. Skempton, A. W.: Discussion: Piles and Piled Foundations, *Proceedings, Third International Conference on Soil Mechanics and Foundation Engineering,* 1953.
64. Meyerhof, G. G.: The Design of Franki Piles with Special Reference to Groups in Sand, *Proceedings, Symposium Design of Pile Foundations, Sixth International Conference on Bridge and Structural Engineering,* Stockholm, 1960.
65. Yang, N. C.: Condition of Large Caissons during Construction, National Research Council, *Highway Research Board Bulletin* 74, 1965, pp. 68–84.
66. Mohan, D., G. S. Jain, and V. Kumar: Load Bearing Capacity of Piles, *Geotechnique,* vol. 13, March, 1963.
67. Mansar, C. I., and R. I. Kaufman: Pile Tests, Low Sill Structure, Old River, Louisiana, *Transactions of the American Society of Civil Engineers,* vol. 123, 1958.
68. D'Appolonia, E., and J. A. Hribar: Load Transfer in a Step-taper Pile, *Proceedings of the American Society of Civil Engineers,* vol. 89, no. SM6, November, 1963.

CHAPTER 7
Pile Capacity-Penetration Resistance

In Chap. 6 it was noted that pile capacity is determined from three basic indicators: (1) calculations based on theory or analogy, (2) empirical correlations with penetration resistance, and (3) load-test data. It also was noted that, generally, each of the different indicators will suggest a different result and that selection of a design value requires interpretation of these differing results. Such interpretation requires familiarity with the variables and nuances influencing each.

This chapter discusses the variables related to determining and evaluating the relation between capacity and penetration resistance. Discussion of the variables relating to the conduct and interpretation of the results of load tests is contained in Chap. 9. The matter of estimating capacity by calculation and the problems of interpreting the results of such calculations was discussed in Chap. 6.

Section 7A MEASURES OF PENETRATION RESISTANCE

7A.1 Blow Count

For piles installed by use of an impact hammer, penetration resistance is measured as the number of blows of the hammer required to penetrate the pile for a certain distance. This is known as blow count, is expressed in terms of blows per foot or blows per inch,* and is measured by the

* The distance over which the penetration resistance is measured is important. Resistance expressed in terms of blows per inch has not the same significance as resistance expressed in blows per foot. For example, assume that friction piles are to be installed in sandy soil to a capacity of 50 tons each and that, for the hammer being

inspector who marks increments of length on the pile and counts the number of blows required for the given length increment to pass a fixed point related to the ground.

7A.2 Rate of Penetration

For piles installed by use of a vibratory hammer, there is no blow count, per se, and penetration resistance is measured, instead, by the rate of penetration, usually expressed as inches of penetration per second. The thought is that, under a fixed set of conditions (fixed type of pile and soil), the rate of penetration for a given energy input and driving mechanism would be constant and that any variation in said conditions would affect the rate of penetration, a reduction in rate, presumedly, representing the equivalent of harder driving.

Rate of penetration is measured by observing the time required for a given increment of pile length to pass a fixed point related to the ground, and converting to a rate in inches per second.

7A.3 Jacking Pressure

Under certain circumstances (underpinning, or where vibration would be a severe problem, for example) piles are jacked into the ground. In

---

used, the penetration resistance corresponding to this capacity is 60 blows per foot of penetration. In theory, the pile can be driven to 60 blows per foot, or 5 blows per inch, or 1 blow per ⅕ in. all to the same effect. In practice, however, the pile driven to 60 blows per foot usually will perform better; i.e., it will settle less under the same applied load than that terminated at a resistance of 5 blows per inch, or 1 blow per ⅕ in. The reason is that cited on page 202, namely, that pile capacity largely is a function of the amount of energy expended in installing the pile and not just of the recorded resistances, so that the pile which is driven to a sustained resistance performs better than one in which driving is terminated the instant that resistance is attained.

This principle applies, however, only so long as the pile is not driven so hard that it is damaged. It does not apply, for example, where end-bearing piles (timber or concrete, in particular) are being driven to bearing on an abrupt unconformity (say, rock). For such case, driving must not be sustained for 1 ft, or even 1 in., but must be terminated directly when bearing is attained, as once the bearing is reached, additional driving is likely to be expended in progressively crushing or crippling the pile tip rather than in producing an increase in capacity. In this connection, it is noted that, in the general case, it is exceedingly difficult to determine just when bearing is reached, so that the installation of concrete or timber piles to end bearing on rock is a job requiring supervision by an experienced person having good judgment.

An intermediate case would be a friction (or end-bearing) pile where the penetration resistance builds up rapidly (but not abruptly). Again, driving must be terminated before the pile (or driving equipment) is damaged, but in this case not the instant that resistance is attained. As a rule of thumb, such piles should be driven to a sustained resistance for a penetration of 1 ft, but terminated at lesser penetration if the resistance, in blows per inch, reaches double the required value or reaches the point where there is no sensible penetration of the pile under the hammer blow.

such case penetration resistance is measured in terms of the jacking pressure, either gauge pressure or gauge pressure translated into total load. The pressure used as a measure of resistance must be the sustained pressure, i.e., the force which can be held on the jack for a given interval of time without excessive displacement. The required time interval and the amount of displacement which is not "excessive" are matters of opinion and would depend on the type of soil being penetrated. As a touchstone criterion, it is noted that the building code of the City of New York specifies, for underpinning piles, that:

> The working load for each permanent underpinning pile shall not exceed the larger of the following values: two-thirds of the total jacking force used to obtain the required penetration if the load is held constant for seven hours without measurable settlement; or one-half of the total jacking force at final penetration if the load is held for a period of one hour without measurable settlement.

7A.4 Augered or Screwed Piles

In theory, penetration resistance of augered or screwed piles could be expressed in terms of the torque requirement or in terms of rate of penetration for a given torque and load. In practice, this is too difficult to do consistently and reliably, and little credence is given to any correlation between capacity and penetration resistance for such types of piling.

Section 7B CORRELATIONS OF CAPACITY WITH PENETRATION RESISTANCE—PILES INSTALLED BY IMPACT HAMMER (DRIVING FORMULAS)

Much attention has been given to the derivation of correlations between capacity and penetration resistance for piles installed by impact hammer. A large number of formulas* (some simple, some erudite) have been proposed, and at least one general method of evaluation (wave equation) has been derived. All the formulas either have an empirical basis or require the use of empirically derived factors. As a result, opinions on the validity and applicability of one formula versus another vary, and there has been a general trend in modern practice toward rising disenchantment with driving formulas in general and increasing application of the rational approach (Sec. 6B) to pile design. Some of the reasons for this will be apparent from the following discussion.†

* Smith [13] refers to 450 such formulas.

† Despite their many deficiencies, the history of performance of piling installed according to formula (with some exceptions) has been relatively satisfactory. However, this is not necessarily because the formulas represent a proper correlation between capacity and penetration resistance, but rather because they incorporate a large safety factor which tends to mask the deficiencies.

7B.1 Driving Formulas

When the hammer strikes the pile, the pile is driven into the ground for some distance. As the hammer weight is withdrawn, a part of the displacement is recovered, and part remains as a permanent set. The total energy expended by the hammer thus is dissipated partly in the work of providing a set (useful work) and partly in losses (temporary, elastic, and recoverable displacement, heat, etc.). The relation may be written

$$WH = R_u s + R_u L \tag{7-1}$$

where W = weight of impacting parts
H = effective height of fall
WH = energy of hammer = energy input
R_u = ultimate pile capacity
s = set or final penetration per blow
L = some factor or expression representing the energy losses, units of length
$R_u s$ = useful energy
$R_u L$ = energy lost

Equation (7-1) may be transposed into

$$R_u = \frac{WH}{s + L} \tag{7-2}$$

This general expression [Eq. (7-2)], modified to incorporate different constants, loss factors, and the relation between weight of pile and hammer, provides the commonly used pile-driving formulas. Some of the better known and more popular formulations are

a. Engineering News Formula

The formula is

$$R_u = 6\left(\frac{2E}{s + 0.1}\right)^* \quad \text{(single- and double-acting hammers)} \tag{7-3a}$$

$$= 6\left(\frac{2E}{s + 1.0}\right)^* \quad \text{(drop hammers)} \tag{7-3b}$$

where R_u = ultimate pile capacity, lb
E = energy of hammer, ft-lb
s = set or final penetration per blow, in.

The energy E conventionally is taken as the manufacturer's rated energy for the hammer. This is not necessarily the energy delivered to the pile, and the results indicated by Eq. (7-3) must be interpreted in regard to the probable energy losses in the system (Sec. 7B.2).

* If the coefficient of 6 is omitted from this formula, one obtains the conventional form, expressed in terms of *working* load. This implies a factor of safety of 6.

The formula originally was derived when piling consisting primarily of timber sections and design loads of, perhaps, 10 to 20 tons each were used. It has many theoretical shortcomings and is, in fact, notoriously unreliable. Nevertheless, it works well for the lightweight piling and relatively low loads for which it was derived and, at least in the United States, remains the most widely used basis for evaluating pile capacity. For example, in a survey of practice conducted by the Building Research Advisory Board [2] and directed to 130 federal and state agencies, almost all respondents indicated that the *Engineering News* formula (or some modification thereof) was used in one way or another as an aid to establishing pile capacity.

b. Hiley Formula

Chellis [5] suggests the use of this formula when the piles are being installed by use of double-acting, differential-acting steam and diesel hammers, as follows:

$$R_u = \left(\frac{12e_fE_n}{s + \frac{1}{2}(C_1 + C_2 + C_3)}\right)\left(\frac{W_r + e^2W_p}{W_r + W_p}\right) \tag{7-4a}$$

and for piles installed by use of drop or single-acting steam hammers

$$R_u = \left(\frac{e_fW_rh}{s + \frac{1}{2}(C_1 + C_2 + C_3)}\right)\left(\frac{W_r + e^2W_p}{W_r + W_p}\right) \tag{7-4b}$$

Chellis further suggests use of the following for evaluation of the factors in Eqs. (7-4*a*) and (7-4*b*).*

R_u = ultimate carrying capacity of pile in pounds, before applying any factor of safety.
W_r = weight of falling mass, in pounds.
E_n = rated energy of hammer per blow, in foot-pounds, as published by manufacturer.
h = height of free fall of ram, in inches, for drop hammers; normal (shortest) stroke of ram, in inches, for single-acting steam hammers; and normal stroke of ram for double-acting steam hammers.
e_f = efficiency. The following percentages are suggested for use in computing bearing capacity (when computing fibre stresses in piles it is well to increase values under 100 percent by 10 percent):
100 percent for drop hammers released by trigger.
75 per cent for drop hammers actuated by rope and friction winch, but bearing in mind that this figure may decrease when the drop is small or the drag considerable, and increase somewhat

* The following definitions and the data in Tables 7-1 to 7-3 are from R. D. Chellis, "Pile Foundations," 2d ed., McGraw-Hill Book Company, New York, 1962, by permission.

if the drop is very large or the drag not great. The haul on the line and drum, the friction in the guides, and the friction-band effect (if the operator does not fully release it or if he "picks up" the hammer before the fall is complete in order to keep the line taut), all reduce the energy of the blow.

85 percent for McKiernan-Terry single-acting steam hammers.
80 to 85 percent for Raymond single-acting and differential-acting hammers.
75 percent for Warrington-Vulcan single-acting steam hammers.
65 percent for Vulcan-California double-acting hammers.
85 percent for McKiernan-Terry Series *B* double-acting hammers.
85 percent for Industrial Brownhoist, National and Union double-acting hammers.
75 percent for McKiernan-Terry double-acting hammers, sizes 0 to 7.
75 percent for differential-acting steam hammers.
100 percent for diesel hammers.

W_p = weight of pile, in pounds, including shoe and driving cap for drop hammers and single-acting steam hammers; weight of pile including shoe and weight of anvil in case of double-acting and differential-acting steam hammers. Also include weight of follower if used. In the case of driving to refusal in end bearing on rock, use only half the actual weight in the formula. The weight of earth wedged into the spaces between the flanges of *H* piles, or into open-end pipe piles, during driving should also be included. It is probable that the weight of earth clinging to the pile should also be included, which may be assumed as ½ to 1 in. thick, in the absence of definite information, and may equal 50 to 100 percent of the weight of a wood pile.

e = coefficient of restitution:
= 0.80 for micarta cushion when driving Raymond piles.
= 0.55 for no cushion. Steel on steel when driving pipe piles.
= 0.50 for oak cap blocks when driving Raymond piles.
= 0.50 for well-compacted cushion when driving pipe piles.
= 0.50 for ram of double-acting hammers striking on steel anvil and driving steel piles or precast concrete piles.
= 0.40 for medium-compacted wood cushion when driving pipe piles.
= 0.40 for ram of double-acting hammers striking steel anvil and driving timber piles, also for striking steel helmet containing wood and driving steel piles.
= 0.40 for ram of single-acting or drop hammers striking directly on head of precast concrete piles not fitted with driving cap.
= 0.32 for ram of single-acting hammers striking on steel plate cover of wood cap of steel piles.
= 0.25 for fresh wood cushion when driving pipe piles.
= 0.25 for ram of single-acting or drop hammers striking on well-conditioned wood cap of driving cap in driving precast concrete piles or directly on wood pile heads.

= 0.0 for deteriorated conditions of heads of timber piles or of wood cap and for excess packing in driving cap.

s = final set of pile (using average of last five blows for drop hammers and of last 20 blows for other types), in inches.

C_1 = temporary compression allowance for pile head and cap, in inches (see Table 7-1).

C_2 = temporary compression of pile, in inches (see Table 7-2).

C_3 = temporary compression allowance for ground for average cases where pile is driven into penetrable ground, in inches (see Table 7-3).

E = modulus of elasticity for pile material.

The factor of safety to be applied to this formula is 3.0.

TABLE 7-1 ***Temporary Compression Allowance C_1 for Pile Head and Cap****

Materials to which blow is applied	Easy driving, $p_1 = 500$ psi on cushion, or pile butt if no cushion, in.	Medium driving, $p_1 = 1{,}000$ psi on head or cap, in.	Hard driving, $p_1 = 1{,}500$ psi on head or cap, in.	Very hard driving, $p_1 = 2{,}000$ psi on head or cap, in.
Head of timber pile	0.05	0.10	0.15	0.20
3–4-in. packing inside cap on head of precast concrete pile	0.05 + 0.07†	0.10 + 0.15†	0.15 + 0.22†	0.20 + 0.30†
½–1-in. mat pad only on head of precast concrete pile	0.025	0.05	0.075	0.10
Steel-covered cap, containing wood packing, for steel piling or pipe	0.04	0.08	0.12	0.16
3/16-in. red electrical-fiber disk between two ⅜-in. steel plates, for use with severe driving on monotube pile	0.02	0.04	0.06	0.08
Head of steel piling or pipe	0	0	0	0

* Largely from A. Hiley, Pile Driving Calculations with Notes on Driving Force and Ground Resistance, *The Structural Engineer*, vol. 8, July and August, 1930.

† The first figure represents the compression of the cap and wood dolly or packing above the cap, whereas the second figure represents the compression of the wood packing between the cap and the pile head.

The following notations relate to Tables 7-1 to 7-3:

p_1 = stress per square inch on driving cushion, or on pile head if no cushion is used.

$$p_1 = \frac{R_u}{\text{area of pile head}}$$

p_2 = stress per square inch on average cross section of wood or concrete piles, or on shell area of tubular steel pile shells, or on area of steel sheet piling, or on average area of mandrels, or on net areas of shell and core for cast-in-place piles.

$$p_2 = \frac{R_u}{A}$$

p_3 = stress per square inch on horizontal projection of pile tip, including driving points under steel pipe casings, and on bounding area under H piles (for end-bearing piles and piles of constant cross section).

$$p_3 = \frac{R_u}{\text{area of tip}}$$

or:

p_3 = stress per square inch on gross area of pile at ground surface in case of tapered friction piles.

$$p_3 = \frac{R_u}{\text{gross area at ground surface}}$$

TABLE 7-2 ***Temporary Compression Values of C_2 for Piles***

Type of pile	Easy driving, p_2 = 500 psi for wood or concrete piles, 7,500 psi for steel, net section, in.	Medium driving, p_2 = 1,000 psi for wood or concrete piles, 15,000 psi for steel, net section, in.	Hard driving, p_2 = 1,500 psi for wood or concrete piles, 22,500 psi for steel, net section, in.	Very hard driving, p_2 = 2,000 psi for wood or concrete piles, 30,000 psi for steel, net section, in.
Timber pile (E = 1,500,000)	$0.004 \times L$*	$0.008 \times L$	$0.012 \times L$	$0.016 \times L$
Precast concrete pile (E = 3,000,000)	$0.002 \times L$	$0.004 \times L$	$0.006 \times L$	$0.008 \times L$
Simplex tube, pipe pile, monotube shell, Raymond steel mandrel (E = 30,000,000) . . .	$0.003 \times L$	$0.006 \times L$	$0.009 \times L$	$0.012 \times L$

* L is in feet, measured from head to center of resistance to driving.

TABLE 7-3 ***Temporary Compression or Quake of Ground Allowance C_3****

All values of p_3 to be taken on projected area of pile tips or driving points for end-bearing piles and piles of constant cross section, on gross area of pile at ground surface in case of tapered friction piles, and on bounding area under H piles

	Easy driving, p_3 = 500 psi, in.	Medium driving, p_3 = 1,000 psi, in.	Hard driving, p_3 = 1,500 psi, in.	Very hard driving, p_3 = 2,000 psi, in.
For piles of constant cross section........	0–0.10	0.10	0.10	0.10

* Largely from A. Hiley, Pile Driving Calculations with Notes on Driving Force and Ground Resistance, *The Structural Engineer*, vol. 8, July and August, 1930.

c. Modified Engineering News Formula

The formula is

$$R_u = 6\left[\frac{2E}{s + 0.1}\left(\frac{W_r + e^2W_p}{W_r + W_p}\right)\right]^* \tag{7-5}$$

where R_u, E, s = same as in Eq. (7-3)
W_r = weight of hammer ram, lb
W_p = weight of pile, including driving appurtenances, lb
e = coefficient of restitution

This formula is a compromise between the *Engineering News* formula and the Hiley formula. As indicated by Housel [1], for the normal range of e factors and pile-ram weight ratios, the modification, in effect, reduces the energy E to about 60 percent of the value (usually manufacturer's rating) used in the unmodified formula.

d. Pacific Coast Uniform Building Code Formula

The formula is

$$R_u = \frac{12E\left(\dfrac{W_r + KW_p}{W_r + W_p}\right)}{s + \dfrac{12R_uL}{AE_{\text{pile}}}} \tag{7-6}$$

where R_u = ultimate pile capacity, lb
E = energy of hammer, ft-lb
s = set or final penetration per blow, in.
A = area of pile section
E_{pile} = modulus of elasticity of pile material

* If the coefficient of 6 is omitted from this formula, one obtains the conventional form, expressed in terms of *working* load. This implies a factor of safety of 6.

K = a coefficient equal to 0.25 for steel piles and 0.10 for other piles
W_r = weight of falling mass, tons
W_p = weight of pile, tons
L = length of pile, ft

The factor of safety to be applied to this formula is 4.0.

e. Canadian National Building Code Formula

The formula is

$$R_u = \frac{12E\left(\dfrac{W_r + 0.5e^2W_p}{W_r + W_p}\right)}{s + \dfrac{R_u}{2A}\left[\dfrac{12L}{E_{\text{pile}}} + 0.0001\right]} \tag{7-7}$$

where e equals the coefficient of restitution, as for the Hiley formula. Other symbols are the same as for Eq. (7-6). The factor of safety to be applied to this formula is 3.0.

f. S_0 Formula

The formula is

$$R_u = \frac{e_f E_n}{s + S_0/2} \tag{7-8}$$

where e_f = efficiency of hammer
E_n = rated energy of hammer
s = penetration per blow
$S_0 = \sqrt{\dfrac{2e_f E_n L}{AE_{\text{pile}}}}$
A, E_{pile} = area and modulus of elasticity of pile material

7B.2 Validity of Formulas

A number of studies have been made to evaluate the relative accuracy and applicability of the various driving formulas indicated. The criterion used to evaluate accuracy is the ratio between capacity as indicated by load test and capacity as indicated by formula. This criterion is not a particularly good one as it is sensitive to the criterion used as a measure of capacity in the load test; also, for those formulas where empirical constants (efficiency, coefficient of restitution, C_1, C_2, C_3, etc.) must be established, almost any answer can be obtained by judicious juggling of constants. The following discussions must be considered with this in mind.

a. Engineering News Formula

I. *From Michigan Study of Pile-driving Hammers and Piles* [1]. Test results indicate that the *Engineering News* formula generally overestimates pile capacity (under the subject test conditions and based on ultimate load capacity) by a factor of 2 to 6; that is, the safety factor inherent in the formula varies from 1 to 3 and is not 6, as intended.

II. *Chellis* [3] indicates that the factor of safety obtained using the *Engineering News* formula may vary from 0.5 to 16 or more. The formula indicates unduly high capacities when light hammers or heavy piles are used, and the results are sensitive to a variety of factors. Chellis also indicates that the formula was designed for driving timber friction piles and notes that its original derivation was for use with drop hammers. Tests using steam hammers always indicate capacities, especially with small sets, as much as six times higher than the apparent capacity as indicated by the formulation applicable to drop hammers. He concludes that the formula is inherently inaccurate.

III. *Hough* [6] notes the increasing unreliability of the *Engineering News* formula for small sets (high capacities), but notes that there is a natural limitation (values of set at which damage to the pile or to the driving equipment begins to occur) on use of the formula which tends to prevent obtaining unsafe bearing values. In the authors' experience, this limiting value is about 60 tons; i.e., the *Engineering News* formula is not sufficiently reliable for pile loads in excess of 60 tons.

IV. *Agerschou* [14] has made a statistical analysis of failure loads from 171 load tests on piles with their tips in sand or gravel and concludes:

> The ENR pile formula is unreliable. It has the highest standard deviation; 96% of the allowable loads determined by this formula will have actual safety factors varying from 1.1 to more than 30. There is no way of knowing what the safety factor is in the individual case.

V. *Flaate* [17] on the basis of 116 tests concludes that:

> The variations obtained with the *Engineering News* formula are so great that it should not be used, especially since better formulae are available.

VI. *Hensey and Petus* [15], on the other hand, suggest that the *Engineering News* formula is satisfactorily reliable for friction piles, loaded to 40 tons or less.

VII. *Terzaghi and Peck* [8] indicate that ". . . bearing capacity . . . computed on the basis of the widely used *Engineering News* formula is so poor and so erratic as to offer no justification for its continued use."

b. Hiley Formula

The endorsement of the Hiley formula by Chellis, a widely recognized authority, is significant, and the formula is understood to have wide acceptance in Great Britain.

Flaate [17] also suggests that this formula gives good results. So does

Agerschou [14].* Nevertheless, perhaps simply because of difficulties in selecting appropriate constants, many authorities disagree as to its validity. Terzaghi and Peck [8], for example, indicate that the formula, although conveying the impression of greater reliability, has been shown by experience to be almost as inaccurate as the *Engineering News* formula. This seems to be confirmed by the Michigan Studies [1] which show a variation of safety factors for the Hiley formula greater than that for the *Engineering News* formula, although the actual safety factors for the Hiley formula exceeded those for the *Engineering News* so that a smaller design factor could be used.

A similar conclusion is contained in report *B* of the Committee on the Bearing Value of Pile Foundations of the ASCE [9] which states: "The use of a complicated formula is not recommended since such formulas have no greater claim to accuracy than the more simple ones."

c. Other Formulas

The Michigan Studies [1] show that *all* the 11 formulas studied, which included, in addition to the *Engineering News* and Hiley, the Uniform Building Code, Redtenbacher, Eytelwein, Navy-McKay, Rankine, Canadian National Building Code, Modified *Engineering News*, Gates, and Rabe formulas, basically were unreliable.

Agerschou's work, in addition to studying the *Engineering News* and Hiley formulas, considered the S_0, Janbu, and Weisbach formulas and the wave equation of Smith [13]. He found the standard deviation on log (P failure/P formula) for the Hiley, Janbu, Weisbach, and S_0 formulas and for the wave equation to be approximately the same. There appeared to be no special merit to selecting among these several relations.

7B.3 Factors Which Influence the Apparent Penetration Resistance (Piles Installed by Impact Hammer)

Even the act of evaluating the penetration resistance itself is not a simple matter. It involves more than just counting blows. The significance of the blow count depends on many variables, as listed below.†

* "Good," however, is a relative term. In the analysis by Agerschou, the safety factors derived from the test piles using the Hiley formula varied from 1.0 to 4.2, which is quite a wide range, and in Flaate's work the safety factor required to insure against failure was about 3.

† In any particular case, it should be understood that for a given type of pile and soil stratification (and except for the special case where the piles are to be hung up in a firm stratum overlying a softer one) the longer the pile, the greater the capacity, and anything which changes the total pile penetration required to achieve the presumptive necessary penetration resistance correspondingly changes the capacity of the pile as installed.

TABLE 7-4a ***Minimum Driving Resistance and Minimum Hammer Energy for Steel H Piles, Pipe Piles, Precast and Cast-in-Place Concrete Piles, and Composite Piles (other than timber)*** [*From N.Y.C. Code (1968).*]

Minimum Driving Resistance*‡§¶						
Pile capacity, tons	Hammer energy,† ft-lb	Friction piles, blows per ft	Piles bearing on hardpan, blows per ft	Nondisplacement piles bearing on decomposed rock, blows per ft	Displacement piles bearing on decomposed rock, blows per ft	Piles bearing on sound, intermediate, or medium rock
Up to 20...	15,000	19	19	48	48	5 blows per ¼ in. (minimum hammer energy of 15,000 ft-lb)
	19,000	15	15	27	27	
	24,000	11	11	16	16	
30.........	15,000	30	30	72	72	
	19,000	23	23	40	40	
	24,000	18	18	26	26	
40.........	15,000	44	50	96	96	
	19,000	32	36	53	53	
	24,000	24	30	34	34	
50.........	15,000	72	86	120	120	
	19,000	49	54	80	80	
	24,000	35	37	60	60	
	32,000	24	25	40	40	
60.........	15,000	96	5 blows per ¼ in. (minimum hammer energy of 15,000 ft-lb)	240	240	
	19,000	63		150	150	
	24,000	44		100	100	
	32,000	30		50	50	
70 and 80..				5 blows per ¼ in. (minimum hammer energy of 19,000 ft-lb)		
100........						
Over 100...						

See page 315 for footnotes.

a. Weight of the Hammer

It will be apparent that it makes no sense to try to drive a spike with a tack hammer. The same thing applies to driving piles. Unless the hammer is heavy enough, one can pound on a pile all day, with very low apparent set, but without attaining any significant penetration or capacity. It is for this reason that formulas such as that of Hiley are popular and that minimum weights of hammer ram (or minimum energy) often are specified. For example, Tables 7-4*a* and *b* appear in the proposed New York City Building Code (1968).

b. Weight of the Pile

In a similar manner to the effect of the weight of the hammer, the weight of the pile influences the set under a given blow. A heavy pile will show a lesser set under a given impact energy than will a light section. This is reflected in the Hiley and similar formulas, but again, there is a limiting value beyond which the inertia due to the weight of the pile is overbalanced by the increased stiffness of the section which permits more of the energy of the hammer blow to reach the tip of the pile. For such case the dynamic formulas do not give reasonable results. For the Hiley formula, there is some experience to indicate that the limiting weight of pile beyond which this occurs is about 200 lb per ft. It also appears to occur when comparing mandrel-driven piles to corresponding sections driven without a mandrel.

---

* Final driving resistance shall be the sum of tabulated values plus resistance exerted by nonbearing materials. The driving resistance of nonbearing materials shall be taken as the resistance experienced by the pile during driving, but which will be dissipated with time.

† The hammer energy indicated is the rated energy.

‡ Sustained driving resistance—where piles are to bear in hardpan or soft rock, the minimum driving resistance shall be maintained for the last 6 in., unless a higher sustained driving resistance requirement is established by load test. Where piles are to bear in gravels, sands, clays, or silt soils, the minimum driving resistance shall be maintained for the last 12 in. unless load testing demonstrates a requirement for higher sustained driving resistance. No pile need be driven to a resistance to penetration (in blows per inch) more than twice the resistance indicated in this table, nor beyond the point at which there is no measurable net penetration under the hammer blow.

§ The tabulated values assume that the ratio of total weight of pile to weight of striking part of hammer does not exceed 3.5. If a larger ratio is to be used, or for other conditions for which no values are tabulated, the driving resistance shall be as approved by the commissioner.

¶ For intermediate values of pile capacity, minimum requirements for driving resistance may be determined by straight line interpolation.

TABLE 7-4b ***Minimum Driving Resistance and Hammer Energy for Timber Piles*** [*From N.Y.C. Code (1968).*]

Pile capacity, tons	Minimum driving resistance,*‡ blows per in.	Limitations on hammer energy, ft-lb†
Up to 20.	See formula§	7,500–12,000
20 to 25.		9,000–12,000 (single-acting hammers) 14,000–16,000 (double-acting hammers)
25 to 30.		12,000–16,000 (single-acting hammers) 15,000–20,000 (double-acting hammers)

* Final driving resistance shall be the sum of values indicated by formula plus resistance exerted by nonbearing materials. The driving resistance exerted by nonbearing materials is the resistance experienced by the pile during driving, but which will be dissipated with time.

† The hammer energy indicated is the rated energy.

‡ Sustained driving resistance—where piles are to bear in hardpan or soft rock, the minimum driving resistance shall be maintained for the last 6 in., unless a higher sustained driving resistance requirement is established by load test. Where piles are to bear in gravels, sands, clays, or silt soils, the minimum driving resistance shall be maintained for the last 12 in. unless load testing demonstrates a requirement for higher sustained driving resistance. No pile need be driven to a resistance to penetration (in blows per inch) more than twice the resistance indicated, nor beyond the point at which there is no measurable net penetration under the hammer blow.

§ The minimum driving resistance shall be determined by the following formula:

$$P = \frac{2W_hH}{s + 0.1} \qquad \text{or} \qquad P = \frac{2E}{s + 0.1}$$

where:

P = Allowable pile load (lb)
W_p = Weight driven (lb)
W_h = Weight of striking part of hammer (lb)
H = Actual height of fall of striking part of hammer (ft)
E = Rated energy delivered by the hammer per blow (ft-lb)
s = Penetration of pile per blow (in.), after the pile has been driven to a depth where successive blows produce approximately equal net penetration

The value $\frac{W_p}{W_h}$ shall not exceed 3.

c. Speed of the Hammer

For impact hammers, and all other things being equal, the energy delivered is a function of the speed of the hammer. In general the faster the succession of blows the more energy delivered to the pile, and hence the greater the capacity for a given indicated set. The hammer speed must be checked during driving by a timed count of blows. The manufacturer's rated speed should be adhered to. In addition, to the extent that a succession of blows keeps the pile moving and the interface between pile and soil lubricated, the capacity which will correspond to given resistance depends on the speed of the hammer being in excess of some critical value, below which the pile would "set up" between blows. In general, this is a problem only if a drop hammer is to be used or if driving is interrupted.

d. Where Driving Is Interrupted

For piles in cohesive (or mixed) soils, if driving is interrupted, there is a tendency for the pile to set-up or freeze; i.e., upon the resumption of driving the set will be much less than that recorded before the interruption. Under such circumstances, the assumed correlation between capacity and penetration resistance is not meaningful, and the pile must be driven a sufficient distance to break the freeze, before readings of penetration resistance can be interpreted. This distance may be a few inches or several feet. A value of 5 ft commonly is used.

e. Cushion or Cap Block

The hammer blow usually is delivered to the pile through a cushion or cap block. This block absorbs part of the energy of impact and reduces that part which is delivered to the pile.

The cap block must be kept in good condition, although readings of penetration resistance should not be made immediately after a new block or cushion is installed or until the new material has an opportunity to be properly set. It must be of reasonably firm materials and must be replaced when crushed. Unless these things are tended to, the recorded sets will be too low, often drastically so. Special care should be taken that foreign energy-absorbent materials (wood chips, tires, etc.) are not used in the cap-block recess. A suggested specification is:

> The cushion or cap block shall be a solid block of hardwood with its grains parallel to the axis of the pile and enclosed in a tight-fitting steel housing, or shall be an equivalent assemblage. If laminated materials are used, the type and construction of these materials shall be such that their strength is equal to, or greater than, hardwood. Wood chips, pieces of rope, hose,

shavings, or automobile tires and similar materials shall not be used. Cap blocks shall be replaced if burned, crushed, or otherwise damaged.

f. Followers

When a follower is used, again the apparent sets will be too low. There are two reasons: (1) the effective weight of the pile is increased and (2) if the follower does not fit tight, there is an energy loss due to "chatter" and to separation between pile and follower under the impact wave.

The effect of the weight of the follower can be considered by use of the Hiley (or similar) formulas. The remedy for relative movement or chatter is to provide the follower in a single-length section, stiff enough to prevent whip during driving and to insist that the follower fit tightly to the pile.

Followers should be of steel, hardwood, or similar material which will minimize absorption and cushioning of the hammer blow in the follower.

g. Jetting

The use of jetting sometimes is required in order to obtain sufficient penetration to get the piles to bear in satisfactory material below compressible strata which may underly the site.

The jetting operation disturbs the soil around and below the pile (this being its function), and readings of penetration resistance should not be taken while jetting is in progress or until the pile has been reseated after jetting has been completed.

h. Condition of Equipment

It will be apparent that the driving equipment must be in good condition if the hammer blow is to be properly delivered to the pile. For example, the hoses must not be kinked or crimped; valves should be fully opened; the hammer must not bind in the leads; and the cap should bear fully on the head of the pile, which should be trimmed to fit.

i. Sequence of Installation and Too-close Spacing

As the piles for a group are driven, the surrounding and underlying soil is compacted, both by the volumetric displacement of the piles and by the vibrations incident to driving. As a result, the ground tightens up, and the last piles driven tend to run shorter, for a given penetration resistance requirement, than the first piles installed. This is particularly true when the piles are "boxed in" by driving; i.e., the perimeter piles in the group are driven first and the center piles then are filled in (Fig. 7-1*b*), or when the pile spacing is too small so that succeeding piles must be driven through progressively more compact soil (Figs. 7-1*c* and *d*).

In Chap. 6, it was indicated that the basic function of piling, as conventionally used for support of buildings, is to form a footing deep in the ground. It will be apparent that the pile groups indicated in Figs. 7-1*b* to *d* fail to satisfy this function. The pseudofootings, as created in the bearing stratum, are not of the desired size and/or do not bear uniformly in the requisite stratum.

On the other hand, if there were no distinct bearing stratum to be reached (the soil at the site being, instead, a uniform deposit increasing

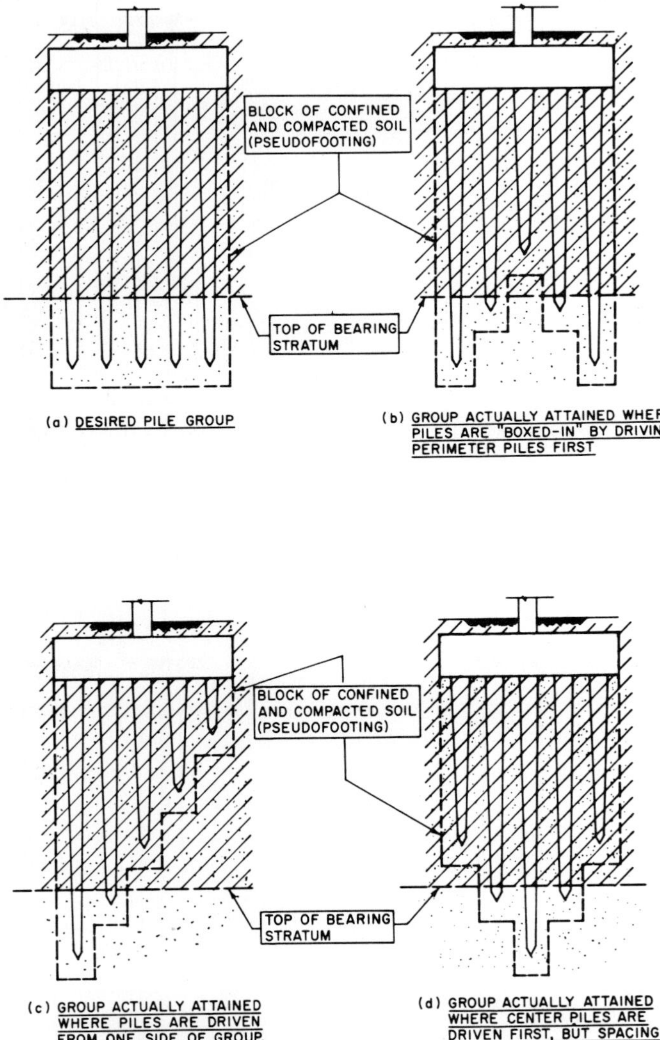

FIG. 7-1 Effect of Sequence of Pile Installation (layered soil profile-distinct bearing stratum).

uniformly in density with depth), the function of the piles would be only to spread the load. In such case, the variation in pile length might be acceptable, depending on circumstances. For example, refer to Figs. 7-2*b* and *c*. In Fig. 7-2*b*, the full load of the pile group devolves on section 2-2. The bearing area at this level is as required (Fig. 7-2*a*). At section 1-1, the bearing area in the soil is less, but the underlying soil is reinforced

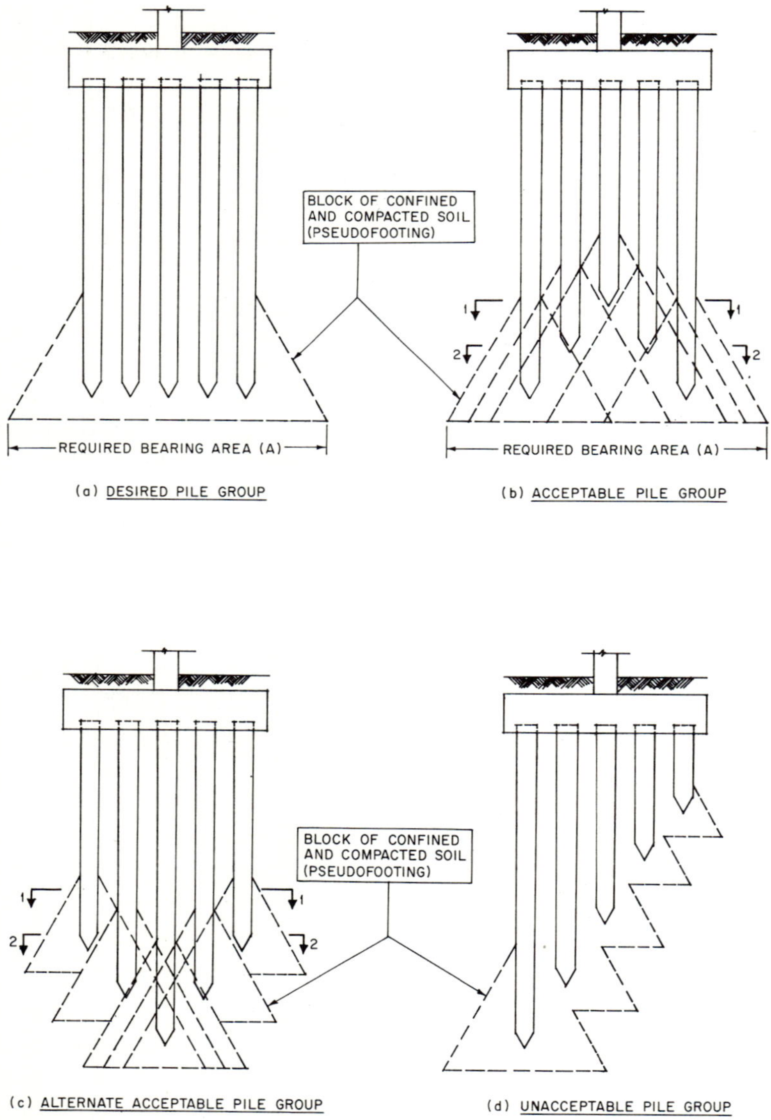

FIG. 7-2 Effects of Sequence of Pile Installation (Uniform Soil Profile).

by the piles which penetrate below section 1-1, the piles, in effect, carrying part of the applied load below this level. The same is true with regard to Fig. 7-2*c*. The situation shown in Fig. 7-2*d* is unacceptable because the pseudofooting will tend to tilt.

The conditions shown in Figs. 7-2*b* and *c*, while potentially acceptable, are not desirable. The strength of the reinforced layer is uncertain. In addition, the shorter piles are underlain by more compressible material, carry less than their pro rata share of the load, causing the longer piles to be somewhat overloaded, by comparison, and requiring that the cap (Fig. 7-2*b*) be provided with extra strength to resist this uneven distribution of reaction. For these reasons, and because soil conditions seldom are uniform, designers properly prefer to have the piles in a group terminate more or less at the same level. Where such is not realized, the adequacy of the group becomes suspect.

Since it seldom is feasible to drive all the piles in a group to the same tip elevation without underdriving some and overdriving others there always is some question on how much variation can be accepted and when the condition should be investigated. As a rule of thumb for friction piles, if the variation of length within a group (ratio of length of longest pile to length of shortest pile) exceeds 1.30, the adequacy of the group should be analyzed as indicated in Figs. 7-2*b* and *c*. For end-bearing piles, larger variations can be accepted without concern.

j. Heaved Piles

Another possible detrimental consequence of improper sequence of installation and/or too-close spacing of piles is that the piles installed first in the sequence may be heaved by the ground displacements resulting from installation of the subsequent piling. In addition, the lateral pressures developed in the soil may displace (or break) adjacent piles or damage nearby buildings or utilities.

Most engineers consider that heave is detrimental to pile capacity and that the heaved piles will settle under load by an amount approximately equal to the original heave plus the normal settlement otherwise anticipated. Accordingly, it is concluded that heaved piles must be redriven to bearing, although a nominal amount of heave (say, ¼ in.) generally is considered to be acceptable without bothering to redrive.

Heave is not *necessarily* detrimental, however, depending on circumstances, as follows (see Fig. 7-3).*

CASE 1 When piles are anticipated to derive their principal support from end bearing, are shallowly embedded in the bearing stratum, and are lifted from bearing by the adhesion and heave of the overlying strata, under load, the pile can be expected to settle back into bearing; i.e., the

* The indicated discussion is based, partly, on Klohn [16].

settlement under load will be increased over and above the normally anticipated displacement, by the amount of heave.

CASE 2 On the other hand, if the pile is expected to derive its resistance almost entirely from friction along the sides, the heave may not pose a future settlement problem.

The general criterion is that, if the pile and the soil from which it will derive its support under load heave equally and together, no discontinuity

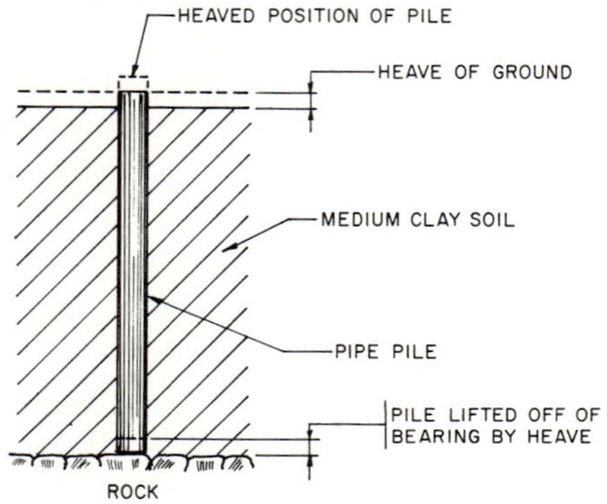

CASE I - END BEARING PILES

PILE WILL SETTLE BACK INTO BEARING UNDER LOAD. THEREFORE SETTLEMENT WILL BE APPROXIMATELY EQUAL TO AMOUNT OF HEAVE. PILES SHOULD BE REDRIVEN TO BEARING.

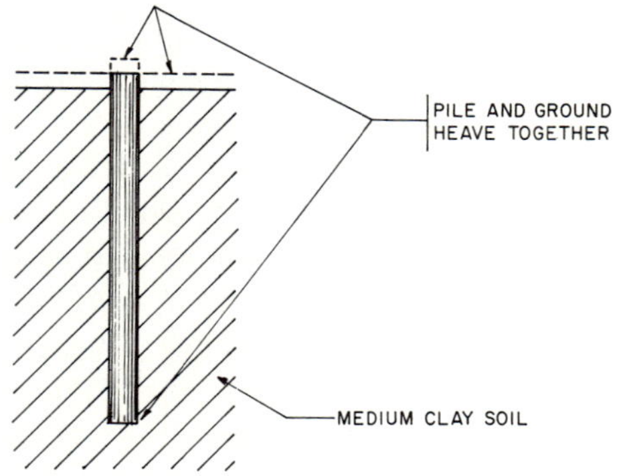

CASE II - FRICTION PILES

PILE AND GROUND HEAVE TOGETHER. LOAD CAPACITY OF PILE IS NOT REDUCED NOR ARE POTENTIAL SETTLEMENTS INCREASED, EXCEPT BY VIRTUE OF ANY CHANGE IN SOIL CHARACTERISTICS DUE TO REMOLDING. IN INSENSITIVE SOILS, PILES PROBABLY NEED NOT BE RETAPPED.

FIG. 7-3 Capacity of Heaved Piles.

is created. Further, if the strength and consolidation (or settlement) characteristics of the soil are not adversely affected by the remolding due to the heave (insensitive soils), then the settlements under load will not materially exceed those anticipated under the conditions existent prior to heave. If an actual discontinuity is created; i.e., the pile is pulled up and away from its seat of bearing, resettlement approximately equal to the amount of heave will occur. For intermediate cases, some increased settlement (but less than the amount of heave) should be anticipated.

The best solution to problems of pile heave, as with other problems, is prevention. Preventive measures, where required, include use of low displacement piles (generally H sections), increase of pile spacing, and proper attention to sequence of installation.

7B.4 Conclusions

The use of dynamic pile formulas for estimating capacity of driven piles and the problem of selecting which formula to use may be summarized by the following quotations:

1. From report of ASCE committee on the bearing value of pile foundations [9]:

> All dynamic pile-driving formulas are subject to definite limitations and any dynamic pile-driving formula is nothing more than a yardstick to help the engineer secure reasonably safe and uniform results over the entire job.

2. From Michigan Studies [1]:

> The estimates of supporting capacity by dynamic pile formulas, considering the entire group of . . . formulas, varied through a wide range and could not be used to predict pile capacity with any degree of certainty. The greatest variability occurred in the high capacity range, and except for this deficiency several of the formulas gave fairly consistent results insofar as the range of variation was concerned. However, even with good grouping of points, only a few of the formulas showed reasonably good agreement between the actual safety factor found in the tests and that prescribed in the formulas.

The practice of the authors (who are among those disenchanted with driving formulas) is to use the principle of similitude (Sec. 7C) as a criterion for capacity (supplementing analysis) wherever economically feasible. For small jobs where the cost of load tests incident to use of the similitude approach would be unreasonably burdensome, the piles are driven to the *Engineering News* formula, and the capacity is limited to 40 tons or less. Of course many exceptions are made as judgment dictates, particularly where analysis (usually on the basis of the equivalent-footing analogy) appears to be a reliable approach, i.e., mainly in granular soils.

Section 7C CORRELATIONS OF CAPACITY WITH PENETRATION RESISTANCE—PILES INSTALLED BY MEANS OTHER THAN IMPACT HAMMER (PRINCIPLE OF SIMILITUDE)

Where piles are not installed by use of impact hammer (thus, jacked piles, screwed piles, augered piles, or piles installed by use of vibratory hammer), no driving formulas have been developed, and correlation between capacity and penetration resistance is predicated on either jacking pressure or similitude.

Jacking-pressure criteria (see Sec. 7A.3) apply only to jacked piles. For other types, the principle of similitude is followed.

The principle of similitude has been described in general terms in Sec. 6B.1*d*. Briefly, it consists of installing one or more index piles which are load tested. If the indicator pile supports the design load (and an overload) satisfactorily, the other nearby piles are driven in like manner on the presumption that they will behave comparably to the test pile. As indicated in Chap. 9, this is a large presumption. Accepting the presumption, however, the principle of similitude is a powerful tool. Moreover, in the cases indicated, it is pretty much the only practical way in which the assumptions of the design computations can be verified.

Application of the principle, however, is not so simple as might first appear. For example, are the nontested piles to be driven to the same level as the tested pile, or to the same penetration resistance, or to the same penetration into the bearing stratum (or main bearing stratum, if there is a succession of capable strata)? Generally, it is not practical to satisfy all three of these criteria at the same time.

The problem is not easily solved. Soils can be variable in character even over short distances, the level of the bearing stratum may vary widely, and, to make matters worse, the character and strength of the soil often is progressively altered as succeeding piles are driven.

The following procedures and criteria are suggested as a possible solution:

STEP 1: Check the borings and investigate the site conditions to be certain that load tests on single piles will provide a meaningful representation of the performance of piles installed in the actual building foundation (see Chap. 9).

STEP 2: Divide the area under the building into areas of apparent uniform conditions as indicated by the soils explorations made prior to construction. If necessary, make additional soil explorations as required to identify areas of similar conditions.

STEP 3: Install an index pile in each area of uniform conditions* using whatever type of pile and/or method of installation and/or equipment desired. Record all driving data for the index piles.

STEP 4: Load-test the index piles. If the load-test results on any pile are unsatisfactory, install a new index pile to replace it and drive the new pile longer, or harder, or whatever is called for so that a better performance will be attained. Repeat the load test until all areas of uniform conditions are represented by a satisfactory index pile (as indicated by load test).

STEP 5: All building piles within the area of influence of a given, satisfactorily tested index pile shall be installed to the same or greater penetration resistance (equal or lesser set for driven piles; equal or lesser rate of penetration for piles installed by vibratory hammer; equal or greater torque for piles installed by auger; equal or greater jacking pressure for jacked piles), as for the successful index pile. The same (or heavier) equipment (of same type), identically operated (rpm, manifold pressure, weight, hydraulic pressure, steam or air pressure, etc.) and the same type, external dimension, equal or greater cross section, and same shape of pile shall be used to install the building piles as was used for the index pile. All building piles within the area of influence of a given satisfactorily tested index pile shall bear in or on the same main bearing stratum as the index pile. Note that this is not nearly so easy as it sounds. For piles installed by impact hammer, there are pressure gauges, hose lengths, binding of the hammer in the leads, condition of head of pile, fit of cap and cap block, and other variables which have to be simultaneously observed. With the vibratory hammer there are several gauges and dials to keep track of. This gets to be a difficult task.

ADDITIONAL PRECAUTIONS: The following optional precautions may be considered.

1. *Comparison Piles.* A correlation between the index piles installed by proposed means and like piles conventionally driven may be obtained by installing comparison piles. The comparison piles are installed by impact hammer to driving resistance as determined by formula. The index piles then are installed near (within 4 to 6 ft) the comparison pile and to the same tip elevation (or if the comparison piles bear on hard

* Site conditions may be sufficiently uniform that only one test pile would be required for an entire site. As a rule of thumb, however, it is suggested (as provided in the New York City code) that at least one index pile be provided for each 15,000 sq ft of building area.

material, to bearing in or on the same material). Steps 4 and 5 (load test and installation of building piles) then are executed.

The purpose of the comparison piles is (*a*) to assist in establishing the probable tip elevations required for the index piles to be adequate (judgment or experience is likely to serve as well, or better, however) and (*b*) to drive, in effect, the index piles to penetration resistance as indicated by driving formula so that the building piles likewise are installed to the equivalent of formula penetration.

2. *Proof Testing.* Following or during installation of the building piles, a certain number (say, 5 or 10 percent) can be proof-tested by load test, by redriving using conventional equipment (air or steam hammer), or by other means. If proof testing is by redriving, the piles must be moved far enough to break the "freeze" before their proof value can be judged. If the proof testing is satisfactory, the installation is judged to be satisfactory. If not, there is trouble because all the piles have to be redriven.

Section 7D CIRCUMSTANCES WHERE PENETRATION RESISTANCE IS INAPPLICABLE AS A MEASURE OF PILE CAPACITY

There are a number of circumstances where penetration resistance is not applicable as a measure of pile capacity and, therefore, should not be used as an indicator.

1. *Piles Driven into Firm Material Overlying Weak or Compressive Substrata.* See Fig. 6-24*a*.

2. *Tapered Pile Driven through Lense of Hard Material.* This situation is illustrated in Fig. 6-24*b*. The pile tends to wedge in the firm crust overlying the bearing stratum, and very little penetration of the bearing stratum is attained. The wedging effect of the crust tends to dissipate as the bearing stratum compacts or consolidates under the pile load. Therefore the recorded penetration resistance is temporary and also will tend to dissipate with time.

3. *Piles Terminating on Boulders.* See Fig. 6-24*c*.

4. *Piles Subject to Negative Skin Friction.* See Fig. 6-24*d*.

5. *End-bearing Piles.* Piles driven to end bearing on rock generally are driven to refusal (4 or 5 blows per $\frac{1}{4}$ in. or to no further sensible penetration), regardless of capacity. Accordingly, penetration resistance has no meaning for such piles. Even for less extreme conditions, the Michigan Studies [1] show conclusively that driving formulas grossly underestimate capacity of end-bearing piles and that the capacity of such piles is more a function of allowable contact pressure than of penetration resistance.

6. *Piles Driven into Fine-grained Soils.* Teng [4] indicates that

"Dynamic pile driving formulas cannot be relied upon for determination of pile capacity in soils containing a fair amount of fine grains." Similarly, Henry [7] and others [8, 12] indicate that the use of driving formulas is inapplicable in soft cohesive soils and should be verified by load test when the piles are installed in hard cohesive soil.

The reasons are: (1) In soils containing a substantial fraction of fine-grained materials, the displaced water cannot readily escape, builds up excess pore pressure around the pile during installation, and tends to reduce the frictional resistance around the periphery of the pile and conversely to create a temporary, dynamic resistance due to the hydraulic (dashpot) effect. (2) The soil itself is disturbed by the driving operation and by the displacement of the pile volume, thereby temporarily modifying its strength value to that of the remolded condition.

In this connection, Terzaghi and Peck [8] point out that in several areas, such as New Orleans, where the use of friction piles in soft clay is prevalent, "no experienced engineer would consider driving according to a pile formula."

REFERENCES

1. "A Performance Investigation of Pile Driving Hammers and Piles," Michigan State Highway Commission, Lansing, Mich., March, 1965.
2. Federal Construction Council, Survey of Practice, Foundation Piling, *Report* 4, National Academy of Sciences, National Research Council, Publication 987.
3. Chellis, R. D.: Pile Foundations in G. A. Leonards (ed.), "Foundation Engineering," McGraw-Hill Book Company, New York, 1962.
4. Teng, W. C.: "Foundation Design," Prentice-Hall, Inc., Englewood Cliffs, N.J., 1962.
5. Chellis, R. D.: "Pile Foundations," 2d ed., McGraw-Hill Book Company, New York, 1961.
6. Hough, B. K.: "Basic Soils Engineering," The Ronald Press Company, New York, 1957.
7. Henry, F. D. C.: "The Design and Construction of Engineering Foundations," McGraw-Hill Book Company, New York, 1956.
8. Terzaghi, K., and R. B. Peck: "Soil Mechanics in Engineering Practice," 2d ed., John Wiley & Sons, Inc., New York, 1967.
9. Committee on the Bearing Value of Pile Foundations, Progress Report, *Proceedings of the American Society of Civil Engineers*, May, 1941.
10. Housel, W. S.: Supporting Capacity of Piles: Estimates and Test Results, unpublished report relating to Michigan study of pile hammers and piles.
11. Parsons, J. O.: Piling Difficulties in the New York Area, *Journal of the Soil Mechanics and Foundations Division, Proceedings of the American Society of Civil Engineers*, vol. 92, no. SM1, January, 1966.
12. Cornfield, G. M.: Discussion of Pile Driving Analysis by the Wave Equation, by E. A. L. Smith, *Journal of the Soil Mechanics and Foundations Division, Proceedings of the American Society of Civil Engineers*, vol. 87, no. SM2, February, 1961.
13. Smith, E. A. L.: Pile Driving Analysis by the Wave Equation, *Journal of the Soil*

Mechanics and Foundations Division, Proceedings of the American Society of Civil Engineers, vol. 86, no. SM4, August, 1960.

14. Agerschou, H. A.: Analysis of Engineering Pile Formula, *Journal of the Soil Mechanics and Foundations Division, Proceedings of the American Society of Civil Engineers*, vol. 88, no. SM5, October, 1962.
15. Hensey, M. D., and R. L. Petus: Discussion of Analysis of Engineering Pile Formula, by H. A. Agerschou, *Journal of the Soil Mechanics and Foundations Division, Proceedings of the American Society of Civil Engineers*, vol. 89, no. SM3, May, 1963.
16. Klohn, E. J.: Pile Heave and Redriving, *Journal of the Soil Mechanics and Foundations Division, Proceedings of the American Society of Civil Engineers*, vol. 87, no. SM4, August, 1961.
17. Flaate, Karl: An Investigation of the Validity of Three Pile Driving Formulae in Cohesionless Material, *Norwegian Geotechnical Institute Publication* 56, 1964.

CHAPTER 8

Design Considerations —Specific Types of Pile

8A Timber Piles

1. *Materials*

The materials standard for timber piling is ASTM D25. This standard contains provisions relating to required straightness and permissible defects (twist of grain, knots, holes, splits, and shakes).

The specification also classifies piles (classes A, B, and C) as to required butt and tip dimension for various lengths. Class A and B piles generally are used for permanent constructions. Class C piles are used for temporary purposes and, sometimes, for light loads in permanent construction. Selection of the class required (for a given species) can be made on the basis of the allowable stress values indicated in Sec. 2 following and assuming load distribution along the length of the pile as described in Sec. 6A.3.

2. *Allowable Stress Values*

In current practice (1968), timber piles are not stress graded. Therefore the stress values assigned in the design standard for timber (National Lumber Manufacturer's Association—NLMA) are not applicable for piling. Subcommittee VII of Committee D-7, American Society for Testing and Materials is working on this problem. However, pending publication of the results of their work, it is recommended (as stipulated by the Uniform Building Code, 1960), that the allowable stress value be taken as

60 percent of the basic *c* stress value for clear material of like species, as set forth in ASTM Standard Specification D245.

3. *Special Capacity-reduction Factor*

Timber piles are not inspectable. Therefore the presence and magnitude of potential distortions of the pile axis during installation are not known, and a special capacity-reduction factor is required to compensate for the potential effects of eccentricity and residual stress.

In addition, timber piles are quite susceptible to damage when installed by impact hammer. The degree of hazard depends on the type of pile (friction or end-bearing), the capacity (hard- or easy-driving), the length (long piles are more fragile than short ones), and the soil conditions. The two extremes would be (1) short friction piles of low capacity (low-hazard conditions) and (2) long end-bearing piles of high capacity (high-hazard conditions). For the low-hazard case, the suggested special capacity reduction factor is 1.2 to 1.3.* For the high-hazard case, an increased value of 1.5 to 2.0 is suggested, that is, a 33⅓ to 50 percent reduction in allowable stress as compared to that for the same section installed in the superstructure.

When piles are installed without impact, the potential for damage is reduced so that a lesser factor would be justified. However, the possibility of distortion of the pile axis remains so that the factor used should not, in any case, be less than 1.2 to 1.3. These values are summarized in Table 8-1.

4. *Maximum Loads*

Theoretically, if the pile is long enough, the section is large enough, and the species is properly selected, there need be no upper limit on the load to be supported by a timber pile, provided that the above indicated stress criteria are satisfied. As a practical matter, however, the load capacity is limited by the exigencies of installation. If the piles are installed by use of an impact hammer (which is the usual procedure), the higher the capacity, the harder or longer the pile must be driven and the greater the danger of damage.

Therefore, practice (which may or may not be justified) has been to establish upper limits for permissible loads on timber piles. The values prescribed by different authorities vary widely reflecting the vagaries of experience, but 30 to 40 tons per pile [5–7] is not an unusual design value for friction piles, with lesser values (say 20 to 25 tons) used for end-bearing piles because of the increased sensitivity of such piles to damage during installation. Even higher loads have been used. Reference 6

* This includes a factor of 1.1 to 1.2 for the effects of eccentricity (see footnote on page 231), plus an additional factor of 1.1 to compensate for the effects of potential residual stress.

TABLE 8-1 ***Summary of Recommended Special Capacity-reduction Factors*** (ϕ')*

Type of pile	Conditions	ϕ'*	Nominal allowable stress predicated on "short-column" formulas
Timber	Low-hazard installation—short friction piles; low to medium capacity (30 tons or less)	1.2–1.3	$0.45c$–$0.50c$†
	High-hazard installation—end-bearing piles and long friction piles of high capacity; installation by impact hammer	1.5–2.0	$0.30c$–$0.40c$
	High-hazard installation; installation without impact	1.2–1.3	$0.45c$–$0.50c$
Cast-in-place concrete	Inside permanent casing	1.0	$0.25f'_c$ (working stress design)‡
	Without permanent casing	1.3–1.5	$0.17f'_c$–$0.20f'_c$ (working stress design)‡
Precast concrete	Short piles; light loads	1.0	$0.225f'_c$–$0.25f'_c$ (working stress design)‡
	Long piles; heavy loads; hard driving	1.3	$0.17f'_c$–$0.20f'_c$ (working stress design)‡
Steel	H sections—short friction piles; favorable soil conditions	1.2–1.3	$0.45f_y$–$0.50f_y$
	H sections—friction piles; heterogeneous or questionable soil conditions:		
	Installation by impact hammer	1.5–2.0	$0.30f_y$–$0.40f_y$
	Installation without impact	1.2–1.3	$0.45f_y$–$0.50f_y$
	H sections—end-bearing piles:		
	Installation by impact hammer	1.5–2.0	$0.30f_y$–$0.40f_y$
	Installation without impact	1.2–1.3	$0.45f_y$–$0.50f_y$
	Pipe sections	1.2–1.3	$0.45f_y$–$0.50f_y$
Compacted concrete	Inside permanent casing	1.0	$0.25f'_c$ (working stress design)‡
	Without permanent casing	1.3–1.5	$0.17f'_c$–$0.20f'_c$ (working stress design)‡

* See also Sec. 6A.5, step 5 for discussion of Special Capacity-reduction Factors (ϕ').

† c value is value for clear material.

‡ Generally, higher allowable loads will result if design is predicated on ultimate-strength criteria.

describes the use of timber piles at 50-ton design loads. Reference 7 discusses cases in which capacities up to 65 tons were used.

5. *Details*

I. *Peeling.* Not required unless bark is patently loose (inspect for same), or pile is to be given a preservative treatment.

II. *Lagging.* There always is a question on the amount of load for which the connection between lagging and the basic pile material should be designed. This would be a function of the amount of friction developed along the lagging as compared to the friction and end bearing developed by and along the basic pile. No definitive means of analysis have been developed for this problem. The authors suggest the following rule of thumb: Proportion connection between double lagging and basic pile material for full design load on pile. Proportion connection between single lagging and basic pile material for one-half design load on pile.

8B Cast-in-place Concrete Piles

1. *Types*

Some common types of cast-in-place concrete piles are shown in Fig. 8-1.

2. *Allowable Stress Values*

See discussions of Sec. 6A.

3. *Special Capacity-reduction*

When installed using a permanent casing, cast-in-place concrete piles may be inspected for defects, and deviations of the pile axis from a straight line can be estimated, or measured.* Thus, except for the possibility of voids in the concrete fill due to some inadvertency in placement, the unknown factors of incidental bending moment and damage are eliminated, and a minimum value of the ϕ' factor 1.0 would apply.

When installed without a permanent casing (casing withdrawn or where the concrete is cast into a bored, uncased hole), a new and unknown hazard is introduced. This is the possibility of necking of the pile section, or the creation of discontinuities, due to intrusion of the surrounding soil, slumping of the hole during filling, or lateral pressures engendered by the installation of adjacent piles. This hazard is very real (see Refs. 1 and 2, for example) and merits a larger, special capacity-reduction factor. A value of 1.3 to 1.5 is suggested, the exact value to be dictated by judgment and the specific circumstances.

* Where the axis deviates from specified tolerances, action must be taken to reinforce or strengthen the pile or to reduce the load.

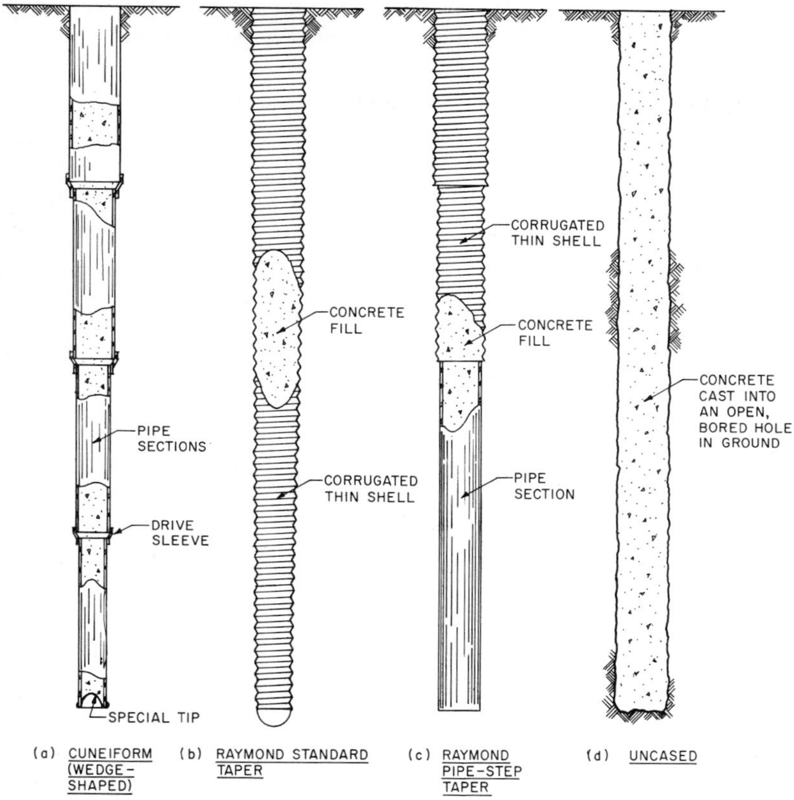

FIG. 8-1 Some Sample Types of Cast-in-place Concrete Piles.

4. *Minimum Dimensions*

Regardless of stress considerations, the diameter of a cast-in-place pile must be sufficient to permit proper placement of the concrete therein. The criteria stipulated in the New York City code are a minimum tip diameter of 6 in. for tapered piles, and 8 in. for piles of constant section. In this connection, the concrete must be sufficiently loose to prevent arching between the walls of the pile section during placement. At least a 5- to 6-in. slump should be specified, and the concrete design strength should be established accordingly.

8C Precast Concrete Piles

1. *Allowable Stress Values.* See discussions of Sec. 6A.

2. *Special Capacity-reduction Factor.* The sensitivity of precast concrete piling to damage during installation and to the occurrence of stress due to distortion is intermediate between that of timber and cast-in-place

piles. Accordingly, an intermediate value of special capacity-reduction factor (1.0 to 1.3) is recommended. The higher value should be used for longer piles, heavier loads, and hard driving.

3. *Handling Stresses.* The design of precast piling must provide for temporary stress conditions during handling of the pile in the casting yard and into the leads. These stresses are due to the weight of the pile being lifted (see Fig. 8-2), weight which should be increased 50 percent as an allowance for impact.

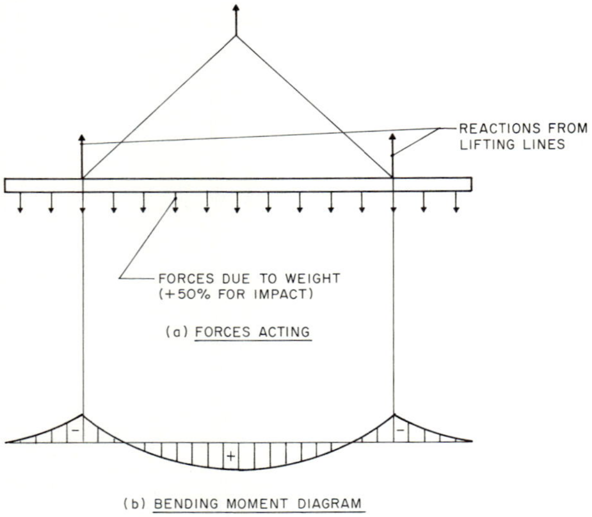

FIG. 8-2 Precast Concrete Piling—Handling Stresses.

4. *Driving Stresses.* Precast concrete piles are subject to some unusual problems of stress during driving.

When the hammer hits the top of the pile, a compression stress wave is created which travels down the pile at a velocity

$$V = \sqrt{\frac{E}{\rho}}$$

where V = velocity of stress wave
E = modulus of elasticity of pile material
ρ = mass density of pile material

The length of the stress wave depends on the length of time that the hammer ram remains in contact with the cushion or pile head. A heavy ram will stay in contact longer than will a light one, thus producing a longer stress wave. Similarly, if the ram strikes a thick soft cushion, it will produce a longer wave than when it strikes a relatively thin hard cushion.

When the compression stress wave reaches the tip of the pile it is partly reflected back up the pile, as another stress wave. The length, sign, and magnitude of the reflected stress wave depends on the amount of resistance offered by the soil at the pile tip. The less the resistance the greater the wave length and tensile stress component. If the pile tip is completely free, the length and magnitude of the reflected tensile wave would be equal to that of the compression stress wave initiated by the hammer blow. If the pile tip is immovable, the reflected wave is compression rather than tension.

The incident and reflected waves combine, algebraically, under proper conditions producing net tensile stresses in the pile. These tensile stresses cause transverse cracking of the pile section. Spiral cracking due to a torsional component of the incident and/or reflected wave also occurs. In addition, possible spalling of the head of the pile or of the tip due to large compressive stresses at these points of discontinuity must be prevented.

These conditions can, with difficulty, be evaluated by analysis as the stress magnitude varies depending on the pile length, speed of the hammer, condition of the cushion block, and resistance encountered at the tip. Therefore, in usual practice no attempt is made at numerical evaluation, but recourse is made to standards of practice as promulgated by the PCA [3] and PCI [4] standards which represent the "state of the art." The provisions as they relate to design of the pile section may be summarized as follows:

a. Precast Piles

1. Reinforcement including ties or spirals should be provided as for column sections. The minimum longitudinal reinforcement is not specified but is suggested to be 2 percent of the concrete section, placed in a symmetrical pattern of not less than four bars.
2. Protective cover—1½ in. clear.
3. Special reinforcement at head and tip—provide number 2 ties at 3-in. centers for a minimum distance from the head and tip equal to the diameter (or side dimension) of the pile. For spiral reinforcement, close up the pitch to 2-in. centers for a minimum distance from the head and tip equal to the pile diameter.
4. Chamfer head of pile to prevent damage and, unless reinforcing bars are to protrude from head, keep bars at least 3 in. back from head. Head of pile to be perpendicular to longitudinal axis.
5. Tip—Use blunt point, 6-in. side or diameter, with tapered length equal to 1½ times side dimension of pile for piles driven into clay or silt soils and equal to 3 times side dimension for piles driven into sand or gravel soils.

6. Minimum size—Conform to ACI 318 [15]. Use a minimum 10 by 10 in. or 8 by 12 in.

b. Precast-prestressed Piles

1. Minimum residual compression—700 psi.
2. Maximum residual compression—$0.2f'_c$.
3. Minimum concrete strength—f'_c = 5,000 psi.
4. Splices—Use cored or drilled dowel holes on both sides of splice. Minimum area of dowels = 1.5 percent of gross concrete section. Dowels to be grouted into holes.
5. Allowable handling stress—$6.0 \sqrt{f'_c}$ (tension).
6. Special reinforcement at head and tip—Not specified (see Precast Piles).
7. Cover—2 in., clear.

To these provisions, the authors would add:

c. Piles with Voids

1. Extend special end and tip tie or spiral reinforcement for a distance of 12 ft or one-third the length of the pile, whichever is smaller.
2. Minimum wall thickness of the concrete in any section to be 4 in.
3. Venting—If void extends through one or both ends of the pile, the head end should be vented to prevent the buildup of hydraulic pressure in the void during driving.

d. Miscellaneous

1. Where reinforcement is to be carried through head of pile, grind projecting ends flush with pile head to prevent direct loading under impact of the ram.
2. Chamfer all top and bottom edges and corners.
3. For driving, use a heavy ram with a low-impact velocity.
4. Slow the ram until the pile develops resistance. Do not use full ram stroke when the driving is easy.
5. Tolerances—Pile must be straight to minimize eccentricity of the hammer blow. Suggested tolerances—Sweep not to exceed (*a*) ⅛ in. in 10 ft, (*b*) ⅜ in. in 40 ft, or (*c*) 0.1 percent of pile length.*

8D Steel Piles

1. *Corrosion*

The first problem to be resolved before dealing with the design of steel piling is the potential for corrosion.

Many authoritative works have been published on this matter (Refs. 8

* Adapted from ACI 347 [16].

through 14, for example). Opinions run the full gamut from "no problem" to presages of inevitable doom and disaster.

The authors have had occasion to review and to give careful consideration to the matter. Bearing in mind that the problem of corrosion of steel piles is different from that of corrosion of, say, pipe lines in that percentage loss of section and not pitting is the important consideration, it was concluded that the weight of evidence, as published and as reported in discussion, including reported evidence from hundreds of H, pipe, and shelled cast-in-place piles which have been exhumed or extracted,* plus the absence of any recorded instance known to them of failure of a structure attributable to corrosion of steel piling, is that steel piles driven in upland areas, away from salt water and fully embedded in soil material, do not experience appreciable loss of section, except under the following circumstances:†

1. Where the soil contains appreciable amounts of decomposing organic materials. This includes garbage or other organic fills.
2. In cinder or slag fills where a residual content of acid or sulfur is present.
3. Where a condition of chemical seepage exists (coal piles, chemical plants, and like circumstances).
4. Where a bona fide condition of electrolytic action occurs. In this connection, it is noted that a search of available literature failed to reveal more than a handful of reports of actual problems of this type involving piles used to support buildings and fully embedded in the ground, under the limiting conditions above described.

Where one or more of these limiting conditions exists, unless there is some compelling economic or other reason, steel piles should not be used. If they must be used, protective measures must be taken. Indirect devices (for example, discounting $\frac{1}{16}$ or $\frac{1}{8}$ in. of metal, reduced values of allowable stress, minimum thickness provisions motivated by considerations of corrosion, etc.) are not satisfactory. Protective coatings appear to be of questionable value because they tend to be damaged in driving.‡ Encasement appears to be the most effective preventive measure.

* Verbal communications—discussions regarding proposed New York City building code.

† Except in the listed circumstances, in those few cases where a problem has occurred, it has been confined to the portion of the pile embedded in the loose, upper few feet of the soil above the water table. The tendency toward corrosion appears to dissipate with depth as the soil becomes tighter and the circulation of air around the pile is impeded and becomes largely obviated if the tops of the piles are cut off 4 or 5 ft below grade, a condition which usually is satisfied by the necessity for embedding the pile cap below grade and below frost depth.

‡ An exception may be the use of coal tar epoxy coatings with which some success has been reported. A relatively heavy coating (15 mils) is used, and the pile is sandblasted before coating.

2. *Allowable Stress Values*

See discussions of Sec. 6A.

3. *Special Capacity-reduction Factor*

Two cases must be considered—H piles and pipe piles.

I. *H piles* are subject to the same hazards and unknowns as are timber piles. Therefore, the same basic values of special capacity-reduction factor would apply. However, for the reasons given in Chap. 6, residual stresses may be neglected for steel piles, and so the above factors could be reduced 10 percent. On the other hand, some concession might be made to the possible occurrence of some corrosive action, so that the same values proposed for timber piles are proposed for H sections, that is, 1.2 to 1.3 for low-hazard installations and 1.5 to 2.0 for high-hazard conditions. The results related to short-column stress values are indicated in Table 8-1. As may be noted, for the more restrictive cases, the allowable stress values vary from $0.30f_y$ to $0.40f_y$ and bracket the criterion of $0.35f_y$ adopted by several prominent codes. For the case of lesser hazard, the allowable stress values are much higher, approaching (for A7 steel) the values of 18,000 to 20,000 psi used under some of the national codes (BOCA, for example).

II. *Pipe piles* are inspectable, as for cast-in-place sections and, ideally, the same special capacity-reduction factor 1.0 would apply. However, again in the spirit of conservatism and as a concession to the possible occurrence of some corrosive action, a value of 1.2 to 1.3 is suggested.

4. *Details*

I. *Minimum Thickness of Metal.* As noted above, provisions relating to required minimum thickness of metal, to the extent that these are motivated by fear of corrosion, are unjustified. If there is a serious corrosion problem, an extra ⅛ in. or so of metal is no panacea, and if there is not a problem, the extra metal is wasted. However, there is a sound basis for such provisions in terms of preventing damage due to excessive driving stresses encountered during installation. Heavy sections are less prone to such damage than are light ones.

This is the reason that *BP* sections are used for piling and that *WF*, *I*, *B*, or *M* shapes are not. There is no reason that these latter types of section (or built-up sections) could not be used, however, if they conformed to the essential stiffness criteria for *BP* sections, i.e., square proportions and width-thickness ratio of projecting flanges not to exceed 14. A number of sections satisfy these criteria, including the 6 by 6, 5 by 5, and 4 by 4 sizes, which sizes are, in fact, being used for lightly loaded piling, in competition with timber.

For pipe piles, there also is a need for minimum thickness criteria, if the piles are installed by impact hammer. Again the reason is the need to prevent damage during installation. The following criteria are suggested in Ref. 17:

1. Pipe installed open-ended and having a nominal outside diameter of less than 14 inches should be at least 0.25 inch thick. For diameters of 14 inches or more, the minimum thickness should be 0.375 inch. Pipe of lesser wall thickness can be used if a cutting shoe is provided.
2. Pipe installed closed-end should have a minimum nominal wall thickness of 0.10 inches.

Where installation is by use of a vibratory hammer, minimum thickness provisions need not apply, provided that the energy input of the hammer is not so great as to produce excessive heat in the pile material.

II. *Materials.* Again because of the need to provide adequate stiffness and to prevent damage during installation, the material for steel piles must be properly selected. For H sections, any of the usual structural alloys (ASTM A7, A36, A242, A373, A440, A441) can be used. For pipe piling, the applicable specification is ASTM A252. Generally, it is well to use grade 2 steel, or better.

8E Compacted Concrete Piles

1. *Description*

Compacted concrete piles (also called bulb piles) include those where the concrete is compacted, either by a falling weight or under hydraulic pressure, during driving. A bulb of plastic concrete is extruded from the base. The concrete in the shaft also may be extruded (from a temporary casing) as the casing is withdrawn, or a permanent casing may be left and the concrete in the shaft not compacted. The essential scheme is shown in Fig. 8-3.

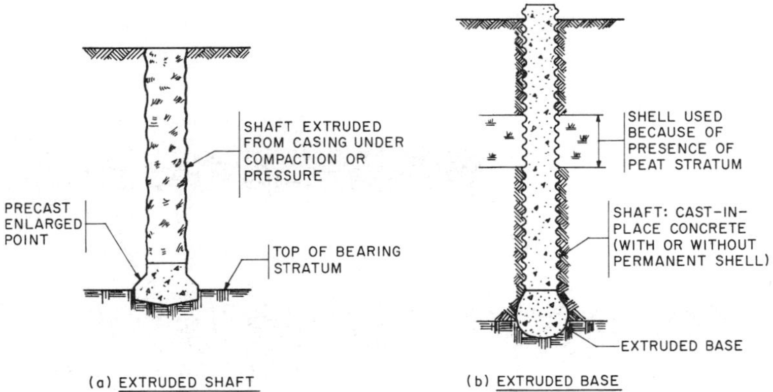

FIG. 8-3 Compacted Concrete Piles.

2. *Design of Shaft*

In the United States, bulb piles generally are designed on the basis of end-bearing resistance alone. Any resistance developed along the shaft is neglected.* Under this circumstance the pile is not a pile at all, but a footing, with the shaft functioning as a pier. Design may be made on the basis of the equivalent-footing analogy as described in Sec. 6B.1*b*, with the particular criteria stipulated on page 245 used when the pile tips are founded in sandy material.

3. *Special Capacity-reduction Factor*

Despite its analogy (with regard to function and behavior) to a footing, the compacted pile differs from a footing in that the shaft and base cannot be inspected after installation. Therefore, the special capacity-reduction factors and allowable stress criteria stipulated for cast-in-place concrete piles would apply.

8F Caisson Piles

1. *Description*

A caisson pile is a concrete-filled pipe pile which is driven to, and socketed in, bedrock.

2. *Design of Shaft*

A caisson pile can be constructed to close tolerances on straightness. The alignment can be checked visually. The juncture between pile and supporting soil is positive. The supporting material is inspected, either by using a drop light or a television camera or by observing the washings from the drilling of the sockets. In addition, all the load is assumed to be supported by the socket, the resistance along the sides of the pile shaft being neglected. Accordingly, this type of pile is the closest thing (in piling) to a direct extension of the building column and may be designed accordingly, as follows:

The core section could, in theory, be designed for $0.60f_y$, although in practice, a lesser value is used (say $0.50f_y$).

The shell may be designed as for a pipe-pile section (special capacity-reduction factor = 1.2 to 1.3).

The concrete fill is designed as for a cast-in-place pile (special capacity-reduction factor = 1.0).

The assemblage is designed on the basis of each unit (shell, core, fill) being stressed to the indicated allowable value, simultaneously.

* It is understood that in European practice both resistances are considered.

3. *Design of the Rock Socket*

The design of the rock socket is predicated on simultaneous development of ultimate bearing pressure on the bottom of the socket and of bond along the sides of the socket. The justification for this assumption is that the displacements required to develop either component are so small that they are compatible. The bearing capacity of the shell and the core, when they bear on the rock, conventionally is disregarded, and the assemblage treated as though the entire load were resisted by the aforementioned two components only (bearing on bottom of socket and perimeter shear).

Allowable bearing values at the bottom of the socket derive from Chap. 5, with the values increased for embedment as indicated therein. This increase is important as the socket usually is several feet deep, and the increase of 10 percent per foot of embedment quickly doubles or triples the allowable bearing pressure, as compared to the basic value.

The strength of the concrete fill, computed as the fully confined (flexural) strength of $0.45f'_c$ and not as the column strength value of $0.25f'_c$, must be of comparable magnitude and must be checked.

The allowable bond stress between the sides of the socket and the concrete fill is a matter of debate. The Boston building code prescribes a value of 100 psi. The New York City code prescribes a value of 200 psi. Even larger values are used. Both of the values noted appear to be conservative, if the rock is sound. If the rock is not sound, it probably would be best to neglect the perimeter shear resistance entirely.

4. *Maximum Loads*

There are no load limitations applicable to caisson piles. Design loads of several thousand tons per pile have been used.

5. *Details*

a. Minimum Diameter. This is controlled by the working space required to drill the socket and place the concrete. A conventional figure is 16 in. but there does not appear to be any reason why substantially smaller sections could not be used.

b. Minimum Shell Thickness. Same as for pipe piles.

c. Minimum Depth of Socket. Equal to diameter of caisson, but not less than, say, 16 in.

8G Composite Piles

Several types of composite pile are shown in Fig. 8-4. Each component section is designed to the criteria cited in Secs. 8A through 8F for like type of section. The only special problem is the joints, which

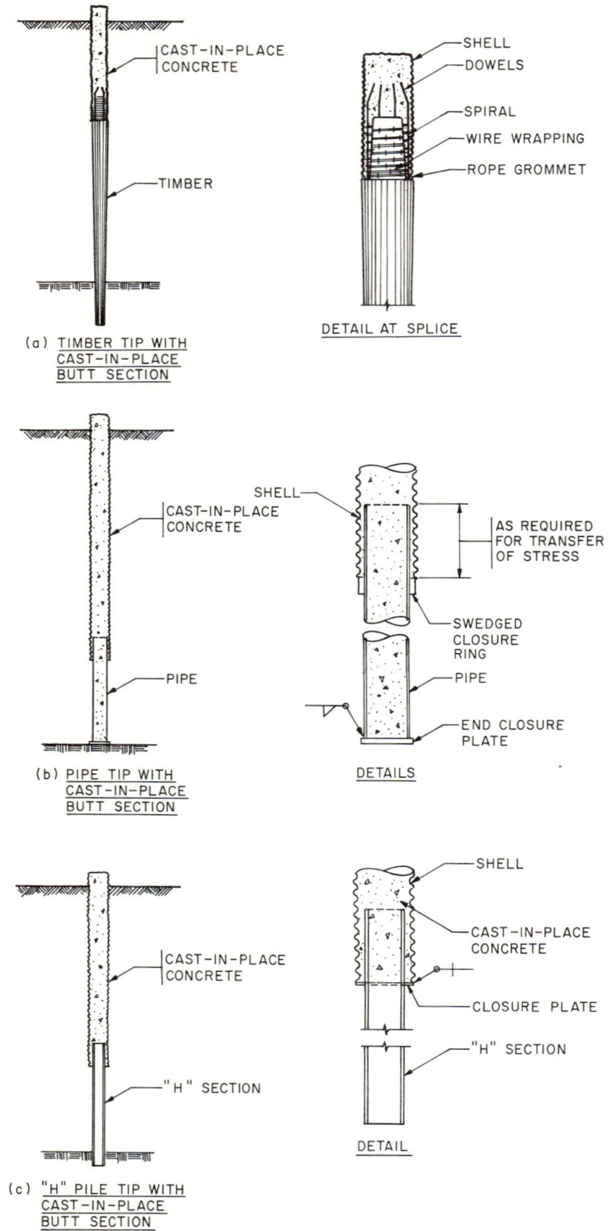

FIG. 8-4 Some Types of Composite Piles.

must be detailed to have some moment capacity (even though no moment theoretically exists in the section), must develop the handling stresses, if such exist, and must resist uplift, if any. No general rules are available. Generally, though, if the joint will stand handling and driving and is not located within the upper 10-ft section of the pile, it probably is adequate.

REFERENCES

1. Feld, Jacob: Failure of Extrusion Pile Foundation, *Proceedings of the Technical Conference on Design of Foundations for the Control of Settlements,* American Society of Civil Engineers, June, 1964.
2. *Engineering News-Record* on Chicago caissons with clay seams.
3. "Concrete Piles, Design, Manufacture, Driving," Portland Cement Association, 33 West Grand Ave., Chicago.
4. "Standard Prestressed Concrete Piles," Joint Committee of AASHO, Committee on Bridges and Structures, and Prestressed Concrete Institute.
5. Capozzoli, L. J.: Current Status of Timber Foundation Piles, *Wood Preserving News,* November, 1966.
6. High Rise Apartment Buildings Built in Low-cost Timber Piles, *Wood Preserving News,* February, 1965.
7. "Timber Foundation Piles," American Wood Preservers Institute, 1707 L St., NW, Washington, D.C.
8. "Corrosion of Steel Pilings in Soils," monograph 58, U.S. Department of Commerce, National Bureau of Standards.
9. Sudrabin, L. P.: Foundation Piling Corrosion: Mechanisms and Cathodic Protection, *Materials Protection,* October, 1963.
10. Long Exposure Underground Has Little Effect on Steel Piling, *Materials Protection,* December, 1962.
11. *Federal Construction Council Technical Report* 32, publication 741, National Academy of Sciences, National Research Council, 1959.
12. Steel Shells in Service 39 Years Still Sound and Serviceable, *Engineering News-Record,* April 12, 1928.
13. Greulich, G. G.: Extracted Steel H Piles Found in Good Condition, *Engineering News-Record,* August 24, 1950.
14. Bjerrum, L.: Norwegian Experience with Steel Pile Foundations to Rock, *Journal of the Boston Society of Civil Engineers,* 1957.
15. "Building Code Requirements for Reinforced Concrete," ACI 318-63, American Concrete Institute.
16. "Recommended Practice for Concrete Formwork," ACI 347, American Concrete Institute.
17. "Pile Foundations," a reference bulletin promulgated by the American Iron and Steel Institute, New York.

CHAPTER 9

Load Tests

Two types of load test are used in foundation design:

1. *Soil load tests* are used to aid in estimating allowable soil-bearing capacity.

2. *Pile load tests* are used as a guide to estimating pile-load capacity.

There is a general tendency to accept load tests as the preeminent, or ultimate, indicator of capacity. Whatever the other indicators (calculation or driving formulas, for example) may suggest, the load test is taken as the "proof of the pudding."

This faith is grossly misplaced. The fact is that load tests are not an especially reliable indicator. It is not that the tests, per se, are unreliable, but that there are serious, fundamental problems in conducting and interpreting load tests results in general. For example:

CASE 1 The simplest load test is the tension test of a steel rod (ASTM A370). In this test, a rod of prescribed diameter, length, and configuration is put into a testing machine and loaded at a prescribed rate to fracture. The fracture load (converted to stress) is deemed to be the ultimate strength of the material, in tension.

This sounds simple, but it is not. If the size, length, or configuration of the bar are changed, the results are changed. If the speed at which the load is applied is changed, to very fast, for example, extraordinarily high apparent strengths appear. Even the conversion from ultimate load to ultimate stress is not certain. Shall it be based on gross section, on net section after deducting the "necking down" of the specimen, or on net section after deducting for the microcracking incident to the failure process?

Finally, the conversion from the strength of the test specimen to the strength of full-size members is not a simple extrapolation. Failure loads for full-size members are much different from what would be computed from failure stresses determined from the test specimens because of the following factors:

1. Increasing the size of specimen increases the statistical probability of occurrence of latent flaws.
2. Increased size makes it more difficult to assure the straightness of the member axis, thereby increasing the probability and amount of eccentricity. Connections to other members also introduce eccentricities.
3. Residual stresses exist in full-size members but may be absent in the test specimens. These stresses are due to the differential cooling of the rolled shape after forming. This matter has been studied at length by the Column Research Council. Residual stresses in the amount of, perhaps, 50 percent of the yield strength have been shown to exist in wide flange and similar rolled shapes.
4. The behavior of either full-size member or test specimen is dependent on the stress history of the specimen (in the case of steel, previous cold working).

CASE 2 Next consider the concrete compression test (ASTM C33). Again, the size and proportions of the test specimen are prescribed, and again, if size or proportions are changed, the results are changed, often drastically. As with the tension test, the speed of loading is important. Very slow application of load introduces creep effects and low apparent strengths. Very rapid loading will indicate inordinately high strength.

There are additional complications:

1. The apparent strength is a function of age.
2. The apparent strength also depends on such variables as whether the specimen is tested wet or dry and on the size of the coarse aggregate as related to the size of the test specimen.
3. The point at which the specimen fails is not so easily established as with the tension bar. Failure of the bar is fairly well defined by a sudden fracture. In the concrete test cylinder, failure is progressive and poorly defined so that some arbitrary failure point must be established.

Finally, the problem of correlation of test strength to the behavior of the actual structure (which was difficult in the case of the steel bar) becomes almost overwhelming with concrete, as it is a matter of common experience that load tests of concrete structures consistently show the structure to have substantially greater strength than is indicated by computation based on strengths indicated from tests on materials samples. The reasons for this include:

1. The structure is *not elastic,* except at very low loads. This is due to nonlinearity of the stress-strain curve and to other factors.

2. *Time-dependent deformations* occur from influences other than applied load. These include creep, shrinkage, and temperature change and are not reflected in the testing of the laboratory specimens.

3. *Three-dimensional behavior.* The structure is designed as a two-dimensional entity; its behavior as a three-dimensional unit is substantially different.

4. *Redistribution of load.* The development of areas of plastic deformation and the redistribution of load begins well before ultimate load is reached. Some tests indicate that this redistribution starts at one-half, or less, of ultimate load capacity. The capacity of the structure, once redistribution of the load starts, would not be anywhere near the simple relations assumed for design.

CASE 3 Finally, soil and pile load tests represent perhaps the highest level of complexity. *All* of the variables indicated as affecting evaluation of the relation between steel and concrete materials tests and behavior of the actual structure similarly affect the soil and pile load tests. These include:

1. The *size and proportions* of the loaded area (test plate)
2. The *rate* of load application
3. The *wet or dry* condition
4. The presence of *coarse particles* in the stressed zone
5. *Inelasticity* of the material being tested
6. *Time-dependent deformations*
7. *Three-dimensional behavior* versus two-dimensional design practice
8. *Redistribution of load* due to plastic deformation
9. *Failure point* not well defined
10. The *stress history* of the soil

In addition, the following new factors apply:

11. *Type of loading* (dead, live, ambient)
12. *Influence of adjacent loaded areas*

It will be no surprise, then, that if it is so difficult to find a reliable correlation between load tests and actual behavior of the completed structure in relatively simple things like steel and concrete members, it is infinitely more difficult to find such correlations when dealing with a heterogeneous, inelastic, nonisotropic material like soil. Because of these difficulties, the use of soil load tests and pile load tests is limited to the role of another aid to judgment; they are *not* a proof of performance, a panacea, or a substitute for careful analysis, experience, and good sense.

Section 9A SOIL LOAD TESTS

9A.1 Procedure*

a. Soil Load Test

A soil load test is a model test of a prototype foundation. It consists of loading a set of plates of varying size and extrapolating the results to the probable performance of the full-size structure.

Usually, three plates (12-, 24-, and 30-in. diam) are tested to establish the necessary extrapolation. Under some circumstances the testing of only two plates (any combination of 12-, 24-, or 30-in. size) or even of a single plate (24- or 30-in. size) may suffice (see Sec. 9A.2*a*).

b. Test Setup

Many different arrangements of test setup may be used, depending on job conditions, testing requirements, and available equipment. A typical arrangement is shown in Fig. 9-1.

c. Test Locations

Sets of test loadings should be made in areas representative of soil conditions at the site, but with the exact locations within the limits of the building area.

The plates for each test should be set as close together (Fig. 9-1*c*) as the following criteria will permit:

1. All plates (when loaded) must test essentially the same soil stratum.
2. The plates must be set sufficiently far apart that the bulbs of pressure under each do not overlap, since, once the soil is stressed, its stress-strain characteristics are changed, and the load test results will be changed.
3. If the plate loadings are to be carried to failure, the spacing must be sufficiently large that the failure planes do not overlap. For the large (30-in.-diam) plate, a distance of 12 to 15 ft to the adjacent plate, or even more, may be required.
4. In no event should the spacing of plates be less than 8 ft 0 in. center to center.

The ground surface at the locations of the proposed tests should be leveled at the elevations of the proposed test for a clear distance of at least two and one-half times the width of the test plate but not less than 5 ft all around the test plates.

* The indicated procedures and criteria are predicated on ASTM standard D1194-57, modified to reflect the authors' views.

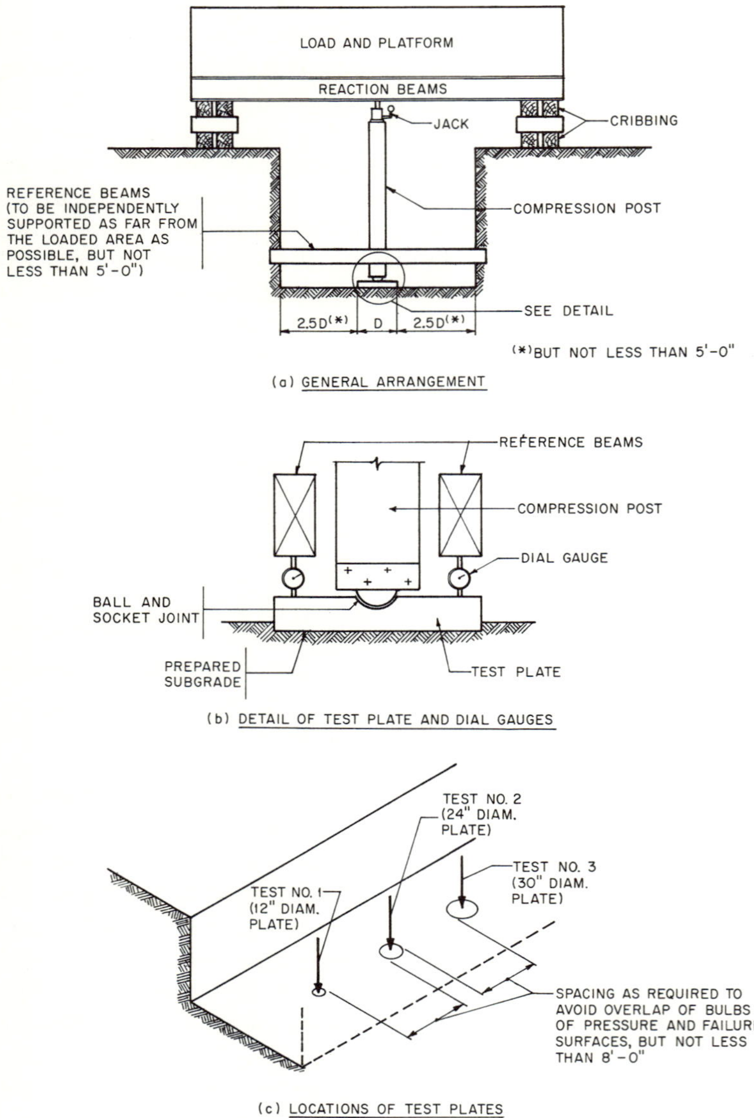

FIG. 9-1 Soil Load Test—Test Setup.

d. Load Platform

The load platform providing support for the test load should be supported on timber or other cribbing, which must not bear within the leveled area around the edge of the test plate.

e. Plates

The plates on which the loads are applied should be made of steel having a minimum thickness of 2 in., be round shaped having a 12-, 24-, and 30-in. diam, and be fitted with a ball and socket depression as indicated in Fig. 9-1. The plate for each test should be set and centered in a depression in the bottom of the pit or loading area about 6 in. larger in diameter than the plate and 4 to 8 in. deep, the least embedment being used for the smallest plate. The area on which the plate is placed should be scraped to remove any disturbed soil and covered with a thin layer of fine clean sand, smoothed to a completely level surface.

f. Settlement Observations

Settlement observations should be made at four points on the test plate (90° intervals) by means of dial extensometers. The extensometers should provide readings to the nearest 0.001 in., readings which should be checked by settlement observations taken using an engineer's level reading to 0.001 ft and properly referenced to a well-established benchmark.

g. Loading of the Soil

I. Load may be applied to the test plates by direct weight or by means of a hydraulic jack. Care should be observed that the load is concentric on the plate. If direct weight is employed, the loading increments should be applied without impact. The weight of the compression post placed on the plate and of the loading platform or box should be obtained prior to the test, and this weight should be considered as the first increment of load. If a hydraulic jack is employed, facilities for maintaining each increment of desired load constant under increasing settlement should be provided. The jack and the gauge for the jack should be calibrated as a unit not more than 2 weeks prior to the test and should represent actual loads within an accuracy of ± 2 percent.

II. The load should be applied to the soil in cumulative, equal increments of not more than 1 tsf, to a total loading intensity which produces failure of the soil (as measured by a rapid, progressive increase in settlement), but not less than six increments. When failure of the soil is not produced with a reasonable total load, the loading procedure may be terminated when the total cumulative settlement has reached 1 in. However, loading to failure is most desirable, if it is feasible.

III. After completion of the loading cycle, the load should be removed from the plate, and the rebound recorded until there is no further, sensible movement.

h. Time Interval of Loading and of Settlement Readings

I. Each load increment should be maintained until the settlement is 0.001 in. or less, over a period of 5 min, but for a time interval of not less than 1 hr.

II. Under each load increment, settlement observations should be made and recorded at intervals of ½, 1, 2, and 4 min, and each 5 min thereafter, until the settlement limitation is reached, the required 1-hr interval has elapsed, and the load is increased or removed.

III. Rebound observations should be made in the same manner as for the loading increments.

i. Precautions

I. When load tests are performed on materials affected by drying, suitable methods shall be used to prevent evaporation from the material.

II. In the event groundwater is present immediately below, at, or above the level required to be tested, dewatering facilities shall be installed to maintain groundwater at a minimum of 4 ft (5 ft for the 30-in. test plate) below the level of the test plate during the preparation and duration of the test or tests. Otherwise, the apparent cohesion imparted to the sand by the soil moisture may introduce an error on the unsafe side.

III. Trenches and other provisions at the ground surface shall be made to prevent inflow of surface water, to remove water that may drain into the test area, and to prevent water from flowing upward into the bottom of the pit, thereby reducing the effective weight of the soil by seepage pressure. The entire test area shall be protected against weather and rainfall for the entire duration of the test.

IV. The test area must be free of frost and protected against freezing during the test.

9A.2 Determining the Design Bearing Pressure from the Test Results

As indicated in Chap. 5, allowable soil bearing pressure is the lesser value indicated by two considerations: (1) ultimate strength, divided by a suitable safety factor and (2) allowable settlement. These two values are evaluated from the load test, as follows:

a. Ultimate Load Capacity

This can be determined only if the test load is carried to actual failure (a rapid, disproportionately large increase in settlement for a fixed increment of load).

Determination of the precise failure load is a matter of judgment. As indicated in Fig. 5-10, the relation between settlement and load, generally, is one of gradually increasing steepness with no well-defined break to establish the failure condition. Convention is to establish the failure load as the intersection of the tangents to the two straight portions of the curve.

The failure load for the full-size foundation is extrapolated from the load test using Eqs. (5-1), (5-2), or (5-3), i.e.,

$$q_{\text{ult}} = K_1 c N_c + \Sigma \gamma_d D_f N_q + K_2 \gamma_d B N_\gamma$$

where K_1 and K_2 are constants representing the shape of the bearing area. In this equation the cohesive strength, c, and the bearing-capacity factors N_c, N_q, and N_γ are unknown. However, the bearing-capacity factors are related to the friction angle ϕ (see Fig. 5-1) so that there are, in effect, only two unknowns c and ϕ. Pairs of simultaneous equations are needed to solve for these unknowns.

If the specified *three*-plate–test series has been conducted, two such pairs of simultaneous equations may be written and solved (by trial and error, since the relation between the bearing-capacity factors and the ϕ value is not constant) for c and ϕ. Two sets of answers will result which may or may not be the same, and one is used as a check on the other.

If only *two* different sizes of plate are loaded, one set of equations and one answer will be derived. If the approximate answer is known before the test is made; i.e., the purpose of the test is to decide whether to design for 3 or 4 tsf, then this one answer (and the two-plate test) is sufficient. If, however, the test is completely exploratory, or if there will be a very large difference in size between the footing and the largest test plate, then the three-plate series should be conducted so that the desired check can be obtained.

In cases where the proposed foundation will bear at shallow depth on purely cohesive or purely cohesionless soils, testing of a *single* plate is sufficient as there is only one variable (ϕ *or* c) and only one equation is needed. For example:

For a surface footing founded on a cohesionless soil, bearing capacity varies directly as the width (b) of the bearing area so that

$$q_{\text{fndn}} = q_{\text{plate}} \left(\frac{\text{width of foundation}}{\text{width of plate}} \right) \tag{9-1}$$

For the same surface footing on a purely cohesive soil, the relation becomes

$$q_{\text{fndn}} = q_{\text{plate}} \tag{9-2}$$

In the case of the single-plate test, it should be noted that, since the test is conducted on the ground surface, the value of q_u indicated by the test (provided that the plate is no larger than the foundation) always will be less than the q_u value for the actual foundation.

b. Factor of Safety

The discussions of Sec. 5A.2 apply.

c. Settlement

The probable settlement of the actual foundation may be extrapolated from the load-test data using the perimeter-shear relation proposed by Housel [Eq. (5-13)], which is

$$P = AP_c + P_rS \tag{5-13}$$

where P = total load on the bearing area required to produce a given settlement, lb
A = contact area, sq ft
P_c = an equivalent unit compressive stress, psf (an experimental constant)
P_r = perimeter of contact area, ft
S = unit shearing stress per foot of perimeter, lb per ft (another experimental constant)

P_c and S are the unknowns and are evaluated by simultaneous solution of the pairs of equations derived from loading the three different sizes of plates (see Sample Computation 9-1). As with evaluating the ultimate bearing capacity, the two sets of values may or may not be the same.

Alternatively, the relation indicated by Taylor [Eq. (5-14)] may be used, to the same purpose.

SAMPLE COMPUTATION 9-1

STATEMENT OF PROBLEM

From plate load test data, predict bearing pressure under a 10- by 10-ft footing which will cause a settlement of 1 in.

DATA

Test plates of 12-, 24-, and 30-in. diameter are load tested. The loads corresponding to 1 in. of settlement are:

12-in. plate = 6,000 lb
24-in. plate = 21,000 lb
30-in. plate = 32,000 lb

PROCEDURE

From Eq. 5-13:

Eq. 1 (12-in. plate):

$$6{,}000 \text{ lb} = 0.785P_c + 3.14S$$

Eq. 2 (24-in. plate):

$$21{,}000 \text{ lb} = 3.14P_c + 2 \times 3.14S$$

Eq. 3 (30-in. plate):

$$32{,}000 \text{ lb} = 4.91P_c + 2.5 \times 3.14S$$

Solving Eqs. 1 and 2:

$P_c = 5{,}720$ psf
$S = 478$ lb per ft

Solving Eqs. 2 and 3:

$P_c = 5{,}780$ psf
$S = 469$ lb per ft

Use: $P_c = 5{,}750$ psf
$S = 470$ lb per ft

For 10- by 10-ft plate, 1 in. of settlement corresponds to:

$$P = 100 \text{ sq ft} \times 5{,}750 \text{ psf} + 40 \text{ ft} \times 470 \text{ lb per ft}$$
$$= 575{,}000 + 18{,}800 = 593{,}800 \text{ lb}$$
$$\frac{P}{A} = 5{,}938 \text{ psf, say 3 tsf}$$

Where the soil being tested is a truly cohesionless deposit, a single-plate test may be used, in lieu of the three-plate series, as the recorded settlements of the test plate may be related to the settlements of the completed structure, using the relation of Terzaghi and Peck [8]

$$S_0 = S_1\left(\frac{2B}{B+1}\right)^2 \tag{5-12}$$

However, use of this equation (and of single-plate load tests) specifically is limited to tests conducted on cohesionless soils.

The allowable bearing pressure (as limited by settlement) is determined by establishing the allowable settlement for the actual foundation and estimating, as described, the soil bearing pressure corresponding to that settlement.

d. Design Soil Bearing Pressure

As was previously indicated, the design soil bearing pressure is determined from the load test as the lesser of the two values limited by (1) ultimate capacity, divided by a factor of safety and (2) settlement.

In most instances the settlement consideration governs. In such a case it is not necessary to carry the load test to failure or to consider the ultimate strength criterion, except indirectly, and some time and money

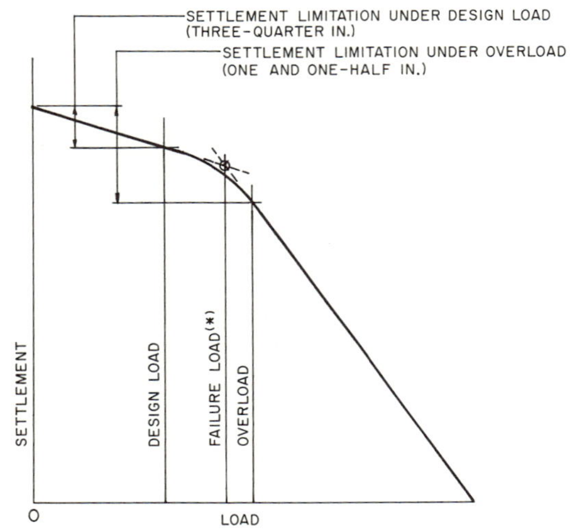

(*)FOR THE PARTICULAR SOIL BEING TESTED (A VERY STIFF AND BRITTLE SOIL), THE FAILURE LOAD (AS DEFINED BY THE INTERSECTION OF THE TANGENT PORTIONS OF THE CURVE) MAY OCCUR AT DEFORMATION VALUES LESS THAN THE STIPULATED LIMITATION ON SETTLEMENT UNDER OVERLOAD (OR EVEN UNDER THE DESIGN LOAD).

FIG. 9-2

may be saved by using rules of thumb, such as the following: The soil load test shall be considered as substantiating the ability of the soil to support the proposed unit intensity of loading, if the gross settlement of the test plate under the proposed unit intensity of loading does not exceed some fixed amount (say, ½ to 1 in.), and the total gross settlement under (some percentage overload) does not exceed some fixed amount (say, 1 in. or more.)*

The problem with this approach is that, unless the limits on settlement (under both design load and overload) are established very conservatively, there is no assurance that the failure load (as previously defined) may not be less than the test overload (or even the design load). Refer to Fig. 9-2, for example. In this case, the requirements of test perform-

* From New York City building code.

ance would have been met, but the desired safety factor would not have been realized.

For this reason load tests carried to failure are more desirable in that they permit a rational interpretation of the probable safety factor and of the design requirements. On the other hand, they also are likely to be more expensive and time-consuming to perform so that where there has been sufficient local experience to establish that the shortcut rule-of-thumb procedures give consistent and reasonable results, the shortcut procedures are popular.

e. Effect of Groundwater Level

Under the prescribed test procedure, the groundwater level is lowered well below the level of the test plates, and the soil in question is tested dry. In theory, if the groundwater level, at some time in the future, will rise to the footing level, or higher, the bearing capacity, as indicated by the load test, should be corrected for the decrease in confining pressure, in accordance with Eqs. (5-1) to (5-3). This correction applies only with regard to ultimate-strength capacity, however. Contrary to common impression, it does not appear to apply with regard to settlement considerations. The reasons are discussed in Sec. 5D.7*a*.

Since the settlement consideration usually controls, the effect of the variation in groundwater level between test and "as built" conditions, generally, is not significant.

9A.3 Limitations on Use of Soil Load Tests

The general limitations and difficulties of conducting and interpreting load tests were discussed in the introduction to this chapter. The following specific limitations apply to the use of soil load tests as criteria for establishing allowable bearing capacities.

a. Effects of Embedment, Limited Depth of Bearing Stratum, and Relief of Overburden Pressure

All these factors increase allowable bearing pressure by reducing the amount of settlement of the bearing area. None of the effects is reflected in the load-test results, however. Therefore, in the general case, the load-test results are too conservative and should be corrected (as described in Chap. 5) to include the effects of these three factors.

b. Load Tests of Fine-grained Soil Materials

The soil load test involves a relatively short-term loading condition. The recorded settlements will not show the long-term displacements which will occur under the permanent load of the building. Soil load tests, therefore, are inapplicable in cohesive soil materials, except as a means for estimating safety against shear failure in the soil. They do not provide valid prediction of probable settlements, except in purely cohesionless (sand and gravel) soils.

c. Load Tests of Rock

Except in the rare case of a massive isotropic, crack-free material, load tests of rock are of no value.

The bearing capacity of rock results from its cementation and, for a given geological entity, is controlled by the manner and degree to which this cementation has been affected or directionalized by jointing, bedding, weathering, differential weathering, etc. These effects are so variable that a load test on a small plate does not permit meaningful extrapolation to the performance of the large bearing areas under the actual footings or pile groups.

d. Load Tests of Gravel and Boulder Soils

Load tests of gravel and boulder soils are valid only so long as the maximum size of the gravel is small compared to the size of the test plate. Load tests of boulder-strewn soils are not likely to give proper results.

e. Where Test Stratum Is Underlain by Weaker Strata

The soil under the test plate is stressed, significantly, to a depth of, perhaps, twice the diameter of the plate. For the large (30-in.) plate, this means that the soil actually tested is limited to a layer 5 ft deep below the test level. Any strata below this level are not tested. The load from the building, on the other hand, stresses the soil to a much greater depth (perhaps 10 to 20 times as deep). If there are weaker, or compressible, strata within this depth, the load-test results do not provide a valid measure of bearing capacity for the actual building foundation. Sowers [4], for example, cites a case where a load test on a crust deposit of firm material overlying a weak stratum indicated a safe bearing capacity of 6,000 psf. The foundation failed during construction under a load of 2,000 psf.

f. Borings

It is impossible to properly interpret the results of a soil load test without adequate borings as described in Chap. 2.

g. Extrapolation

Errors or inaccuracies in the test results will be magnified in extrapolating said results to the full-sized footing. Accordingly, a limit should be placed on the extrapolation, a limit which may be expressed as some percentage of the bearing-capacity value as indicated by analysis. For example, assume that analysis has been made of probable bearing capacity and settlement for the proposed construction (see Chap. 5). The designer is not certain of the validity of said analysis, either because the soil is not one of the types described in Chap. 5 or because some special conditions prevail. A load test is conducted and indicates a design capacity three or four times as great as that indicated by analysis. Despite this indication however, the revised design capacity might be limited to, say, 150 or 200 percent of the capacity as indicated by analysis, so that the extrapolation is kept within reasonable bounds.

Section 9B PILE LOAD TESTS (AXIAL COMPRESSION)

9B.1 Procedure*

a. Pile Load Test

Like a soil load test, a pile load test represents a limited-sized model of the proposed foundation. It may consist of loading a single pile or a group of piles.

b. Test Setup

Again, many different arrangements of test setup may be used. A typical arrangement is shown in Fig. 9-3.

c. Test Locations

See Sec. 7C.

d. Load Platform

The load platform providing support for the test load should be balanced on cribbing which bears at least 8 ft away from the test pile.† When anchor piles are used, they should be located similarly remote

* The indicated procedures and criteria are predicated on ASTM standard D1143-61T, modified to reflect the authors' views and experience.

† Unless load is added to the platform in successive increments, so that the total load just equals the desired jacking load, that part of the platform load in excess of the jacked reaction will bear on the soil under the cribbing. This creates a bulb of pressure and an effective confining pressure in the soil under, and spread out from, the point of reaction. At depth, this confining pressure reacts on the pile, increasing the normal pressure on, and frictional resistance along, the pile shaft. The intent of setting the reaction points away from the test pile is to limit the effect of the confining pressure to the lower portion (and as small a length, as possible) of the pile. Within the limits of practical spacing (say, 8 ft, as indicated), the confining-pressure effect is eliminated only for a very short length of pile (perhaps 10 ft, at most), so that, for longer piles, the test results, particularly at low load values, can be very sensitive to the presence of these points of reaction. The best solution is elimination, i.e., to put the load on the load platform in increments and only as much as is needed to develop the desired pile reaction.

from the test pile. The details of the load platform should be such that no lateral load is imparted to the pile and that the vertical load will be applied concentrically and without impact. Caution must be observed

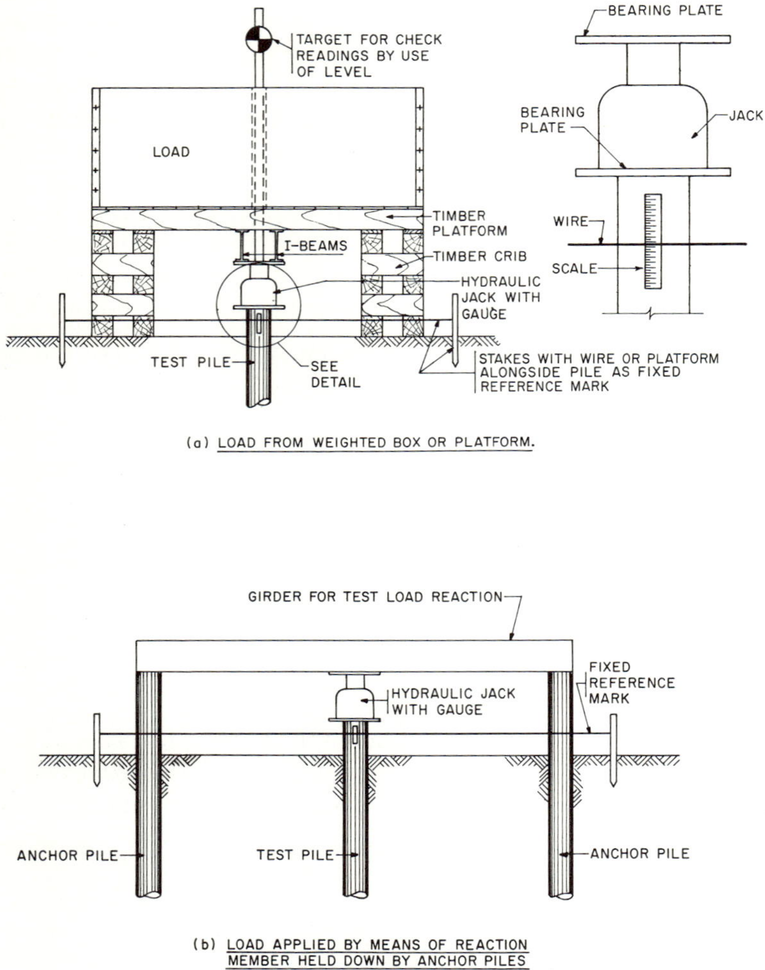

FIG. 9-3 Pile Load Test—Typical Setup.

that the loaded platform does not tilt during the test, and suitable wedges should be provided for this purpose.

e. Settlement Observations

Settlement observations may be made using a target rod as shown in Fig. 9-3, using the scale and string line arrangement there indicated, or dial extensometers may be provided. Readings to 0.001-ft accuracy should be obtained, and a properly referenced bench mark should be provided.

f. Loading

I. Load may be applied to the test pile by direct weight or by means of a hydraulic jack, and care should be observed that the load is concentric on the pile. If direct weight is employed, the loading increments should be applied without impact or jar. The weight of the loading platform or box should be obtained prior to the test, and this weight should be considered as the first increment of load. If a hydraulic jack is employed, facilities for maintaining each increment of desired load constant under increasing settlement should be provided. The jack and the gauge for the jack should be calibrated as a unit not more than two weeks prior to the test and should represent actual loads within an accuracy of ± 2 percent.

II. The load should be applied to the pile in cumulative, equal increments of 50, 75, 100, 125, 150, 175, and 200 percent of the proposed working load. If economically feasible, further loading increments should be added until failure (as measured by a rapid, progressive increase in settlement) occurs.

III. After completion of the loading cycle, the load should be removed, and the rebound recorded until there is no further sensible movement.

g. Time Interval of Loading and of Settlement Readings

I. Increments of load less than the proposed working load should remain in place for a period of not less than 1 hr. The proposed working load, and each increment thereafter, should remain in place until there is no measurable settlement in a 2-hr period. The total test load shall remain in place until settlement does not exceed 0.001 ft in 48 hr. The total load shall be removed in decrements not exceeding 25 percent of the total load at 1-hr intervals or longer. The rebound shall be recorded after each decrement is removed, and the final rebound shall be recorded 24 hr after the entire test load has been removed.

II. Under each load increment, settlement observations should be made and recorded at ½, 1, 2, and 4 min, and at each 4 min thereafter after application of load increment, except in the instance of the total load when, after the 4-min reading, the time interval should be successively doubled until the final settlement limitation is reached and the load is removed.

h. Recorded Data

The following data should be recorded:

1. Soil profile at location of test pile
2. Description of pile

3. Complete driving record (hammer, speed, air or steam pressure, weight and frequency of vibrator, blows per foot, and final resistance)
4. Plot of load-settlement and time-settlement curves
5. Remarks concerning any unusual occurrences during driving or loading of the test pile

i. Precautions

I. The pile head should be carefully squared to receive the bearing plate so that no eccentricity is induced in pile section. The bearing plate should be secured in place by a drift pin, by welding, or by use of a dowel or collar.

II. In testing friction piles in cohesive soil, a time period of not less than 24 hr should elapse between driving and loading of the test pile in order to permit a stable condition to develop, any temporary excess pore pressures to dissipate and any necessary pile "freeze" to develop. For sensitive soils, substantially longer periods (perhaps 2 weeks, or more) may be required. Do not rush. The building will be in place for many years, and the pile-load test is supposed to reflect behavior of the actual building piles, at least to the extent practical.

III. If the load test is to be performed in cohesionless soils, use a deadweight loading and not a test setup wherein reaction is taken against tension piles. The reason is that the uplift reaction on the tension piles reduces the confining pressure around the test pile, thereby tending to indicate a reduced bearing capacity.

9B.2 Determining Working Load from the Test Results

As with the soil load test, the working load is the lesser value indicated by (1) ultimate load capacity, divided by a suitable factor of safety and (2) limitations on settlement. These two values are evaluated from the load test, as follows:

a. Ultimate Load Capacity

In some pile load tests, the plot of load versus settlement shows a sharp break (Fig. 9-4*a*) so that a clearly defined failure load is indicated. More often, the slope changes so gradually (Fig. 9-4*b*) that the failure load is not clearly defined.

A similar situation prevails with the soil load test (Fig. 5-10). In that case, however, the failure, arbitrarily, was defined as the point of intersection of the tangents to the two straight portions of the curve, and, as was discussed in Sec. 9A.2*a*, this criterion has general acceptance (as indicated by published expressions of opinion and known practice).

In the case of the pile load test, there is no such generally accepted criterion for failure. Opinions and practice differ widely. For example, various sources [1, 3, et al.] suggest the following criteria:

1. Maximum load carried without exceeding a net settlement (gross settlement less rebound) of ¼ in. (AASHO).

2. Maximum load at which *total* settlement, including elastic deformation of pile, is not over 0.01 in. per ton of test load and at which no settlement has occurred for 24 hr (International Conference of Building Officials).

3. Maximum load which causes a net settlement (gross settlement minus rebound) not exceeding 0.01 in. per ton of test load (New York City building code and Uniform Building Code).

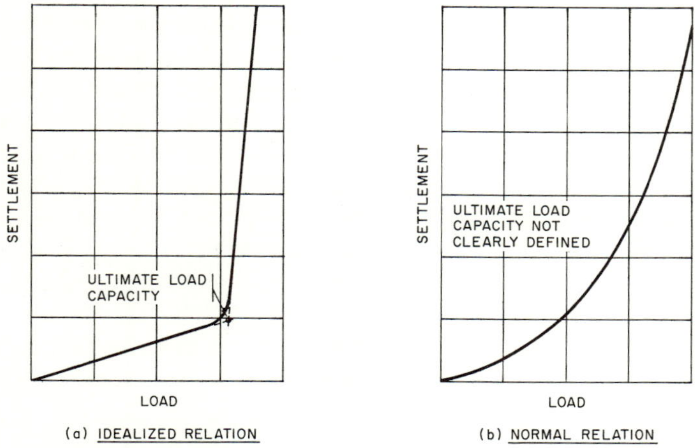

FIG. 9-4 Pile Load Tests—Load-settlement Curves.

4. Point of intersection of tangent lines to upper and lower portions of curve.

5. Point at which slope of gross-settlement curve is four times the slope of the elastic deformation of the pile.

6. Load at which the *gross* settlement exceeds 0.03 in. per ton of additional load (W. H. Rabe, Bureau of Bridges, state of Ohio).

7. Point at which the *gross* settlement exceeds 0.05 in. per ton of additional load or at which the *plastic* settlement (as determined from cyclical loading) exceeds 0.03 in. per ton of additional load (Dr. R. L. Nordlung, Raymond Concrete Pile Company).

8. Point on load-settlement curve where penetration no longer is proportional to load [3].

9. Point at which an increase in load produces a disproportionate increase in settlement (Uniform Building Code).

10. Load at which the penetration of the test pile is equal to 10 percent of the pile diameter (Terzaghi).

11. Plot rate of settlement for the last 30 min of each load increment against the total load applied. Failure load is intersection of two straight lines tangent to the terminal branches of the curve (Fig. 9-5) (Housel).

It is not clear why there should be so much difference of opinion over criteria on failure in the pile load test and relative unanimity of opinion on the same problem as regards the soil load test. Possibly the difference reflects the intensely competitive nature of the piling business and the much more frequent use of pile, as compared to soil, load tests. But it

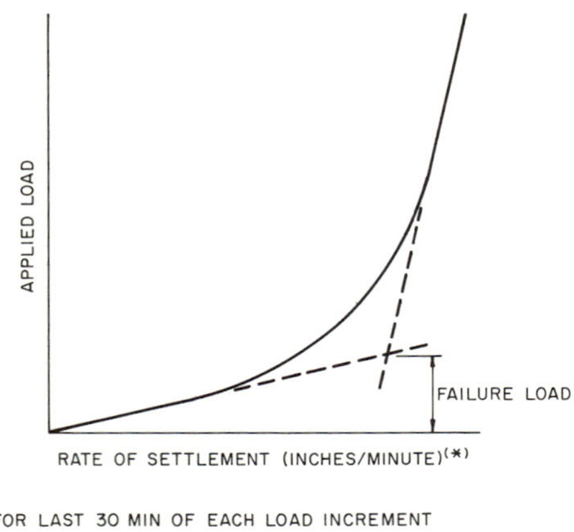

FIG. 9-5 Housel's Method for Estimating Failure Point in Pile Load Test.

makes a considerable difference which criterion is used. Muzumdar and Bose [2], for example, refer to evaluation of a single-load test on the basis of six different criteria. The failure load so determined varied from 34 to 50 tons. Fletcher [6] also has analyzed a specific pile-load–settlement curve in accordance with the procedures stipulated in different building codes and design standards (16, in all). For this same curve, 5 procedures indicated that failure conditions had been developed in the test. The other 11 tests indicated that failure conditions were not developed. Of these, 10 indicated an allowable load from 60 to 75 tons, and 1 indicated an allowable load of 100 tons.

These variations are beyond reasonable design practice, and the reader must interpret according to his own judgment. The authors' personal preference is for criteria 3, 7, or 10 in the above list.

b. Factor of Safety

Recommended safety factors accompanying the above criteria in the indicated sources [1, 3, et al.] range from 1.5 to 2.5 for static load and 3.0 for vibratory loads. Building loads are static. For this case, the great preponderance of opinion favors a safety factor of 2.0. Accordingly this value is recommended; i.e., the working load would be one-half the ultimate load capacity, estimated as described above.

c. Settlement

As with the problem of determining the apparent failure load from the test data, opinions as to criteria for allowable settlement differ. Examples are:

1. One-half the load which causes a net settlement of the pile of 0.75 in., computed as gross settlement (total downward movement) less the amount of elastic shortening in the pile section due to the test load (New York City building code).
2. One-half of the load that causes a gross settlement of 1 in., provided that the load-settlement curve shows no sign of failure within this range [3].
3. One-half of load producing a specified gross settlement after the load has been in place for the specified test period [3].

Criteria 1 and 2 roughly are equivalent and are predicated on limiting total settlement of the building foundation to something under 1 in. Criterion 3 expresses the same thing, except that the limiting settlement value is not specified, permitting the designer to insert whatever settlement he thinks the building can tolerate (see Chap. 5). This might be substantially more than 1 in. Therefore, the authors favor criterion 3.

I. *Elastic Shortening of the Pile.* The stipulation in criterion 1 that the elastic shortening of the pile is to be included in the settlement determination should be noted. There are two schools of thought on this matter. One is that since limitations on total settlement really reflect indirect limitations on differential settlement, they apply only to nonelastic strains in the soil and do not include the elastic strain of the soil and the elastic shortening of the pile; i.e., if the allowable settlement of the pile under the design load (or specified overload) is 1 in. and if the elastic strain in the soil plus elastic shortening of the pile is ½ in., the total allowable settlement under the test load is 1½ in. This approach appears to be based on the thought that the plastic deformation of the soil is the variable and that the elastic strains, particularly the elastic shortening of the piles, would be roughly the same for all the piles.

The other point of view is that the elastic strains will be variable (they will depend on the lengths of the different piles and on the distribution of load along said lengths), and so the proposed limits on settlement apply to the total settlement, i.e., that if the allowable settlement of the pile under the design load (or specified overload) is to be 1 in., the total settlement under the test load (compression of the soil *plus* elastic strains) shall not exceed 1 in.

Both points of view have merit. The elastic strains (particularly the elastic shortening in the pile) are more likely to be uniform than the inelastic compression in the soil. It certainly is not constant for the whole foundation, however. Differences in pile length, alone, would rule against this. Realistically then, the settlement criterion should be established as the total settlement assumed to be tolerable in the building, *plus* some portion of the anticipated elastic strain (shortening of the pile), perhaps one-third to two-thirds the total amount anticipated.

d. Design Load

As was stipulated for soil bearing capacity, the design pile load is the lesser of the two values limited by (1) ultimate load capacity, divided by the indicated factor of safety or (2) the indicated settlement criteria. As with the soil load test, the settlement criterion usually governs so that in theory, it often is not necessary to carry the load test to failure. The problem, however, is the same as with the soil load test; i.e., how to be sure? Therefore, if the load-settlement curve shows any sign of breaking, or the piles are in soft or loose soils, or there is any doubt whatsoever, it is well to carry the test beyond the failure point.

9B.3 Cautions

a. Similitude

It is essential that the test pile be as nearly like the proposed contract piles as is reasonable, especially as regards size, shape, and type of pile; the make and model of pile hammer; the operation of the hammer with regard to speed, height of fall, stroke, accessories, pressure and other variable factors; and equal or greater final resistance to penetration, or, for piles installed by vibratory hammer, equal or lesser rate of penetration.

b. Casing Off

Where the test piles are installed *through* fills, *through* firm strata overlying compressible strata, or under other conditions (including possible occurrence of vibratory loadings) where the resistance that the soil might provide to the pile during a load test would tend to be dissipated with time, said temporary resistance should be eliminated by "casing off" the pile down to the bearing stratum, or by other suitable means.

Casing off consists of driving an open-pipe shell having a larger diameter than the pile, removing the material inside the shell, and installing the pile through the shell.

c. Watchfulness

The assumption of similarity of soil conditions at the location of the test pile and of the other building piles is presumptuous, at best. Accordingly, penetration records should be kept on each pile, and, should a pronounced departure be observed from the pattern of penetration resistance shown by the test pile, additional load tests should be performed.

d. Borings

As with soil load tests, it is impossible to make an adequate evaluation of a pile load test unless borings, carried well below the pile tips, are available for reference. Such borings should be made close to the location of the test pile, but not so close that the boring operation disturbs the soil around the pile. A distance of 10 to 15 ft is suggested.

9B.4 Limitations on Use

a. Tests of Soil Materials Which Are Not Free Draining

As with a soil load test, a pile load test involves a relatively short-term loading condition which does not reflect the long-term settlements which will occur in fine-grained (slow-draining) soil materials under the permanent load of the building. Therefore, pile load tests are inapplicable for use with piles embedded in soils which are not free draining, except as a means for estimating ultimate bearing capacity, exclusive of settlement considerations.

b. Behavior of the Single-test Pile versus Behavior of the Actual Foundation Group

The bulb of pressure created by the actual foundation group is much larger (and deeper) than that created by the single-load–tested pile so that the stress condition created by, and behavior of, the pile group may be much different from that created by the single-test pile (Fig. 9-6). This has two important effects.

1. The ultimate load capacity of the group is likely to be considerably less than the capacity of the single-test pile, multiplied by the number of piles in the group. This is due to the overlap of adjacent bulbs of pressure and, if the pile spacing is less than the spread of the pile load in the soil (which is the usual case), diminution of the size of the analogous footing at each pile (see Fig. 9-6*a*). No rational means for evaluating this effect has been proposed, and it is conventional to assume that the group capacity is equal to the sum of the individual pile capacities multiplied by a group efficiency factor as described in Sec. 6E, Chap. Note 3.

2. The settlement of the group, for the same load per pile, is likely to be much greater than that for the single pile. In the simplest case (Fig. 9-6*b*) this is because a greater depth of soil is strained by the group load. In more critical cases, the pile tip may be underlain by weaker, or more

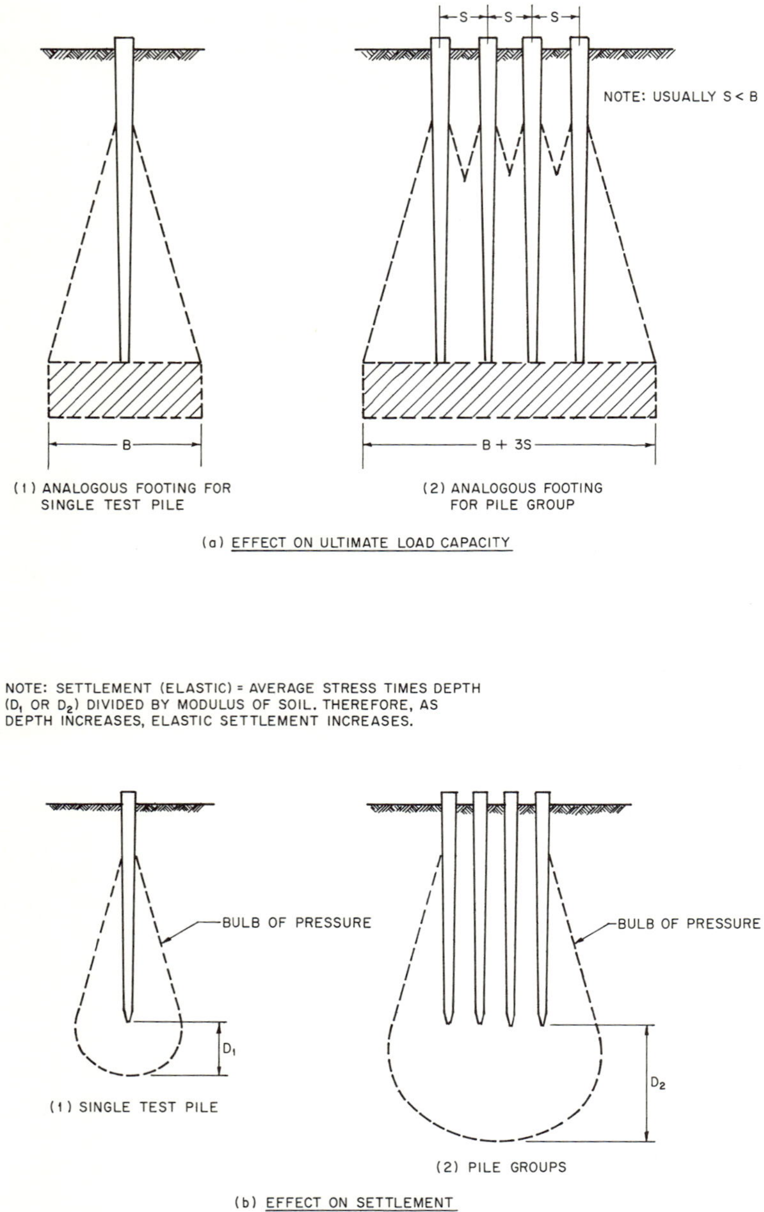

FIG. 9-6 Pile Load Test—Action of Single-test Pile versus Action of Group.

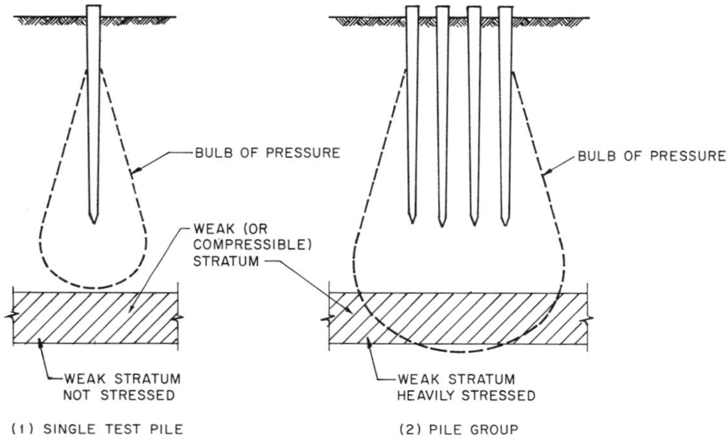

(c) EFFECT OF WEAK (OR COMPRESSIBLE) STRATUM BELOW PILE TIPS

FIG. 9-6 *(Continued)*

compressible strata (Fig. 9-6*c*), strata which are not stressed by the test load but which may be heavily stressed by the group loading.

Group load tests may be conducted to attempt to reflect the above conditions in the test. Unless the size of the test group approximates the size of the actual foundation group, little benefit will be gained. Also, group tests are expensive and seldom are feasible, economically, except on large projects.

Section 9C PILE LOAD TESTS (LATERAL LOAD)

9C.1 Test Setup

Lateral load tests on piles are used to estimate lateral resistance. Procedures and criteria for interpretation of the results have not been standardized. A typical test setup is shown in Fig. 9-7.

9C.2 Design Load

Criteria for interpretation of results which are in use are:

a. Design lateral load equals test load causing 3/8-in. lateral displacement of pile head (New York City building code).

b. Design lateral load is one-half test load causing a deflection of 1/2 in. at the ground line, with 75 percent recovery (Los Angeles building code).

9C.3 Limitation

As with axial load tests, lateral load tests on piling are of limited applicability in estimating the capacity of the actual foundation installation. Relating the action of the single-test pile to the larger, final group

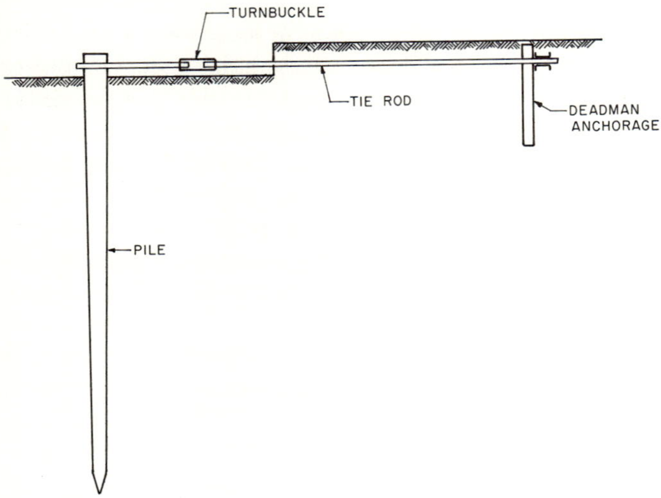

FIG. 9-7 Setup for Lateral Load Test on Pile.

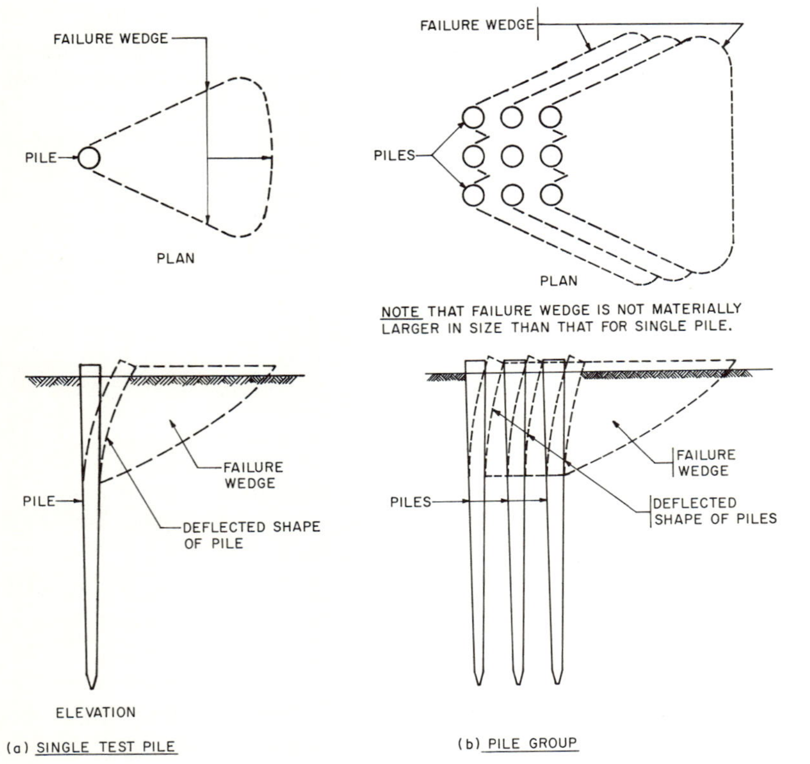

FIG. 9-8 Lateral Pile Load Tests—Action of Single-test Pile versus Action of Group.

is the biggest problem, as illustrated in Fig. 9-8. Conventional pile spacings (2 ft 6 in. to 3 ft 0 in.) are much smaller than the dimensions of the failure wedge developed in the soil by lateral displacement of the single pile. Accordingly, the lateral load capacity of a pile group is likely to be much less than the test-load capacity of a single pile, multiplied by the number of piles in the group.

Section 9D CYCLICAL AND INSTRUMENTED LOAD TESTS

9D.1 Cyclical Load Tests

Cyclical load tests permit an estimation of the distribution of the axial load along the length of the pile. In the simplest case, the purpose is to

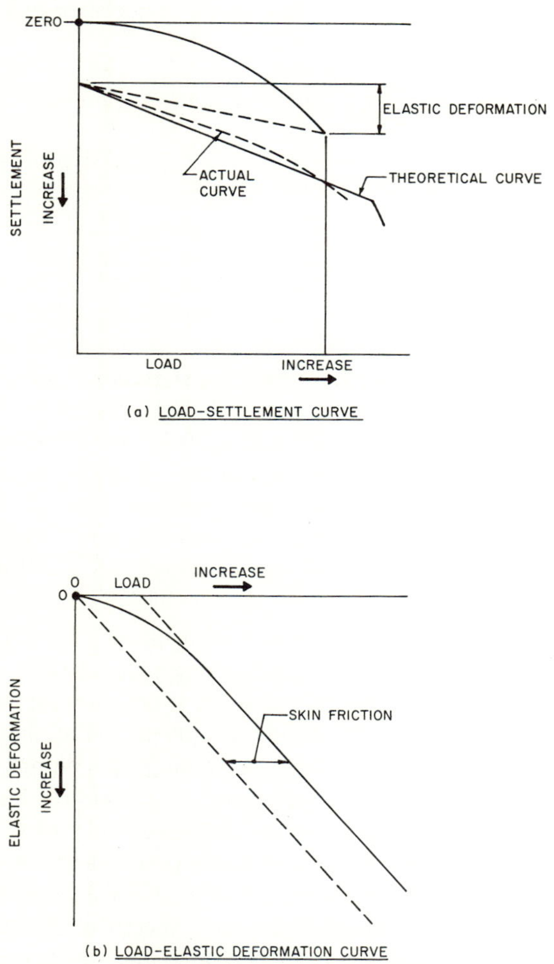

FIG. 9-9 Cyclical Load Test by Method of Van Weele.

differentiate the frictional and end-bearing components of the total resistance. The procedure suggested by Van Weele [7] (see Fig. 9-9) may be used for this purpose. This procedure is as follows:

a. The elastic deformation of the pile plus soil at each load increment is determined. It is presumed to be equal to the difference between settlement under load and settlement under no load during any given cycle (Fig. 9-9*a*).

b. The load on the pile butt is plotted against the elastic deformation (Fig. 9-9*b*). Under higher loads, a straight-line plot should result.

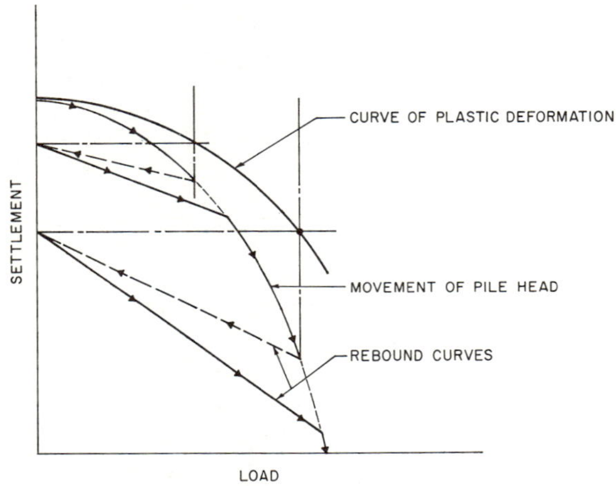

FIG. 9-10 Cyclical Load Test by Method of Chellis.

c. Draw a line through the origin in Fig. 9-9*b*, parallel to the straight-line portion of the elastic-deformation curve. The distance between the curve of elastic deformation and a straight line through the origin is assumed to be the total skin friction acting on the pile. This assumes the skin friction increases with increasing load up to some limiting value, after which it remains constant.

d. The load at the tip equals the test load minus the skin friction.

Chellis [1] describes a somewhat more general procedure. In this test, the load is removed from the pile several times during the process of incremental loading, and the rebounds are plotted (Fig. 9-10). A curve of plastic deformation is plotted by projecting the points representing rebound readings at zero load to intersect the respective loads removed.

The difference in ordinates between the curve of plastic deformation and the curve of movement of the pile head represents the elastic deformation of the pile-soil system. If the elastic deformation of the soil is assumed to be negligible (which it is not), then the effective center of resistance along the pile length can be determined from $e = PL/EA$ by determining the length (L) corresponding to the elastic deformation for a given load (P) on the given pile (area A). Variable values of A (tapered or composite piles) may be considered.

Knowing the location of the center of resistance, the load distribution along the pile can be approximated by assuming a rectangular, triangular, or some other reasonable distribution of the total load about the calculated center of resistance.

Unfortunately, all these assumptions provide a highly problematic solution, so that the principal value of the cyclical test is to establish whether the test load actually devolves on or in the soil stratum which it is desired to test or whether the load is being dissipated at a higher level, so that the critical stratum is not, in fact, being tested.

9D.2 Instrumented Load Tests

Instrumented load tests provide a more reliable estimation of the distribution of load along the pile length. Strain gauges can be installed in the pile at various points along the length and the total load at each level calculated as the product of EA times the recorded unit strain. In theory, this method should provide precise answers, but as anyone will understand who has tried to use strain gauges on full-size test specimens (particularly where the conditions of eccentricity of load application cannot be well controlled and under adverse conditions as obtained in the pile during and after driving), the operation of executing and interpreting an instrumented load test sounds a lot easier than it is. Still, with care, and if enough spare gauges are installed (to compensate for those that do not function, or are torn, or are broken, or whose leads break, or fail to be waterproof), and if the gauges are adequately distributed around the perimeter of the pile so that average readings (compensating for the eccentricities which should not but generally do exist) can be made, an instrumented test pile can provide quite valuable information on the distribution of the load along the length of the pile.

Use of simplified instrumentation, in the form of a series of sounding points along the length of the pile (rods set on shelves set in a hollow tube in, or attached to, the pile), is reported by Chellis [5].

REFERENCES

1. Chellis, R. D.: Pile Foundations, in G. A. Leonards (ed.), "Foundation Engineering," McGraw-Hill Book Company, New York, 1962.
2. Muzumdar, M. N., and C. N. Bose: Evaluation of Safe Bearing Capacity of Friction Piles in Soft Clays, *Proceedings, Symposium on Bearing Capacity of Piles,* Central Building Research Institute, Roorkee, India, February, 1964.
3. Federal Construction Council, Survey of Practice Foundation Piling, *Report* 4, National Academy of Sciences, National Research Council, Publication 987, 1962.

4. Sowers, G. F.: Shallow Foundations in G. A. Leonards (ed.), "Foundation Engineering," McGraw-Hill Book Company, New York, 1962.
5. Chellis, R. D.: "Pile Foundations," 2d ed., McGraw-Hill Book Company, New York, 1961.
6. Fletcher, G. A.: "Pile Load Tests and Their Evaluation," *American Society for Testing and Materials Special Technical Publication* 322, 1962.
7. Van Weele, A. F.: A Method of Separating the Bearing Capacity of a Test Pile into Skin Friction and Point Resistance, *Proceedings, Fourth International Conference on Soil Mechanics and Foundation Engineering,* vol. II, 1957.
8. Terzaghi, K., and R. B. Peck: "Soil Mechanics in Engineering Practice," 2d ed., chap. IX, art. 54, John Wiley & Sons, Inc., New York, 1968.

CHAPTER 10
Some Miscellaneous Matters

Section 10A STRENGTHENING AN EXISTING FOUNDATION

A problem which occasionally arises is that of increasing the load capacity of an existing foundation. The need sometimes is due to deficiencies in, or deterioration of, the existing construction. More often, it derives from a requirement to add load to the building (increased height or change of occupancy).

There are three basic techniques by which the necessary increased strength may be provided: (1) to increase the size of the bearing area (increase the size of the footing or increase the number of piles) so that the *pressure* imposed on the soil by the existing, plus the additional, load is not increased, (2) to permit the imposed pressure to increase, but to carry the load down to a firmer bearing stratum capable of supporting the increase, or (3) again to permit the imposed pressure to increase, but to correspondingly strengthen the soil at the level of the bottom of the footing. These three devices are effected as follows:

TECHNIQUE 1 *Increasing the Effective Size of the Bearing Area*

a. Relieving the Existing Load

Construction used to enlarge a footing is shown in Fig. 10-1. An example of the addition of piles to a pile group is shown in Fig. 10-2. As may be noted, these constructions are elementary and would appear to present no unusual problems. As a matter of fact, however, analysis of the stresses in the final assemblage can be quite complex because the soil does not necessarily behave elastically, and so the principle of

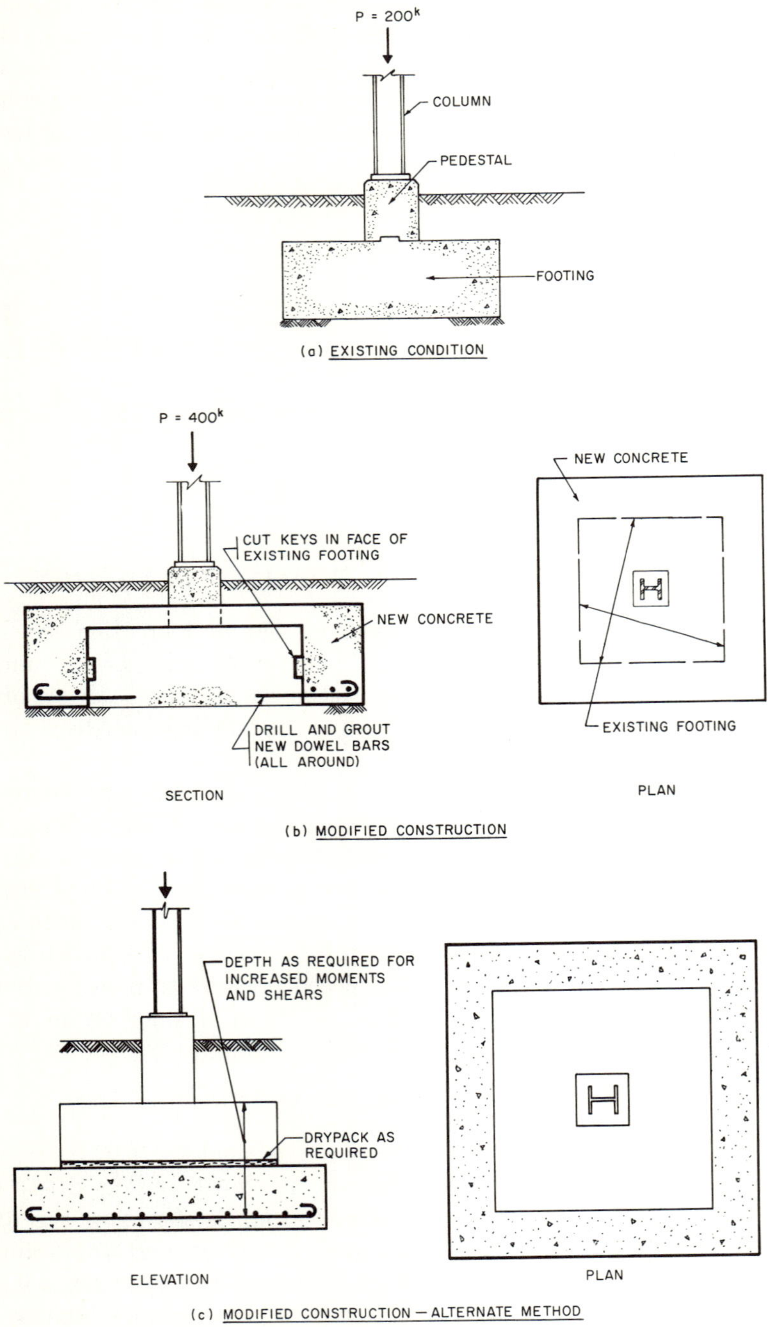

FIG. 10-1 Enlarging on Existing Footing.

superposition does not apply. The effects of the existing and of the newly added load must be considered together. The two components of load interact, and the total effect is not simple. For example, consider the construction shown in Fig. 10-1. Assume that the existing column load of 200 kips is to be increased to 400 kips. The existing footing has

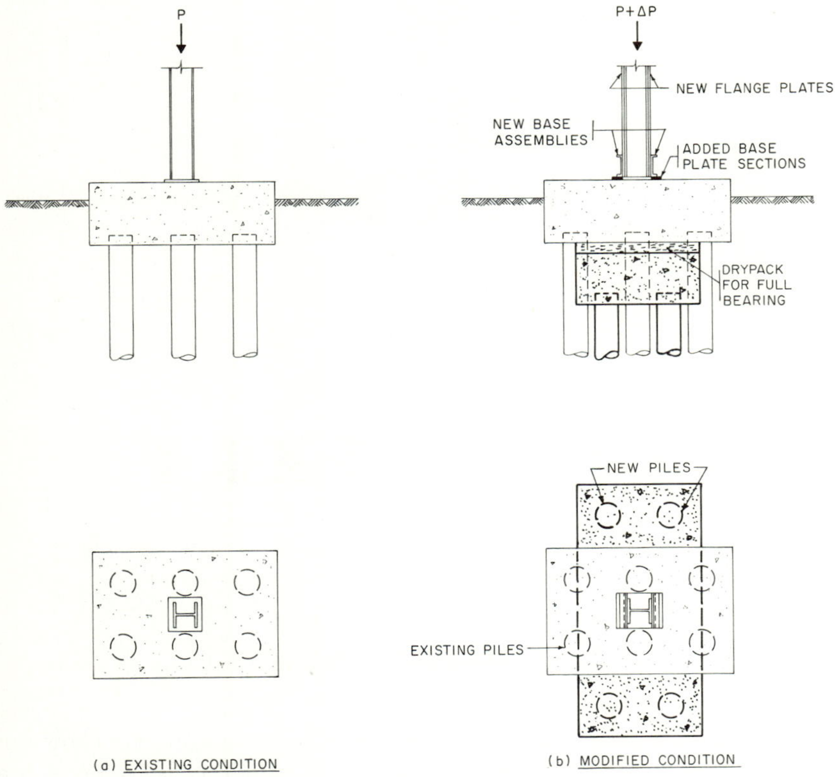

FIG. 10-2

been proportioned for a bearing pressure of 4 tsf (8 ksf), which is the presumptive, or allowable, bearing capacity of the soil.

Obviously, the new construction must be completed before the new load is added to the building, otherwise the footing would be overloaded. Assume that this is done and that the total load of 400 kips now devolves upon the new combined construction. If the soil were to behave elastically, the bearing pressure under the old footing would be 6 tsf and under the new footing only 2 tsf.* If the allowable soil bearing pressure is 4 tsf, as indicated above, the result would be that the soil under the existing

* See Fig. 10-3.

footing would be overloaded, and the design, ostensibly, would be unsatisfactory. In order to provide for this effect, what normally is done is to relieve the load on the existing footing before the new load is added to the building, so that the final bearing pressures under both the new and existing constructions are more nearly the same, and the maximum bearing pressure is more nearly the average value of 4 tsf, which is the presumptive allowable bearing pressure.

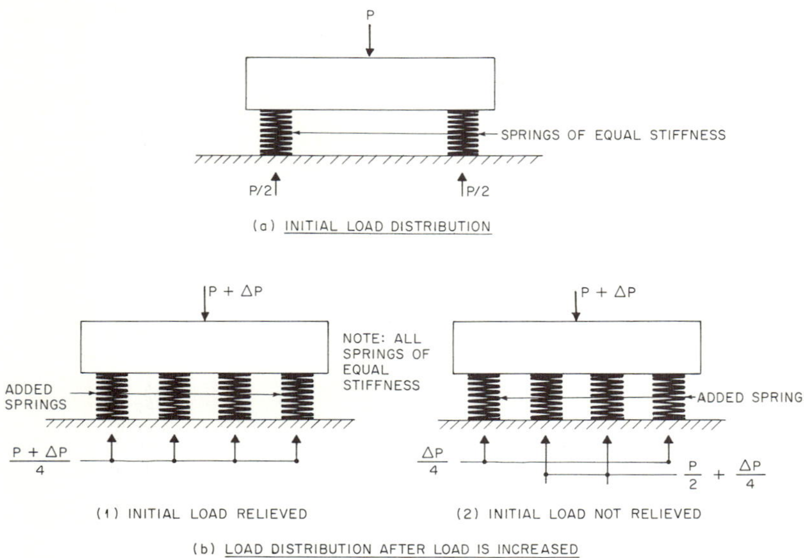

FIG. 10-3 Distribution of Added Load to New and Existing Elements.

Devices used to relieve the load on an existing foundation are shown in Figs. 10-4 and 10-5.* As will be apparent, such constructions are expensive.

The above analysis and requirement are predicated on the assumption of elastic behavior of the soil corresponding, in principle, to working stress design as conventionally practiced in proportioning steel and concrete members. More modern practice is to treat the analysis on the basis of ultimate strength criteria, the object being to eliminate the need for relief of the existing load. To consider the implications of this approach, return to the discussions of Chap. 5.

As indicated therein, allowable bearing pressure is a function of two considerations: (1) potential shear failure in the soil and (2) limitations on settlement. Let us treat first the possibility of shear failure in the soil.

* The figures are schematic. For details see Ref. 1.

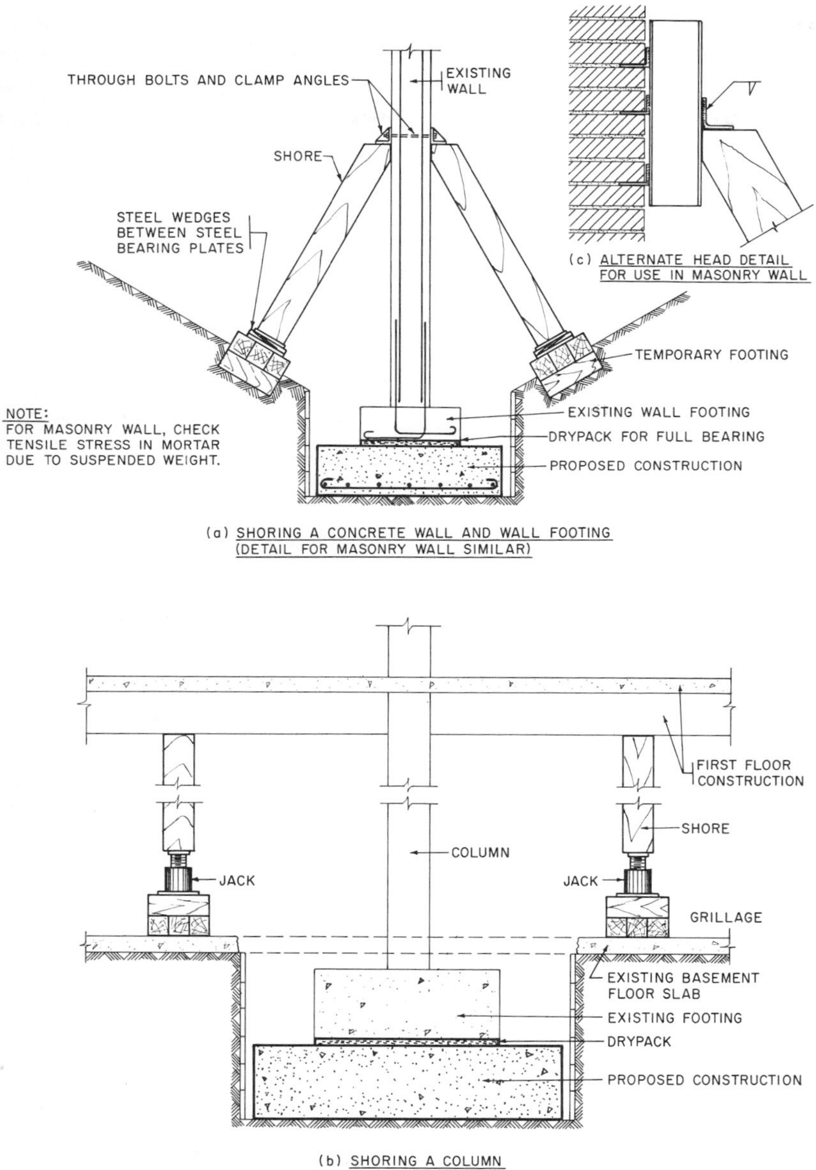

FIG. 10-4 Relieving the Existing Load—Shoring.

Bearing capacity as limited by this consideration is expressed by Eq. (5-1).

$$q_{\text{ult}} = cN_c + \Sigma\gamma_d D_f N_q + \frac{\gamma_d B N_\gamma}{2}$$

Assume that the footing is enlarged without relieving the existing load (but before the new load is added). The value of B in the combined construction (Fig. 10-1) has increased. Therefore, except for purely cohesive soil, the value of average allowable bearing pressure (total load

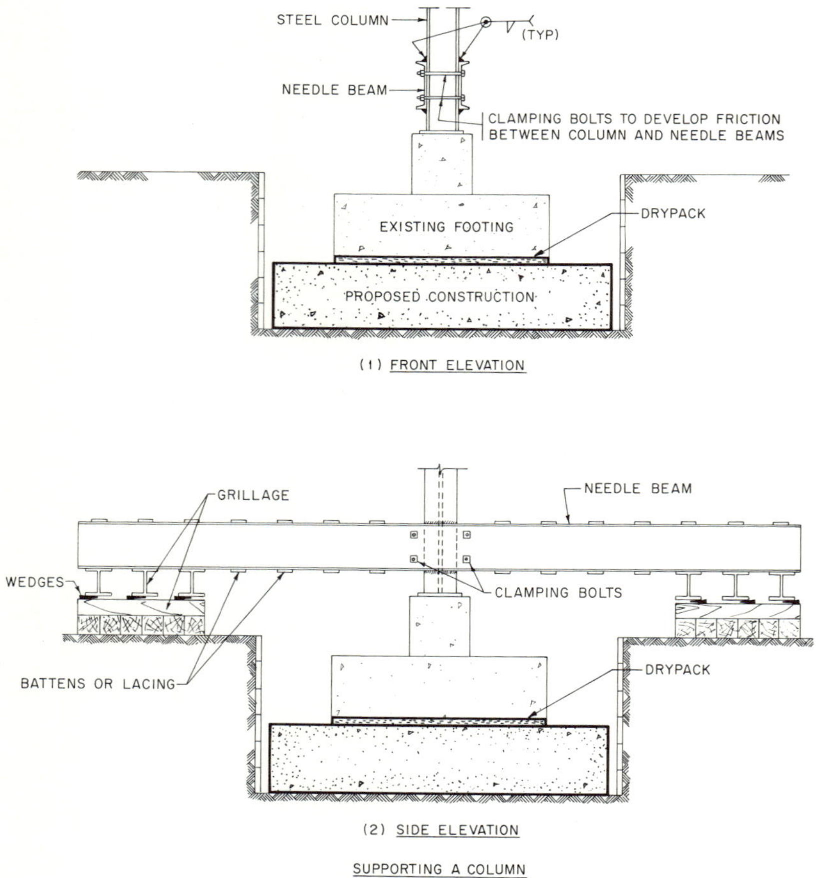

FIG. 10-5 Relieving the Existing Load—Needles.

divided by total area) has been increased. More important, the allowable bearing pressure, locally, under the existing footing also has been increased by virtue of the confining effect of the added footing area. The distribution of bearing pressure under the combined construction, of course, will tend to be much greater under the existing footing than under the added section. The final effect, however, is that the soil (providing that it is not a brittle material subject to progressive fracture) near the center of the bearing area yields (or creeps), causing the pressure to be redis-

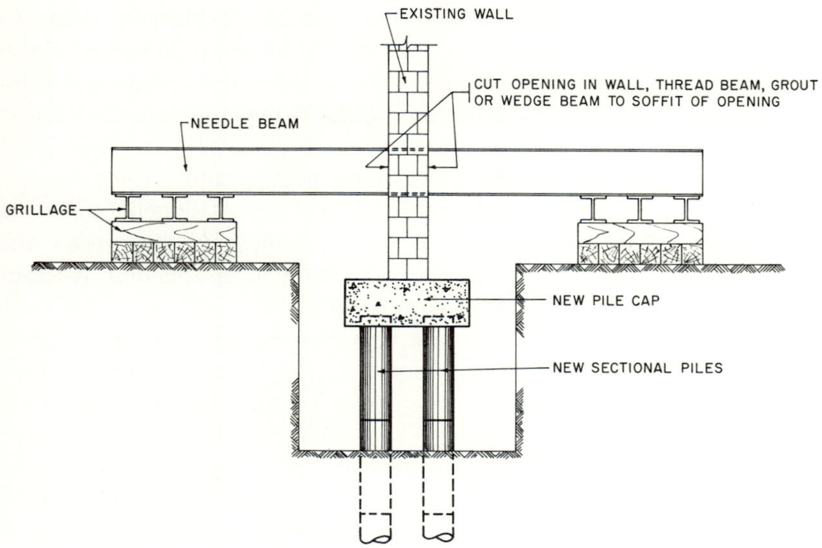

(1) THROUGH BEAMS

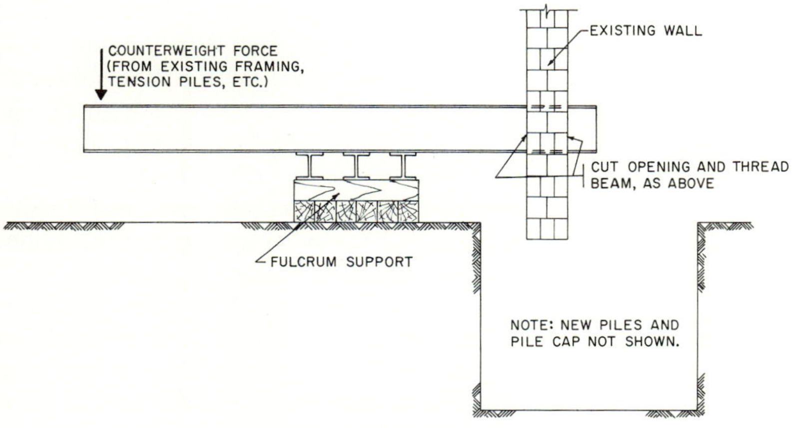

(2) CANTILEVER BEAMS

SUPPORTING A WALL

FIG. 10-5 *(Continued)*

tributed, "ironing out" the differentials, and producing a final pressure distribution which is more or less uniform. This is, in fact, the behavior of conventional footings and is the basis for the usual design assumption of uniform bearing pressure thereunder, an assumption which we know is not strictly true.

Finally, as often as not, and particularly with granular soils, allowable bearing pressure is controlled by considerations of settlement rather than shear failure. Thus, insofar as *shear failure of the soil* is concerned, relief of the existing load is not necessary if the size of the new footing is properly proportioned and if provision is made to resist the considerable shear which occurs at the interface between the new and existing construction, and provided that the underlying soil is not a brittle clay or other type which is subject to progressive fracture rather than to yield (or creep) deformation.

The problem of settlement is a somewhat different matter. If new load is added to the footing without relieving the existing load, the settlement will surely be increased. The amount depends on such things as the relative sizes of the new and existing bearing areas and on the difference in stress-strain characteristics of the soil underlying the two areas.* If the stress-strain curve for the underlying soil material can be established (triaxial-test data are recommended for this purpose), then the amount of settlement can be estimated using the procedures described in Chap. 5.

The only serious uncertainty in the calculation is the amount of settlement which is tolerable in the building. This matter, too, is discussed in Chap. 5. The problem is, even if the allowable settlement can be established (as a criterion on either gross or differential settlement), how much has already occurred? Sometimes this can be established by level readings and the amount of tolerable increase compared to the estimated potential occurrence. Sometimes the existing occurrence cannot be established. Then the decision is strictly a matter of judgment. It must be borne in mind, however, that, as discussed in Chap. 5, tolerable settlement is a function of the time rate of its occurrence and that the total allowable displacement in an old building, occurring as two distinct increments, one during the initial construction and the second occurring after an extended period of time, is likely to be greater than if the settlement occurs during one increment. The reason is that previously discussed; i.e., the building has time to adjust, inelastically, to the incident strains.

If, after considering the above, it still appears that relief of the existing load will be necessary, attention should be directed to the possibility either of taking some of the live load out of the building or, if this is not

* The soil under the existing footing has been prestressed and so advanced along the stress-strain curve.

possible or is not sufficient, of redistributing the column reactions in the building by way of the stiffness of the superstructure.

In general, the walls and interior framing of buildings have considerable capacity to bridge the load from one footing to another and if a few underloaded footings or pile groups can be identified, it may be possible to strengthen the elements adjacent to them without relieving the load. These elements, having been strengthened, will then serve to support the load devolving on succeeding elements while they are reinforced, and so on. If the necessary underloaded footings or pile groups do not exist or are not strategically located, a few footings or pile groups can be relieved and strengthened to start the cycle. Of course, everything has to be right: the footings must to be in the proper place, the loads must not be too heavy, and the superstructure or foundation walls or framing must have enough strength to carry the load.

b. Checking the Structural Strength of the Revised Construction

This is done in accordance with standard design procedures (see Ref. 2). There are a few special points to consider, however.

I. Old foundations conventionally were designed with substantial factors of safety and incorporate a considerable reserve of strength. In particular, the old-style corbelled masonry or plain concrete footings (or where a second footing is placed under the existing footing as in Fig. 10-1*c*) patently are so strong that actual reinforcement of the section seldom is necessary.

II. Technique in design has progressed, permitting greater loads for a given design element. For example, if a footing or pile cap be checked by ultimate strength theory usually it may be found to have some reserve of strength as compared to the same design executed under working stress theory. The same thing applies to the analysis of allowable soil bearing pressures. Most existing designs have been proportioned on the basis of prescription values specified in building codes. Analysis (see Chap. 5) often indicates greater allowable capacity.

III. Allowable stresses in structural elements have increased, drastically in some cases, as a result of our improved knowledge of the action of such elements under load. Increases in allowable stress in steel and timber piles (Chaps. 6 and 8) and in allowable stresses on grillage beams and reinforcement (ACI 318-56 versus ACI 318-63 and AISC-47 versus AISC-63) are examples.

IV. Where concrete stresses (compression, bending, or diagonal tension) control, as they usually do in footings and pile caps, the increase of strength of the concrete with time can be counted on to increase the load capacity. For example, the design doubtless has been predicated on the 28-day cylinder strength. By the time strengthening is required, the concrete may be several years of age and its strength may have increased one-third, or more.

v. In addition to the final loading condition, the temporary loading condition during installation of the underpinning must be considered. Several of the standard underpinning techniques (Figs. 10-4 and 10-5, for example) involve temporary undermining of existing supports. Analysis should be made of this condition, as well as to that of permanent loading. An increase in allowable stress (and soil-bearing capacity) of 10 to 15 percent, or more, would be applicable to such condition. For large, or long footings, this limitation can be met by underpinning in sections. For small footings, this may not be practical, however, because of the expense of the multiple operations involved.

TECHNIQUE 2 *Transfer the Load to a Deeper Stratum*

This is done by the use of underpinning piers or of piling. The piling can be installed in sections, by jacking, or can be driven at a slight batter and either doglegged or pulled into place under the footing.

If a firmer stratum underlies the existing foundation, the piers or piles can be designed to transfer the load to that deeper stratum. If the soil remains substantially the same capacity with increasing depth, the use of friction piles will provide for enlargement of the analogous footing (see Chap. 6).

TECHNIQUE 3 *Increase the Capacity of the Supporting Soil*

This may be done by grout or chemical solidification.

If either technique is used, a point to be considered is that the solidification must be carried deep enough below the footing to ensure that the spread of the load is sufficient to obviate overload of the soil below the grouted zone. Note, however, that this is likely to be a relatively shallow depth. For example, for a 10- by 10-ft footing, a 5-ft-deep grouted zone reduces the bearing pressure on the soil beneath the grouted zone to something less than half the pressure under the existing footing.

Other limitations on the application of grouting or chemical solidifications are that the soil must be sufficiently porous to accept the grout, that the critical load on the foundation must not be controlled by considerations of deep-seated settlement (which is a function of total load and not of unit soil pressures), and that the local building code or building official must recognize and permit the use of the technique. The grouted zone should be drilled and cored to check the efficacy of the work.

Section 10B UNDERPINNING

The previous section (10A) referred briefly to underpinning techniques such as installing jacked piles and underpinning pits and piers. The discussion was limited, however, to consideration of the problem of establishing allowable loads for the revised construction. The following problems were not discussed:

1. Design and detailing of shores, needle beams, and other devices for relieving the existing load (if this is required)
2. Design and detailing of sheeting, bracing, or other devices required incident to undermining or excavating alongside of the existing construction
3. The construction problems incident to the installation of the new work
4. Structural design of the new piers, footings, or pile caps

The last item (structural design of the new piers, footings, or pile caps) is a matter of structural design and is carried out using standard techniques.

The other items (1, 2, and 3) relate to the specialty of underpinning. This is a highly refined and specialized art, techniques of which already have been the subject of an excellent book [1], to which the reader is referred. Because underpinning is such a specialized art and because the satisfactory execution of the work is sensitive to many details, both of design and workmanship, it is suggested that the matters are best left to firms specializing in underpinning work.

Section 10C MINIMUM SIZE REQUIREMENTS

10C.1 Cast-in-place Concrete Piles

A minimum diameter usually is specified to ensure that the pile can be properly filled with concrete. An 8-in. diam appears to be adequate with a 6-in. tip and a 8-in. butt for tapered sections. Even smaller sections probably could be used if a grout mix were pumped into place through a hose inserted into the pile.

10C.2 Foundation Piers

A foundation pier conventionally is designed as a reinforced concrete column, for which the provisions of ACI 318 would apply, i.e., minimum 10-in. diam, or for rectangular columns, 96 sq in. with minimum 8-in. dimension.

These dimensions are only adequate, however, if the fill around the pier is placed uniformly and with reasonable care. If the fill is likely to be dumped or dozed around the pier, substantial lateral forces will be developed during construction and, as a matter of practical necessity, larger sections are required to resist the possible hazards of impact by the bulldozer, dropped boulders, or similar incidents. A minimum 24-in.

section is not excessive under such circumstances, even though it may not be needed for stress, i.e., to support the building.

In no case, of course, should the pier be smaller than the column it supports or, in the case of a steel column, the size of the base plate, plus an allowance for clearance all around.

Section 10D LATERAL SUPPORT CRITERIA FOR FOUNDATION PIERS AND WALLS

10D.1 Piers

As indicated in the previous section, foundation piers conventionally are designed as reinforced concrete columns. This requires that the effective unbraced length, somehow, be estimated.

In theory, the criteria indicated in Chap. 6 (Fig. 6-4) would apply, and these may be used. There is an additional consideration, however, in that piles are driven or bored into the ground whereas the soil around a pier is placed after the pier has been built. The confinement offered by the soil around the pier, therefore, is less reliable, and it is suggested that the equivalent unbraced length not be taken as less than eight times the least dimension of the pier (or the full height, if it is less than eight times the least dimension).

10D.2 Walls

Foundation walls also are designed as compression members (columns). Sometimes, as for basement walls, the column action is combined with bending because of lateral load. In either case, again it is necessary, for purposes of dealing in the design with the compressive load, to estimate an effective unbraced length.

This effective unbraced length may be estimated by analysis using the equations derived for beams on elastic supports [see Hetenyi, Ref. 3]. Alternatively, the criteria indicated for the design of foundation piers (eight times the thickness of the wall, or the full height, whichever is less) would apply, except that where the wall functions as a retaining wall, the unbraced height should not be taken as less than the height of the exposed face.

Section 10E FOUNDATION ELEMENTS OF UNREINFORCED CONCRETE

Foundation elements (piers, footings, pile caps and walls) generally are built of reinforced concrete and are designed in accordance with ACI 318, modified as indicated in this text. Sometimes unreinforced concrete is used. In such case, the requirements of ACI 318 do not directly apply and certain rules of thumb are followed, such as:

1. *Unreinforced Footings:* Thickness should be such that assumed load distribution, at 45° from the limits of the pier or wall, falls entirely within the concrete [4] (see Fig. 10-6). Minimum thickness (thickness of lip for stepped or sloped footings) to be 8 in. (ACI 318).

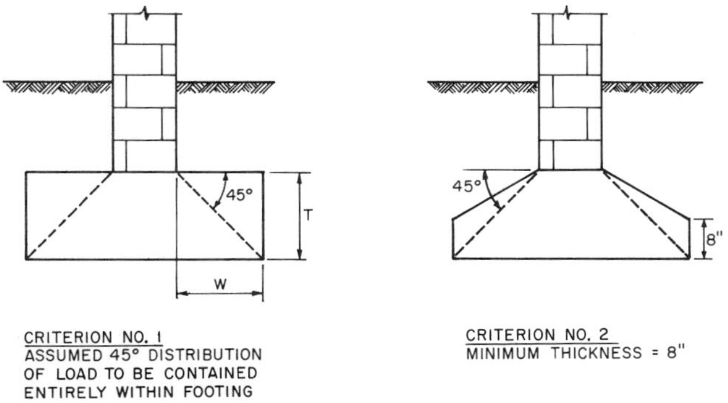

FIG. 10-6 Unreinforced Concrete Footing.

2. *Unreinforced Piers:* (*a*) the eccentricity (deviation of line of action of resultant of forces acting on section from centroid of section) at any level should not exceed one-sixtieth the height of the pier or one-tenth the diameter (Boston building code) and (*b*) the ratio of height to diameter should not exceed 6 (New York City building code).

REFERENCES

1. Prentis, E. A., and L. White: "Underpinning," Columbia University Press, New York, 1931.
2. Johnson, S. M.: "Deterioration, Maintenance, and Repair of Structures," chap. 9, McGraw-Hill Book Company, New York, 1965.
3. Hetenyi, M.: "Beams on Elastic Foundations," The University of Michigan Press, Ann Arbor, Mich., 1947.
4. Tomlinson, M. J.: "Foundation Design and Construction," John Wiley & Sons, Inc., New York, 1963.

Index